PRACTICAL PROBLEMS WITH SOLUTIONS IN WASTEWATER ENGINEERING

AUTHORS

Dr. S.N. Kaul obtained his bachelor and master degrees from IIT, Kanpur and was awarded Ph.D. degree from the University of New Castle Upon Tyne, UK. Dr. Kaul joined CSIR-National Environmental Engineering Research Institute [NEERI], Nagpur in 1970 and worked for over three decades. Dr. Kaul superannuated as Acting Director and Director Grade Scientist from NEERI and was also associated with MIT College of Engineering, Pune as its Principal. Dr. Kaul has published over 350 Papers in National and International journals and authored over 30 books on Environmental Science and Engineering, 8 Patents and 150 Conference papers. Dr. Kaul has designed and operated several waste treatment plants in India.

Dr. Kaul has supervised over a dozen of Ph.D. and 40 M.Tech. students. In addition, Dr. Kaul was responsible for organizing several National and International Conferences and has traveled extensively in India and abroad concerning various R&D programs. Dr. Kaul has received several National and International awards and is expert member of various important National committees on Environmental Science and Engineering.

Dr. D.R. Saini obtained his B.Tech. and M.Tech. from HBTI, Kanpur and Ph.D. (Chem. Engg.) from IIT-Delhi. Presently working as Principal Scientist at NCL, Pune. Dr. Saini has co-authored three books, six chapters and published seventy papers in international journals. Dr. Saini has guided 3 Ph.D. and few M.Tech. students.

Er. Prateek Kaul has obtained his B Tech and M Tech from NIT, Nagpur. He has worked in several multinational industries in India and abroad and has number of patents and publications to his credit.

PRACTICAL PROBLEMS WITH SOLUTIONS IN WASTEWATER ENGINEERING

VOLUME 2

Authors

Prof. (Dr.) S.N. Kaul
Dr. D.R. Saini
Er. Prateek Kaul

2016

Daya Publishing House®

A Division of

Astral International (P) Ltd

New Delhi 110 002

Practical Problems with Solutions in Wastewater Engineering
(6 Volumes Set)
Volume 1: Page 0001-0398
Volume 2: Page 0399-0798
Volume 3: Page 0799-1198
Volume 4: Page 1199-1598
Volume 5: Page 1599-1998
Volume 6: Page 1999-2376

Cataloging in Publication Data—DK
Courtesy: D.K. Agencies (P) Ltd. <docinfo@dkagencies.com>

Kaul, S. N., author.
Practical problems with solutions in wastewater engineering / authors, Prof. (Dr.) S.N. Kaul, Dr. D.R. Saini, Er. Prateek Kaul.
6 volumes cm
Includes bibliographical references (pages) and index.
ISBN 9789351308959 (Vol. 2)
ISBN 9789351243717 (International Edition)

1. Sewage—Purification. I. Saini, D. R., author. II. Kaul, Prateek, author. III. Title.

DDC 628.3 23

Published by : **Daya Publishing House®**
A Division of
Astral International Pvt. Ltd.
– ISO 9001:2008 Certified Company –
4760-61/23, Ansari Road, Darya Ganj
New Delhi-110 002
Ph. 011-43549197, 23278134
E-mail: info@astralint.com
Website: www.astralint.com

Laser Typesetting : **Classic Computer Services,**
Delhi - 110 035

Printed at : **Thomson Press India Limited**

Dedicated to the Sacred Memory

of

Sri Sri Sarda Maa

and

Swami Vivekananda

PREFACE

Solving problems is one of basic function of engineers. Text books in the field of wastewater engineering devote most of their pages to the development and explanation of the theory. This Book is devoted to the pragmatic idea that performance and understanding of calculations should have more proportion in illustrating the principles. In order to help the reader An overview of wastewater treatment processes has been presented in Chapter-1.

This book is organized wherein a problem is stated and the necessary equations are given in the beginning followed by the solution. The present book stresses on the following topics which are of great importance to the country at large, *viz.*:

- Sludges
- Nutrient Removal
- Natural System of Treatment
- Stream Sanitation
- Process Economics
- Elimination of VOCs
- Data Analysis-Statistical Techniques

This book will be useful to students, teachers, practising engineers, and also the researchers in the field of wastewater treatment. Some of the topics bring together information which probably has not been accumulated nor presented in one place before. Ample notes have been given at the end of each problem to assist the reader in comprehending the entire process of the systems calculation, (salient points to ponder). Sufficient references have also been added in the problems.

Towards the end of the book a large number of references have been added to help the user of the book (Bibliography). In addition, two appendices have been given for necessary valuable data, some salient information in Wastewater engineering.

The authors have over three decades of experiences in teaching, research, and application of full scale plants in the field and, therefore a need was felt to pass the experience to all the stake holders. We are sure that the readers will enjoy reading the book as we did while collecting and collating the information for writing this book. The authors would like to thanks Dr. A. Gupta, Director Grade Scientist, CSIR-NEERI, Nagpur for writing a chapter on Statistical Techniques for us in this book.

The trust reposed by the Astral International (P) Ltd. New Delhi is thankfully acknowledged. In addition authors thank all persons who directly or indirectly helped us to bring out this book particularly the students, all our beloved parents, teachers and associates in the CSIR family.

Prof. (Dr.) S.N. Kaul

Dr. D.R. Saini

Er. Prateek Kaul

CONTENTS (VOL. 1-6)

ABBREVIATATION AND ACRONYMS

AACE	American Association of Cost Engineers
Al2(SO4)3.14.3(H2O)	Alum
ASCE	American Society of Civil Engineers
ASME	American Society of Mechanical Engineers
AWWA	American Water Works Association
AE	Aeration Efficiency
BOD	Biochemical oxygen demand
BDOC	Biodegradable Dissolved Organic Carbon
BNR	Biological Nutrient Removal
BPR	Biological Phosphorus Removal
BPT	Best Practicable Technology
BOD/COD	Biological or Chemical oxygen Demand
CBOD	Carbonaceous Biochemical Oxygen Demand at 5 days and 20°C
CFID	Continuous feed and intermittent discharge
$Ca(OH)_2$	Lime
CH_4	Methane
COD	Chemical oxygen demand
DO	Dissolved oxygen
DAF	Dissolved Air Flotation
DBP	Disinfection By-product
DCS	Distributed Control System
DO	Dissolved Oxygen
DOC	Dissolved Organic Carbon
EPEA	Environmental Protection and Enhancement Act
EC	Equipment cost
EC EDR	European Community Environmental Directive Requirements

ENR	Engineering News Record
EPA	Environmental Protection Agency
F/M	Food to micro-organisms ratio
Fe2(SO4)3	Ferric sulfate
FeCl3.6H2O	Ferric chloride
FeSO4.7H2O	Ferrous sulfate
F/M	Food to Microorganism ratio
G	Velocity Gradient
GCDWQ	Guidelines for Canadian Drinking Water Quality
GWUDI	Groundwater under the direct influence of surface water
GAC	Granular activated carbon
GCC	Gulf Cooperation Council
gfd	Gallon per square foot per day
H2S	Hydrogen sulfide
ha	Hectare
HRT	Hydraulic retention time
HPC	Heterotrophic Plate Count
HRT	Hydraulic Retention Time
IX	Ion exchange
IFID	Intermittent feed and intermittent discharge
KOH	Potassium hydroxide
MF	Microfiltration
mgd	Million of gallons per day
MLSS	Mixed liquor suspended solids
MWTP	Municipal waste-water treatment plant
MAC	Maximum Acceptable Concentration
MLSS	Mixed Liquor Suspended Solids
NH3-N	Ammonia nitrogen
N	Nitrogen
Na2S2O5	Sodium metabisulfite
Na2SO3	Sodium sulfite
NaOH	Sodium hydroxide

NF	Nan filtration
NGO	Non-governmental organization
NPDES	National pollutant discharge elimination system
NTS	Natural treatment system
NSF	National Sanitation Foundation
NTU	Nephelometric Turbidity Unit
OTE	Oxygen Transfer Efficiency
OTR	Oxygen Transfer Rate
ORP	Oxidation Reduction Potential
OU	Odour Unit
O&M	Operation and maintenance
OF	Overflow
ORP	Oxidation-reduction potential
P	Phosphorus
P2O7	Polyphosphate
PAC	Powered activated carbon
POTWs	Publicly owned treatment works
PO4-3	Orthophosphate
PVC	Polyvinyl chloride
PLC	Programmable Logic Controllers
QA/QC	Quality Assurance/Quality Control
RBC	Rotating Biological Contactor
RBC	Rotating biological contactor
RI	Rapid infiltration
RO	Reverse osmosis
SBR	Sequencing batch reactor
SCADA	Supervisory control and data acquisition
SO_2	Sulfur dioxide
SR	Slow rate
SRT	Solids retention time
SS	Suspended solids
SAR	Sodium Adsorption Ratio

SBR	Sequencing Batch Reactor
SRT	Sludge Retention Time
TBOD	Total Biochemical Oxygen Demand at 5 days and 20 oC
TOC	Total Organic Carbon
TP	Total Phosphorus
TSS	Total Suspended Solids
TTHM	Total Trihalomethanes
TCC	Total construction costs
TIC	Total indirect cost
TOC	Total organic carbon
TOD	Total oxygen demand
TSS	Total dissolved solids
UF	Ultra filtration
UNEP	United Nations Environment Programme
USAID	United States Agency for International Development
USEPA	United States Environmental Protection Agency
UV	Ultraviolet
UC	Uniformity Coefficient
UV	Ultraviolet
WaTER	Water Treatment Estimation Routine
WEF	Water Environment Federation
WWTP	Waste-water treatment plant

[a]ratio of apparent volumetric mass transfer coefficient in wastewater to clean water

* MLVSS (X) $= \dfrac{Y_t S_i (S_i - S_e)}{(\theta_c^{-1} + k_d)\theta}$

* Aeration basin volume (V):

$$V = \frac{S_i (S_i - S_e) Q}{k'_e S_e X}$$

or $$V = \frac{Y_t S_i (S_i - S_e) Q}{(\theta_c^{-1} + k_d) X}$$

* Other relationships:

$$X_w = VX\left[\frac{Y_t S_i (S_i - S_e)}{\theta X} - kd\right]$$

$$\frac{VX}{\theta_c} = VX\mu .$$

- Activated sludge design expressions: McKinney's model

  * Effluent BOD (S_e) $= \dfrac{S_i}{k_m \theta + 1} = \dfrac{(S_i - S_e)}{k_m \theta}$

  * MLVSS (X) $= \dfrac{Y_t k_m S_e}{(\theta_c^{-1} + k_d)\theta} = \dfrac{Y_t (S_i - S_e)}{(\theta_c^{-1} + k_d)\theta}$

  * Aeration basin volume (V):

$$V = \frac{(S_i - S_e) Q}{k_m S_e}$$

or $$V = \frac{Y_t (S_i - S_e) Q}{(\theta_c^{-1} + k_d) X}$$

  * Other relationships:

$$X_w = VX\left[\frac{Y_t k_m S_e}{X} - k_d\right]$$

and $$\frac{VX}{\theta_c} = VX\mu$$

$$k_m = k_e X.$$

- Activated sludge design expressions: Lawrence and Mc Carty model

  * Effluent BOD (S_e) $= \dfrac{K_s (1 + k_d \theta_c)}{\theta_c (Y_t K - k_d) - 1}$

$$= \frac{K_s\left(\frac{1}{\theta_c}\frac{1}{N}k_d\right)}{Y_t K - \left(\theta_c^{-1} + k_d\right)}$$

* MLVSS (X) $= \frac{\theta_c Y_t (S_i - S_e)}{\theta(1 + k_d \theta_c)} = \frac{Y_t (S_i - S_e)}{\left(\theta_c^{-1} + k_d\right)\theta}$

* Aeration basin volume (V) :

$$V = \frac{\theta_c Y_t (S_i - S_e) Q}{(1 + k_d \theta_c) X}$$

or

$$V = \frac{Y_t (S_i - S_e) Q}{\left(\theta_c^{-1} + k_d\right) X}$$

* Other relationships :

$$X_w = VX\mu.$$

- Activated sludge design expressions: Gaudy's model

* Effluent BOD (X_e) $= \frac{K_s(\mu + k_d)}{\mu_m - (\mu + k_d)}$

$$= \frac{K_s\left[D\left(1 + R - R\frac{Xr}{X}\right) + k_d\right]}{\mu_m - \left[D\left(1 + R - R\frac{Xr}{X}\right) + k_d\right]}$$

* MLVSS (X) $= \frac{Y_t[S_i - (1+R)S_e] + RX_r}{1 + R + k_d \theta_c}$

$$= \frac{Y_t[S_i - (1+R)S_e]}{(\mu + k_d)\theta}$$

* Aeration basin volume (V):

$$V = \frac{QY_t[S_i - (1+R)S_e] + RX_r Q}{k_d X} - \frac{(1+R)Q}{k_d}$$

$$V = \frac{Y_t[S_i - (1+R)S_e] Q}{(\mu + k_d) X}$$

* Other relationships:

$$D = q^{-1}$$

$$X_w = VXm$$

$$R = \frac{Q_r}{Q}$$

$$\mu = D\left[1 + R - R\left(\frac{X_r}{X}\right)\right]$$

- Activated sludge design expressions: Kincanon and Stover's model

  * Effluent BOD (S_e) $= S_i - \dfrac{U'_m S_i}{K_b - \dfrac{S_i}{X\theta}}$

  * MLVSS (X) $= \dfrac{Y_t (S_i - S_e)}{(\theta_c^{-1} + k_d)\theta}$

  * Aeration basin volume (V):

$$V = \frac{Q S_i / X}{\dfrac{U'_m S_i}{S_i - S_e} - K_b}$$

  * Other relationships:

$$R = \frac{1 - \dfrac{Y_t U'_m S_i}{X(K_b + S_i / X)} - k_d}{\dfrac{X_r}{X} - 1}$$

$$\theta_c = \frac{1}{\dfrac{Y_t (S_i - S_e)}{X\theta} - k_d}$$

1.8 Biological Systems

Table 35 : Electron acceptors.

Environment	*Electron Acceptor*	*Process*
Aerobic	O_2	Aerobic respiration
Anaerobic	NO_3^-	Denitrification
	SO_4^{-2}	Sulphate reduction
	CO_2	Methanogenesis

6.2.2 Reaction kinetics

$$\frac{dX_v/dt}{X_v} = \frac{\mu_m S}{K_s + S} = \mu$$

where X_V = Viable or active biomass concentration

μ_m = Maximum specific growth rate constant

K_s = Saturation constant [at which m is equal to $\mu_m/2$]

μ = Specific growth rate

$$\frac{dX_v}{d_t} = -Y\frac{dS}{dt}$$

$$r = \frac{\mu_m S X_v}{YK_s} = -k_b\, S\, X_v$$

where r = Rate of consumption of a substrate

k_b = Apparent first order biodegradation rate constant [= m_m/YK_s]

- First order approximation

$$\mu = \frac{\mu_m S}{K_s}$$

- Minimum substrate concentrations

There is a minimum primary substrate concentration (S_{min}) below which cells have a negative growth rate under steady-state conditions.

Table 36 : Typical values of m_m, K_s and S_{min}

Reaction	*Electron Donor*			*Electron Acceptor*	*Parameters for Electron Donor* m_m (d^{-1})	K_s *(mg/L)*	S_{min} *(mg/L)*
Methanogenesis	BOD_L	COD	TOC	CO_2	0.14	20-400	3.3-6.7
Sulphur reduction	BOD_L	COD	TOC	SO_4^{-2}	2.0	10	1.1
Denitrification	BOD_L	COD	TOC	NO_3^-	1.9	13.5	0.36
Aerobic	BOD_L	COD	TOC	O_2	9	1-100	0.02-2

Primary substrate removals (BOD_L, COD, TOC) to levels less than 1 mg/L (BOD_L) would appear to be impossible for typical biological systems. It is possible to remove organic compounds to below S_{min} levels, if the target compound is not the primary substrate, rather it must be a secondary substrate. The secondary substrate can be removed by secondary utilization that is driven by the intrinsic kinetics of the substrate and the amount of biomass. The amount of biomass generated is determined by the primary substrate utilization. For very low secondary substrate concentrations where $S << K_s$, then

$$r = -k_b X_v S$$

X_v is determined by primary substrate utilization, and is independent of S, therefore, the term $k_b\, X_v$ can be assumed to be constant:

$r = -kS$

[Although X_V is independent of S, it changes as the primary substrate changes, and k_b must be determined experimentally for each waste].

where k_e = Eckenfelders first order substrate removal constant

k'_e = Eckenfelders second order substrate removal constant [= $k_e \times S_i$]

K = Maximum substrate utilization rate [Lawrence and Mc Carty]

m_{max} = Maximum specific growth rate [Gaudy]

K_s = Saturation constant [Lawrence and Mc Carty]

Y_t = True yield coefficient [All models; $Y = Y_t\ (1 + k_d\ q_c)^{-1}$]

k_d = Decay coefficient [Maintenance energy; all models]

U_{max} = Maximum substrate utilization rate [Kincannon and Stover]

k_b = Substrate loading at which the rate of substrate utilization is one half the maximum rate [Kincannon and Stover]

k_m = Mc Kinney's constant [= ke. MLVSS (X)].

Table 37 : Design expression for aerated stabilization basin.

Model (s)	*Effluent BOD (S_e)*	*Rate Constant*
Eckenfelder	$S_e = \dfrac{S_i}{1+K_e\theta}$	$K_e = \dfrac{S_i - S_e}{S_e\theta}$
Exponential	$S_e = S_i \exp[-K'_e\theta]$	$K'_e = \dfrac{\ln(S_i/S_e)}{\theta}$
Lawerence and McCarty	$S_e = \dfrac{K_s(1+bt)}{\theta\left[\dfrac{yK}{N}b\right]-1}$	$K = \dfrac{(K_s+S_e)(b\theta+1)}{S_e y\theta}$

1.9 Secondary Clarifier Expression

- Pflanz : $$X_e = K\frac{Q_o}{A}\text{MLSS}$$
- Modified Pflanz : $$X_e = K_1 + K_2\frac{(Q_o+Q_r)}{A}\text{MLSS}$$
- EPA [Agnew] : $$X_e = 18.2 + 0.0136\frac{(Q_o)}{A} - 0.003\ \text{MLSS}$$

- Ghobrial $$X_e = K_N\left[\frac{Q_o + Q_r}{A}\right]\text{MLSS}\left[\frac{V_s/V}{(V_s/V)\text{CR}}\right]$$

- Chapman :

$$X_e = -180.6 + 4.03\,\text{MLSS} + 133.24\left[\frac{Q_o + Q_r}{A}\right] + \text{SWD}\left[90.16 - 65.24\left(\frac{Q_o + Q_r}{A}\right)\right]$$

- Stephenson and Thompson: $X_e = K_{env}(SF - SF_{min})^2 + X_{avg}$

where X_e = Effluent suspended solids
Q_o = Influent flow rate
Q_r = Recycle flow rate
A = Secondary clarifier area
MLSS = Aeration tank mixed liquor suspended solid concentration
SWD = Side water depth
V_s = Volume of sludge (sludge blanket volume)
V = Secondary clarifier volume
CR = Critical value of parameter
K = Slope of Pflanz equation
K_1 = y-intercept for modified Pflanz equation
K_2 = Slope for modified Pflanz equation
K_N = Constant
Q_o/A = Overflow rate
K_{env} = Envelope constant
X_{avg} = Average effluent suspended solids
SF = Solids flux rate at X_e
SF_{min} = Minimum solids flux rate [at minimum $X_e = X_{min.}$]
$\frac{Q_o\,\text{MLSS}}{A}$ = Solids flux without recycle flow
$\frac{(Q_o + Q_r)\,\text{MLSS}}{A}$ = Solids flux with recycle flow.

Stephenson and Thompson model can be written in terms of solids flux at optimum with $X_e = X_{min}$, *i.e.,*:

$$X_e = K_{env}\left[SF - SF_{opt}\right]^2 + X_{min}$$

and the plot of X_e (TSS) Versus SF is:

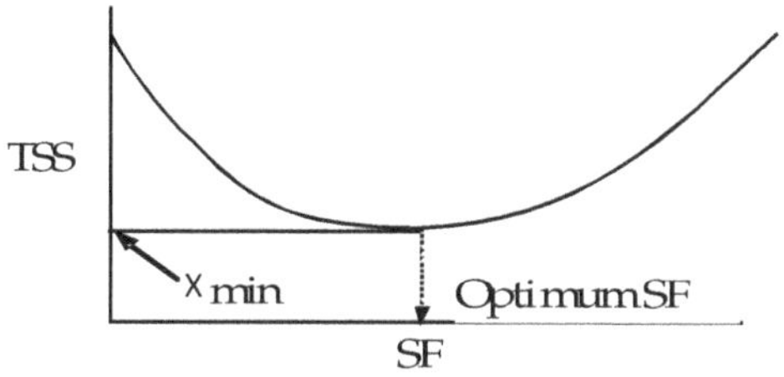

1.10 Parshall flume [Combination of Venturi-Weir principle]

The schematics of Parshall Flume is shown in Figure 26, and the dimensions are presented in Table 42.

Open channel liquid flow, when passed through a prescribed constriction in the channel, will produce a hydraulic head at a specific point upstream of the constriction. The flow is proportional to the head, raised to the 1.56th power times a throat coefficient.

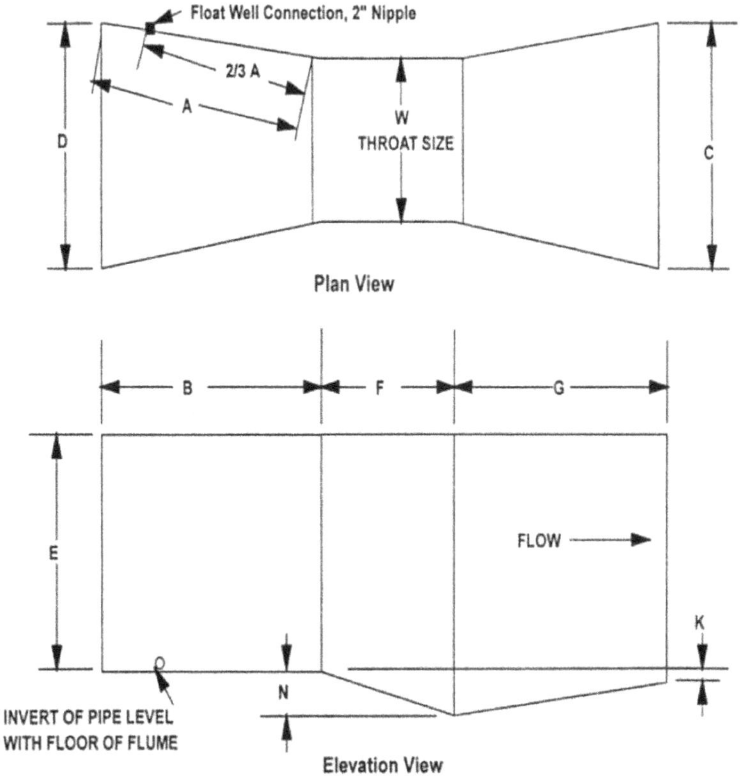

dia. (in.)

Figure 26 : Parshall flume Design.

1.11 Minimum Slopes for Sewers

Table 38 : Minimum Slopes for Sewers

Sewer pipe dia. (in.)	*6*	*8*	*10*	*12*	*15*	*18*	*21*	*24*	*greater than 24*
Min. slope	0.008	0.004	0.0028	0.0022	0.0015	0.0012	0.001	0.0008	0.0008

1.12 Design criteria for PSTs

Table 39 : Design criteria for PSTs.

Parameter	*Range*	*Typical*
Detention time (h)	1.5-2.5	2.0
Overflow rate ($m^3/m^2.d$)		
- Average flow	32-48	
- Peak flow	80-120	100
Weir loading ($m^3/m.d$)	125-500	250
Rectangular system		
- Depth (m)	3-5	3.6
- Length (m)	15-90	25-40
- Width (m)	3-24	6-10
[Must divide into bays of not greater than 6.0 m wide for mechanical sludge removal equipment]		
Sludge scraper speed (m/min.)	0.6-1.2	1.0
Circular		
- Depth (m)	3-5	4.5
- Diameter (m)	3.6-60	12-45
- Bottom slope (mm/m)	60-160	80
Sludge scraper speed (rpm)	0.02-0.05	0.03

1.13 Design parameter for aerated grit chamber

Table 40 : Design parameter for aerated grit chamber.

Parameter	*Range*	*Typical*
Depth (m)	2.5	-
Length (m)	7.5-20	-
Width (m)	2.5-7.0	-
Width-depth ratio	1:1-5:1	2:1
Detention period at peak flow (min.)	2-5	3
Air supply (m^3/min. m of length)	0.15-0.45	0.3
Grit quantity ($m^3/10^3\ m^3$)	0.004-0.200	0.015

1.14 Chlorine dosages for various wastewaters

Table 41 : Chlorine dosages for various wastewaters.

Wastewater Type	*Cl_2 dosage (mg/L) to yield 0.2 mg/L free residual after 15 min. contact time*
Raw	
- Fresh to stale	6-12
- Septic	12-25
- Settled	
- Fresh to stale	5-10
- Septic	12-40
- Effluent after chemical precipitation	3-6
Trickling filter	
- Normal	3-5
- Poor	5-10
Activated sludge	
- Normal	2-4
- Poor	3-8
Intermittent sand filter	
- Normal	1-3
- Poor	3-5

1.15 Aeration Design System

- ***Carbonaceous matter :*** One pound of BOD combined with 0.5 lb of oxygen will produce 0.65 lb of new bacterial cells plus by-products [CO_2, H_2O, NH_3, and energy]. After all the food is used up, the new cells with feed on themselves (endogenous respiration), and will utilize another 0.75 lb of O_2 to produce 0.12 lb of biologically inert residue, and more by-products of oxidation. The schematics of biological oxidation of organic matter is shown in Figure 27.

  In the entire process, 0.15 lb of ammonia are produced which are biologically oxidized to nitrates, consuming 0.30 lb more of O_2. The carbonaceous demand of BOD is (0.5 + 0.75) lb = 1.25 lb of oxygen, and if nitrogenous demand is added, a pound of BOD requires 1.55 lb of oxygen to carryout the process to completion [BOD is the ultimate BOD]. Complete biodegradation required 20 to 50 days for most organics and more for a difficult to biodegrade waste. In activated sludge process, sludge age is less than 20 days, so a value of 1.1 lb of O_2 per lb of BOD is commonly used for the design of aeration systems for biological degradation that have SRTs of less than 10 days.

- ***Ammonia :*** Ammonia consumes 4.6 lb of O_2 per lb of ammonia for bio-oxidation to nitrates.

- ***Sulphide :*** Sulphides consume 2 lb of O_2 per lb of sulphide as sulphur, and occurs in about 20 minutes [chemical oxidation process].

  These values must be added to the oxygen requirements for BOD oxidation. Other reduced compounds may be in the wastewater that require oxygen [iron, etc.].

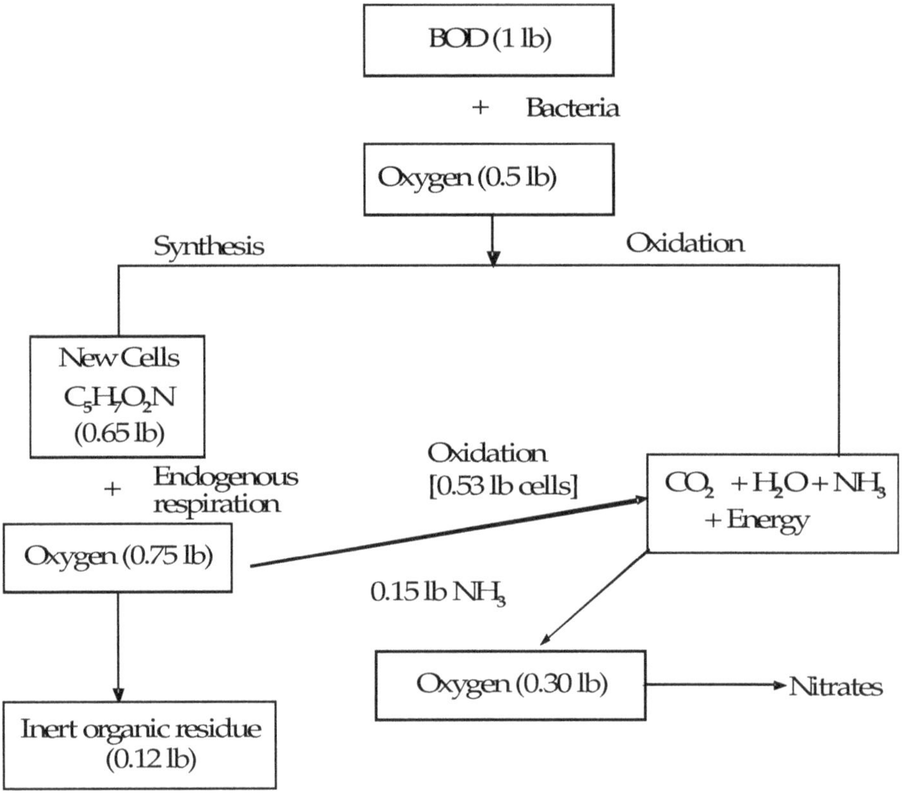

Figure 27 : Biological oxidation of organic matter

Example 3.46

STP Sludge Volume Change on Dewatering

Determine the volume of sludge to be stored.

Assume the following data:

Surplus STP sludge	:	75 m^3/d
Concentration of sludge	:	8000 mg/L
Desired concentration of dewatered sludge	:	4%

Table 42 : Dimensions of Parshall Flumes

W Feet	Flume Flow		A	2/3 A	B	C	D	E	F	G	K	N
	Min.	Max.										
0.25	0.02	1.23	1´ 6 3/8´´	1´ 0 1/4´´	1´ 6´´	0´ 7´´	0´ 10 3/8´´	2´ 0´´	0´ 6´´	1´ 0´´	0´ 1´´	0´ 2 1/4´´
0.5	0.03	2.52	2´ 0 7/16´´	1´ 4 5/16´´	2´ 0´´	1´ 3 1/2´	1´´ 3 5/8´´	2´ 0´´	1´ 0´´	2´ 0´´	0´ 3´´	0´ 4 1/2´´
0.75	0.06	5.75	2´ 10 5/8´´	1´ 11 5/8´´	2´ 10´´	1´ 3´´	1´ 10 5/8´´	2´ 6´´	1´ 0´´	2´ 6´´	0´ 3´´	0´ 4 1/2´´
1	0.07	10.41	4´ 6´´	3´ 0´´	4´ 4 7/8´´	2´ 0´´	2´ 9 1/4´´	3´ 0´´	2´´ 0´´	3´ 0´´	0´ 3´´	0´ 9´´
1.5	0.10	15.90	4´ 9´´	3´ 2´´	4´ 7 7/8´´	2´ 6´´	3´ 4 3/8´´	3´ 0´´	2´ 0´´	3´ 0´´	0´ 3´´	0´ 9´´
2	0.27	21.39	5´ 0´´	3´ 4´´	4´ 10 7/8´´	3´ 0´´	3´ 11 1/2´´	3´ 0´´	2´ 0´´	3´ 0´´	0´ 3´´	0´ 9´´
3	0.39	32.57	5´ 6´´	3´ 8´´	5´ 4 3/4´´	4´ 0´´	5´ 1 7/8´´	3´ 0´´	2´ 0´´	3´ 0´´	0´ 3´´	0´ 9´´
4	0.84	43.88	6´ 0´´	4´ 0´´	5´ 10 5/8´´	5´ 0´´	6´ 4 1/4´´	3´ 0´´	2´ 0´´	3´ 0´´	0´ 3´´	0´ 9´´
5	1.03	55.32	6´ 6´´	4´ 4´´	6´ 4 1/2´´	6´ 0´´	7´ 6 5/8´´	3´ 0´´	2´ 0´´	3´ 0´´	0´ 3´´	0´ 9´´
6	1.68	66.89	7´ 0´´	4´ 8´´	6´ 10 3/4´´	7´ 0´´	8´ 9´´	3´ 0´´	2´ 0´´	3´ 0´´	0´ 3´´	0´ 9´´
7	1.94	78.46	7´ 6´´	5´ 0´´	7´ 4 1/4´´	8´ 0´´	9´ 11 3/8´´	3´ 0´´	2´ 0´´	3´ 0´´	0´ 3´´	0´ 9´´
8	2.26	90.16	8´ 0´´	5´ 4´´	7´ 10 1/8´´	9´ 0´´	11´ 1 3/4´´	3´ 0´´	2´ 0´´	3´ 0´´	0´ 3´´	0´.9´´

- *Flows are MGD, extreme capacities: i.e. the actual measuring range will depend on the type of instrument used and will lie within these limits*
- *Dimensions are feet and inches excepts as noted.*
- *All dimensions are plus or minus 1/8 inch except throat width which is plus or minus 1/16 inch.*

Solution

1. Solids Content

   1 litre of water has a mass of 1 kg (10^6 mg)

$$\text{Solids content } = \frac{8000 \text{ mg}}{10^6 \text{ mg}} = 8 \times 10^{-3}\,(0.8\%)$$

2. Required sludge volume reduction

$$\text{Solids content (S) } = \frac{100 \text{ m}_s}{\text{m}_t}(\%)$$

$$= \frac{100 \text{ m}_s}{\rho_s V}$$

If $\rho_s = 1000 \text{kg/m}^3$

$$S = \frac{100 \text{ m}_s}{1000 \text{ V}} \text{ or } \frac{\text{m}_s}{10 \text{ V}}(\%)$$

where m_s is the mass of solids, and m_t is total mass of a volume, V (m^3) of sludge.

$$(S)_1 = \left[\frac{\text{m}_s}{10 \text{ V}}\right]_1$$

$$(S)_2 = \left[\frac{\text{m}_s}{10 \text{ V}}\right]_2$$

Assume no loss of solids on dewatering:

$$(m_s)_1 = (m_s)_2$$

$$10 \text{ V}_1 \text{ S}_1 = 10 \text{ V}_2 \text{ S}_2$$

$$V_2 = \frac{V_1 S_1}{S_2}$$

$$= \frac{(75)(0.8\%)}{4\%} = 1.5 \text{ m}^3 \text{ (A reduction of 75\% in volume)}$$

Example 3.47

Sludge Dewatering : Centrifuge

1. Determine the feed rate for centrifuge No. 2 based on the performance of centrifuge No. 1 (solid bowl machine). Centrifuge No. 1 was found to operate satisfactory at a flow rate of 0.5 m^3/h. Assume the following data:

Table 43 : Centrifuge data.

Parameters	*Machine No. 1*	*Machine No. 2*
Bowl diameter (D, cm)	20	40
Pool depth (P, cm)	2	4
Bowl speed (W_s, rpm)	4000	32000
Bowl length (L, cm)	30	72

1.1 Required design by scale-up from prototypes

- Sigma (Σ) concept has been widely used in scaling up (or down) between two similar centrifuge machines
  - * Stokes low is given as

$$V = \frac{g(\rho_s - \rho)d^2}{18\mu}$$

  - * Centrifuges:
    - A particle (d) in a centrifuge attains this terminal velocity immediately
    - Particle settling is slow enough to be in laminar flow regime
    - Centrifugal acceleration is w^2r

$$\text{Therefore, } V = \frac{r\omega^2(\rho_s - \rho)d^2}{18\mu}$$

    - Radial distance (dr) travelled by the particle during time (dθ) is :

$$dr = Vd\theta = \frac{r\omega^2(\rho_s - \rho)d^2}{18\mu}$$

Integrating from r_1 (centerline to pool surface) to r_2 (centerline to bowl) gives:

$$\int_{r1}^{r2} \frac{dr}{r} = \frac{\omega^2(\rho_s - \rho)}{18\mu} d^2 \int_0^\theta d\theta$$

or $$\ln\frac{r_2}{r_1} = \frac{\omega^2(\rho_s - \rho)d^2\theta}{18\mu}$$

$$\text{Detention time } (\theta) = \frac{18\mu}{\omega^2(\rho_s - \rho)d^2}\ln\left(\frac{r_2}{r_1}\right)$$

For the particle to be removed, the detention time (q) is:

$$\theta = \frac{\cancel{V}}{Q} = \frac{\text{Volume}\ (\cancel{V})}{\text{Flow rate}\ (Q)}$$

Therefore, $$\frac{\cancel{V}}{Q} = \left[\frac{g}{\omega^2}\ \text{In}\left(\frac{r_2}{r_1}\right)\left(\frac{18\mu}{g(\rho_s - \rho)d^2}\right)\right]$$

or $$Q = \left[\frac{\omega^2}{g\ \text{In}\ (r_2/r_1)}\right]\left[\frac{g(\rho_s - \rho)d^2}{18\mu}\right]$$

= (Machine variables)(Sludge variables)

First term = Machine variables = Σ

Therefore, $Q = \Sigma$

For two geometrically similar machines:

$$\frac{Q_1}{Q_2} = \frac{\Sigma 1}{\Sigma 2} = \frac{(\text{Sigma})_1}{(\text{Sigma})_2}$$

- **β concept :** The machine is said to be solids limiting when its design precludes its ability to discharge a sufficient quantity of solids. The solids throughput is proportional to the ratio of the volumes occupied by solids
  - * Volume occupied by the solids V_s is:

    Vs = P D y L, and the fraction of pool volume occupied by solids is:

    $$\frac{V_s}{V} = \frac{\pi\, D\, L\, y}{\pi\, D\, L\, \cancel{Z}} = \frac{y}{\cancel{Z}}$$

    where y = Depth of solids

    D = Bowl diameter

    L = Bowl length
  - * Assuming no slippage, the travel time (T) for a particle in the bowl is:

    $$T = \frac{L}{\Delta w\, S\, N}$$

    where Δw = Bowl/conveyor differential speed

    S = Pitch of blades

    N = Number of leads
  - * Solids flow into a machine can be approximated as Q_s/ρ_c

    where: Q_s = Solids throughput (PPH)

ρ_c = Specific weight of wet cake

Q_s/γ_c = ft^3/h

* Volume occupied by the solids is:

$$V_s = \left(\frac{Q_s}{\gamma_c}\right) T$$

$$V_s = \frac{Q_s}{\gamma_c} \frac{L}{\Delta w\, S\, N}$$

* Inside wall area is: A = pDL or L = A/pD

Therefore, $\frac{V_s}{A} = \frac{Q_s}{\gamma_c} \frac{L}{\Delta w\, S\, N\, \pi\, D} = y = \text{Cake depth}$

$$\frac{V_s}{\cancel{V}} = \frac{y}{\cancel{Z}} = \left[\frac{Q_s/_c}{\Delta w\ SN\, \pi\, D}\right]\left[\frac{1}{\cancel{Z}}\right]$$

$$= \frac{(\text{Process variables } Q_s \text{ and } _c)}{(\text{Machine variables})}$$

Machine variable is defined as $\beta = \Delta w\ S\ N\ \pi\ D\ P\ (m^3/h)$

where S = Scroll pitch

$\Delta w = \omega_s - \omega_b$

ω_s = Bowl speed

ω_b = Scroll speed

P = Pool depth

D = Bowl diameter

* Scale up of solids throughput between any two solid bowl centrifuge is:

$$\frac{[Q_s/_c]_1}{\beta_1} = \frac{[Q_s/_c]_2}{\beta_2}$$

If specific weight $(\rho_c)_1 = (\rho_c)_2$, then:

$$\left[\frac{Q_s}{\beta}\right]_1 = \left[\frac{Q_s}{\beta}\right]_2$$

2. Estimate the capacity of centrifuge No. 2 based on the performance of centrifuge No. 1 which is operating satisfactory at 60 kg solids/h. Assume the following additional data (section 1):

Table 44 : Performance Data.

Parameters	*Machine-1*	*Machine-2*
Conveyor speed (rpm)	3950	3150
Pitch (cm)	4	8
No. of leads	1	1

- Required volume of sludge in pool

$$V = 2\pi\left[\frac{r_1 + r_2}{2}\right](r_2 - r_1)L$$

$$V_T = 2(3.14)\left(\frac{8+10}{2}\right)(10-8)(30) = 3400 \text{ cm}^3$$

$$V_2 = 2(3.14)\left(\frac{16+20}{2}\right)(20-16)(72) = 32600 \text{ cm}^3;$$

$$\omega = (\text{rpm})(\text{min}/60 \text{ s})(2\pi \text{ radians/revolution})$$

$$\omega_1 = (2 \times 3.14)(4000/60) = 420 \text{ rad/s}$$

$$\omega_2 = (2 \times 3.14)(3200/60) = 335 \text{ rad/s}$$

$$\Sigma = \frac{\omega^2}{g \text{ In}(r_2/r_1)}.V$$

$$\Sigma_1 = \frac{(420)^2}{980}\frac{3400}{\ln(10/8)} = 2.74 \times 10^6$$

$$Q_2 = \frac{\Sigma_2}{\Sigma_1}Q_1$$

$$= \frac{16.8 \times 10^6}{2.8 \times 10^6}(0.5) = 3.2 \text{ m}^3/\text{h}.$$

2.1 Required capacity of centrifuge No. 2

$$Q_2 = \frac{\beta_2}{\beta_1}Q_1$$

$$= \frac{[\Delta w \text{ S N } \pi \text{ D Z}]_2}{[\Delta w \text{ S N } \pi \text{ D Z}]_1}Q_1$$

$$= \frac{[50 \times 8 \times 1 \times 3.14 \times 40 \times 4]_2}{[50 \times 4 \times 1 \times 3.14 \times 20 \times 2]_1}(60)$$

$$= 480 \text{ kg/h}$$

Note :

1. Centrifugal acceleration

$$\text{CA}(\text{X gravity}) = \left(\frac{r_1 + r_2}{2}\right)\omega^2 g$$

where r_1 = Radial distance from the center of the centrifuge to the top of the sludge

r_2 = Radial distance to the bottom of the tube or to inner wall of the bowl

ω = Radial velocity (rad/s)

g = Gravitational constant.

2. Percent recovery (R)

$$R = \frac{C_o - C_e}{C_o} \times 100$$

where C_e = Solids concentration of the centrate

C_o = Feed solids concentration.

3. Filter press scale up equation proposed by Jones

$$T = \frac{0.321 r \mu d^2 (C_o - C_k)^2}{P C_o (100 - C_o)}$$

where T = Length of time to complete pressing (h)

d = Distance between filter plates in the press (in.)

C_o = Feed solids concentration (%)

C_k = Cake solids concentration (%)

$r \times 10^7$ = Specific resistance (s^2/g)

P = Filtration pressure (psi)

μ = Filtration viscosity (cP).

4. Determine the filter yield. Assume the following data from a filter leaf test:

Feed solids concentration = 4%

Cake solids concentration = 20%

Area of filter test = 0.1 ft^2

Total cake dry weight = 0.028 lb

Cycle: 30 seconds submerged

60 seconds drying

30 seconds off the filter

2 minute/cycle

Filtration solids concentration = 500 mg/L

$$\text{Filter yield} = \frac{0.028\ \text{lb}/0.1\ \text{ft}^2}{2\ \text{min} \times 1\ \text{hr}/60\ \text{min}} = 8.4\ \text{lb}/\text{ft}^2.\text{h}$$

5. Determine the solids recovery in any dewatering device from a solids balance

Feed flow rate = Q_o

Feed solids concentration = C_o

Filter cake flow = Q_f

Filter solids concentration = C_f

- Liquid balance is

$$Q_o = Q_f + Q_k$$

- Solids balance

$$Q_oC_o = Q_fC_f + Q_kC_k$$

$$= (Q_o - Q_k)C_f + Q_kC_k$$

or $$Q_o(C_o - C_f) = Q_k(C_k - C_f)$$

$$Q_k = \frac{Q_o(C_o - C_f)}{(C_k - C_f)}$$

- Percent solids recovery (dry), %R

$$\%R = \frac{\text{Mass of dry solids as cake}}{\text{Mass of dry feed solids}} \times 100$$

$$\%R = \frac{C_kQ_k}{C_oQ_o} \times 100$$

Therefore, $$\%R = \frac{C_k\left[\dfrac{Q_o(C_o - C_f)}{C_k - C_f}\right]}{Q_oC_o} \times 100$$

or $$\%R = \frac{C_k(C_o - C_f)}{C_o(C_k - C_f)} \times 100$$

Example 3.48

Centrifuge

It is desired to study the feasibility of using a centrifuge to separate solid particles of specific gravity (1.60) from water with specific gravity of 1.0 and a viscosity of 1.20 centipoise. The pilot plant has a basket that has an inside diameter of 800 mm, is 400 mm in height, and operates at 1000 rpm. The thickness of the liquid layer in the centrifuge (radial distance from the inside wall to the air-water surface) is to be 100 mm. Determine the time required for a particle 20 mm in diameter to settle through the 100 mm of water.

Solution

1. Required terminal velocity

   The terminal velocity of a particle is a constant value of velocity reached when all forces (gravity, drag, buoyancy etc.) acting on the particle are balanced. The sum of all the forces is then equal to zero (no acceleration):

$$K = dp\left[\frac{g(\rho_p - \rho)\rho}{\mu^2}\right]^{1/3}$$

[Centrifuge, replace g of by rw^2]

where K = Dimensionless constant

dp = Particle diameter

$g = rw^2$

r = Radius or distance from centrifuge axis

w = Angular velocity

d_p = Particle diameter

ρ_p = Particle density

ρ = Water density

μ = Water viscosity

$K < 3.3$; Stokes law range

$3.3 < K < 43.6$; Intermediate law range

$43.6 < K < 2360$; Newton's law range

- Inside radius of the centrifuge (outside surface of the water) is:

$$r_2 = \frac{(800/2)}{1000} = 0.40 \text{ m}$$

- Distance from the axis to the air-water interface is:

  [Thickness of the water is 100 mm or 0.10 m, the radius at the inside surface of the water]

  $r_1 = (0.40 - 0.10) = 0.30$ m

- $\rho_p = (1.60)(998) = 1597$ kg/m^3

  $\rho = (1.0)(998) = 998$ kg/m^3

- $\mu = (1.2)(0.001) = 1.2 \times 10^{-3}$ kg/m.s

  $dp = 20\ (10^{-6}) = 2.0 \times 10^{-5}$ m

  $w = 2\ \pi f = (2\pi)\ (1000)/60 = 104.7$ rad/s

$$k = \left(2.0 \times 10^{-5}\right)\left[\frac{0.40(104.7)^2(998)(1597-998)}{\left(1.2 \times 10^{-3}\right)^2}\right]^{1/3}$$

at $r = 0.40$ m

$= 2.44 < 3.3$, at $r_2 = 0.40$ m [Stoke's low applies]

$= 2.13 < 3.3$ at $r_1 = 0.30$ mm [Stoke's laws applies].

- Terminal velocity at Stoke's law values :

$$v = \frac{r\omega^2 dp^2 \rho_p}{18\mu}$$

$$v = \frac{r(104.7)^2\left(2.0 \times 10^{-5}\right)^2(1597)}{18\left(1.2 \times 10^{-3}\right)}$$

$= 0.324$ r.

2. Required time for a particle 20.0 mm in diameter to settle through the 100 mm water

$$v = \frac{dr}{dt}$$

$$dt = \frac{dr}{v} = \frac{dr}{0.324\ r}$$

$$= \frac{1}{0.324} \ln\left[\frac{0.40}{0.30}\right]$$

$= 89$s

[It takes 0.89 s for the particle to settle from the top (inside surface) of the water layer to the bottom (outside surface) in the centrifuge].

Example 3.49

Polyelectrolyte Requirements for Centrifuge Dewatering

Centrifuge performance data in respect of polymer addition is given by the following relationship:

$$R = \frac{C_1(C_2 + P)^m}{Q^n}$$

where R is the percent recovery, P is the polymer dosage (lb/ton of day solids feed), Q is the feed rate (gal/ft^2.min); C_1, C_2, m, and n constants

Calculate the polymer required and size of the centrifuge with respect to surface area for dewatering 5000 lb of sludge with 4% solids. The centrifuge will operate over 6 hours with 95% solids recovery use the following information:

C_1 : 48

C_2 : 0.47

m : 0.37

n : 0.52

Solution

1. Sludge flow

$$Q' = 5000\left(\frac{1}{0.04 \times 8.34}\right) = 15,000 \text{ gal/d}$$

2. Determine the sludge flow for 6 hours

$$Q' = 15000 \times \frac{24}{6} = 60,000 \text{ gal/d} = 61 \text{ gal/min}$$

3. Percent recovery equation

$$95 = \frac{48(0.47 + P)^{0.37}}{Q^{0.52}}$$

4. Centrifuge size (A)

$$A = \frac{62.5}{Q} \text{ ft}^2$$

5. Polymer dosage

$$P' = P\left(\frac{5000}{2000}\right)$$

6. The results can be plotted

6.1 Centrifuge area (A) versus hydraulic loading $\left(\frac{\text{gal}}{\text{ft}^2.\text{min}}\right)$

6.2 Polymer dosage (lb/d) versus hydraulic loading $\left(\frac{\text{gal}}{\text{ft}^2.\text{min}}\right)$

6.3 Capital and operating costs for each combination of required area polymer dosage should be found and the least cost situation determined.

Note :

1. The variables that influence dewatering process are solids concentration, sludge and filtrate viscosity, sludge compressibility, Chemical composition, and the nature of the sludge particles (size, shape, water constant, etc.).
2. Polyelectrolytes are used to increase recovery. The flocculents both increase the structural strength of solids and flocculate fine particles. Because of increased removal of fine particles, chemical addition usually lowers the cake dryness.

Example 3.50

Performance of Centrifuge

1. Percent capture

The performance of a centrifuge is evaluated by percent capture, and is expressed as:

$$\text{Percent capture} = \left[1 - \frac{S_r(S_c - S_f)}{S_f(S_c - S_r)}\right](1000)$$

where S_r is solids concentration in reject wastewater (centrate : mg/L or %), S_c is the solids concentration in cake (mg/L or %), and S_f is the solids concentration in sludge feed (mg/L or %). The two parameters of concern are : (a) Sigma equation (settling of particles), (b) Beta equation (rejection of solids).

- Sigma equation (setting of particles)

  If the centrifuges are to have the same settling effects within the bowl, the relationship in expressed as:

  $$\frac{Q_1}{\sigma_1} = \frac{Q_2}{\sigma_2} \quad \text{(1, and 2 refers to two centrifuges)}$$

  where Q_1 and Q_2 are the liquid flow rates into two centrifuges (m^3/s), s_1 is the parameter related to the characteristics of the first centrifuge (does not depend an the sludge characteristics), and s_2 is the parameter related to the characteristics of the second centrifuge

  The Sigma value for a solid-bowl centrifuge is expressed as:

$$\sigma = \frac{\left[V_r^2 . V\right]}{\left[g.\ln\left(\frac{r_2}{r_1}\right)\right]}$$

where V_r is the rotational velocity of the bowl (rad/s), V is the liquid volume in the pool (m^3), g is the gravity constant (m/s^2), r_1 is radius from the centerline to the surface of the sludge (m), and r_2 is the radius from centerline to inside bowl wall (m).

- Beta equation (rejection of solids)

  The solids movement may be estimated through beta equation for two geometrically similar centrifuges:

$$\frac{W_1}{\beta_1} = \frac{W_2}{\beta_2}$$

where W_1 and W_2 is the solids loading rate to the first and second centrifuges, respectively (kg/hr), and b_1 and b_2 are the beta functions for the two centrifuges. The Beta function is the expressed as:

$$\beta = (V_w)(d_p)(n)\,(\pi)\,(Ƶ)\,(d)$$

where V_w is the difference in rotational velocity between the bowl and the carvey or (rad/s), d_p is the distance between blades or the scroll pitch (m), n is the number of leads, Ƶ is the depth of sludge in the bowl (m), and D is the bowl diameter (m).

2. Find the flow rate at which the new centrifuge will perform as well as the old one. Use the following data

   Sewage sludge solids : 3%

   Sludge rate : 10 m^3/d

Table 45 : Performance data.

Parameters	*Old Centrifuge*	*New Centrifuge*
Bowl length (L) (cm)	25	60
Bowl diameter (D) (cm)	15	30
Bowl spaced (n_r) (rpm)	5000	4200
Bowl depth (Ƶ) (cm)	2.5	5.0
Scroll pitch (d_p) (cm)	5	10
Number of leads	1	1
Convey or velocity (n_w) (rpm)	4950	4150

Due to air increase in sewage sludge that needs dewatering the old centrifuge needs to the scaled up to another similar and large one.

2.1 Determination of r, V_r, and V_w

$$r_2 = \frac{15}{2} = 7.5 \text{ cm (old)}; r_2 = \frac{30}{2} = 15 \text{ cm}$$

$$V_r = \left(\frac{2\pi}{60}\right)(5000) = 524 \text{ rad/s(old)}; \ V_r = \frac{2\pi}{60}(4200) = 440 \text{ rad/s(new)}$$

$$r_1 = (r_1 - \cancel{z}) = 7.5 - 2.5$$

$$= 5.0 \text{ cm (old)} \ ;$$

$$r_1 = (r_2 - \cancel{z}) = (15 - 5)$$

$$= 10 \text{ cm/s (new)}$$

V_w = (5000 – 4950) = 50 rad/s (old);

V_w = (4200 – 4150) = 50 rad/s (new).

2.2 Use of Sigma equation

$$\text{Volume}(V) = 2\pi\frac{(r_1 + r_2)}{2}(r_2 - r_1)L$$

$$\sigma = \frac{V_r^2 V}{g \ln\left(\frac{r_2}{r_1}\right)}$$

Using the values from item 2.1, the Sigma is calculated as:

σ_{old} = 1,693,984 ; σ_{new} = 11,468,055

where V(old)= 2454 cm^3 ; V(new) = 23,562 cm^3

g = 981 cm^2/s.

2.3 Determination of liquid flow rate from the new centrifuge

$$\frac{Q_2}{Q_1} = \frac{\sigma_2}{\sigma_1} \text{ or } Q_2 = Q_1\left(\frac{\sigma_2}{\sigma_1}\right)$$

$$= 10\frac{(11,468055)}{(1,673,984)} = 68 \text{ m}^3/\text{d}$$

[Solids settling will be of equal quantity for dewatering the sludge at a rate of 68 m^3/d for the new centrifuge (Geometrically similar system)].

2.4 Application of Beta equation

$$\beta = (V_w)(d_p)(n)(\pi)(\cancel{Z})(D)$$

Weight of solids (W) flow rate

$$= \left[\frac{10 \times 20{,}000 \text{ mg/L} \times 10^3}{10^6}\right] = 200 \text{ kg/d (old)}$$

Using the Beta equation

β_1 = 29,452 (old)

β_2 = 235,619 (new).

2.5 Required solids movement

$$W_2 = \left(\frac{W_1}{\beta_1}\right)\beta_2$$

$$= \left(\frac{200}{29{,}452}\right)(235{,}619) = 1600 \text{ kg/d}$$

or in flow units : $\left(\dfrac{1600 \text{ kg/d}}{0.02 \times 1000}\right) = 80 \text{ m}^3/\text{d}$ (This governs the system.)

Example 3.51

Dewatering of Wastewater Sludges through Centrifuges

1. Required centrifuge performance data

$$R = \frac{C_1 (C_2 + P)^m}{Q^n}$$

where R is the recovery (%), P is the polymer dosage (lb per tonne dry solids feed or kg/tonne), Q is the feed rate (gal/ft^2 .min or m^3/m^2.min), C_1 and C_2 are the constants, and m and n are the exponents

C_1 = 48; C_2 = 0.47; m = 0.37; n = 0.52.

2. Determine the polymer required and size of the centrifuge (surface area requirements) for dewatering of thickened sludge.

Assume the following data:

Amount of sludge to be handled	: 20,000 lb/d
Operating period for the centrifuge	: 8 hours/d
Solids recovery (R)	: 95%
Thickened sludge	: 4% solids concentration.

2.1 Required sludge flow

$$Q^* = 20{,}000 \times \frac{1}{0.04} \times \frac{1}{8.34}$$
$$= 60{,}000 \text{ gal/d}$$

Total sludge handled in 8 hours per day by the centrifuge:

$$= 60{,}000 \times \frac{24}{8} = 180{,}000 \text{ gal/d}$$

2.2 Required surface area

Sludge solids recovery (R):

$$95 = \frac{48(0.47 + P)^{0.37}}{Q^{0.52}} \qquad (1)$$

Centrifuge area is given by:

$$A = \frac{62.5}{Q} \text{ft}^2 \qquad (2)$$

Plot polymer dosage (P – lb/d) versus Q (gal/ft^2.min), equation-1 and centrifuge area (A) versus Q (gal/ft^2 min) equation-2 and where they intersect, is the optimal value for A and P.

Note :

1. If the feed rate to a centrifuge is increased, the retention time in the unit is decreased, and the recovery decreases (R).
2. Flow rates (Q) are usually limited to 0.5 to 2.0 gal/hp.min [3.65 to 14.6 m^3/kW.d] to obtain satisfactory solids recovery. Since the lower recovery results in the removal of only larger particles, a drier cake is produced. Increasing the feed solids concentration reduces the liquid overflow from the machine, resulting in an increased recovery of solids.
3. Polymers (poly-electrolytes) are used to increase recovery (polymers increase the structure strength of solids and flocculate fine particles).
4. Increased removal of the fine particles, and addition of chemicals usually lowers the cake dryness.

Example 3.52

Scroll-Type Centrifuge for Dewatering of Wastewater Treatment Plant Sludges

Estimate the theoretical capacity factor (ΣT) for the solid bowl-scroll type (or imperforate basket type) centrifuge. Assume the following data:

Quantity of combined primary and secondary sludges form STP : 43000 gal/d

Operation cycle [One hour for startup + work-up + shut down]	: 12 h/d
Influent sludge solids	: 1.5%
Pool depth	: 1.5 in.
Conveyor differential	: 10 rpm
Final sludge solids desired	: 25% at 2100×G

Pilot plant data:

90% recovery can be achieved at 3.5 gal/min with 3 lb/ton of polymer-X or 15 gal/min with 5 lb/tonne with polymer-X.

Solution

1. Required expressions for scroll-type (imperforate basket type centrifuge)
   - Extended Stoke's law with application of centrifugal force:

$$V_s = \frac{d^2\omega^2 r\Delta\rho}{18\mu} = V_g \frac{\omega^2 r}{g}$$

where V_s = Settling velocity of a particle (droplet) in a centrifugal field (cm/s)

ω = Angular velocity of particle in the settling zone (rad/s)

r = Radius at which settling velocity is determined (cm)

V_g = Settling velocity of a particle or droplet in a gravity field (cm/s)

$\Delta\rho$ = Density difference between true mass density of solid particle (liquid droplets) and that of the surrounding liquid medium (g/cm^3)

d = Solid particle (liquid droplet) diameter (cm)

g = Gravity constant (981 cm/s^2)

μ = Surrounding medium viscosity (poises or g/cm.s).

   - Theoretical capacity factor (ΣT) for the solid bowl scroll and imperforate basket type centrifuges
     * Scroll bowl (imperforate basket type) centrifuge:

$$\Sigma T = 2\pi l \frac{\omega^2}{g}\left[0.75\, r_2^2 + 0.25 r_1^2\right]$$

where l = Effective clarifying length of the centrifuge bowl (cm)

r_2 = Radius of the inside wall of the centrifuge bowl (cm)

r_1 = Radius of the liquid surface in the centrifuge bowl (cm)

ω = Angular velocity of the centrifuge bowl (rad/s)

ΣT = Theoretical capacity factor (cm^2)

* Disc-nozzle type centrifuge:

$$\Sigma T = \frac{2\pi n}{3}\frac{\omega^2}{g}\text{Cot}\theta\left(r_2^3 - r_1^3\right)$$

where θ = 1/2 the angle of the cone

n = Number of discs.

- Solid recover (%, SR):

  Quantity of solids removed from the carrier during centrifugation is defined as solid recovery, and is expressed as:

$$SR = \frac{S_s(S_i - S_e)}{S_i(S_s - S_c)}$$

where SR = Solids recovery (%)

S_s = Solids in discharged sludge (% by weight)

S_i = Influent solids (% by weight)

S_c = Solids in concentrate (% weight).

- Sludge characteristics:
  * Type (primary, digested, etc)
  * Flow (Q, gal/min)
  * Solid quantity (lb/h)
  * Solids concentration (%).
- Dewatering requirements:
  * Moisture content of cake (%)
  * Solids recovery over the centrifugation step (%).

2. Required theoretical capacity factor (ST)

$$\Sigma T = 2\pi l\frac{\omega^2}{g}\left[0.75\, r_2^2 + 0.25\, r_1^2\right]$$

$= 0.82 \times 10^7\ cm^2$ at (2100 d× G)

- ΣT at the design flow rate:

  Q = 43000 gal/d = 48 gal/min

  * At 3.0 lb/ton, ΣT is:

$$\Sigma T = \frac{48\text{ gal/min}}{3.5\text{ gal/min}} \times \left(0.82 \times 10^7\ cm^2\right)$$

$= 11.25 \times 10^7\ cm^2$

* At 5 lb/tonne, ΣT is:

$$\Sigma T = \frac{48 \text{ gal/min}}{15 \text{ gal/min}} \times \left(0.82 \times 10^7 \text{ cm}^2\right)$$

$$= 2.625 \times 10^7 \text{ cm}^2$$

In general, centrifuge capital costs are closely related to the ΣT value and, in this case, it is probably economical to use 2 lb/tonne of additional polymer to reduce the capital expenditure to one-fourth (and has to be evaluated).

3. Required centrate and concentrate characteristics

Influent = 43000 gal/d, Solids = 1.5 %

- lb-SS/d = (0.043 MGD)(15000 mg/L)(8.34)

  = 5380 lb/d

- Concentrate :

  Assume : 85% Suspended solid recovery

  30% Suspended solid concentration in cake

$$\text{Suspended solid in concentrate} = 5380 \text{ lb/d}\left[1 - \frac{S_e}{S_i}\right]$$

$$= 5380 \times 0.85$$

$$= 4373 \text{ lb/d at 30\% Suspended solid}$$

$$\text{Concentrate flow} = \frac{4373 \text{ lb/d} \times 10^6}{(300{,}000 \text{ mg/L})(8.34)}$$

$$= 1748 \text{ gal/d.}$$

- Centrate:

  * Flow = (4300 – 1748) gal/d

  = 2552 gal/d

  * Suspended solid in concentrate = 0.1 (5380 lb/d)

  = 538 lb/d

$$= \frac{538 \text{ lb/d}}{(0.002552 \text{ MGD})(8.34)}$$

$$= 25278 \text{ mg/L}$$

$$= 2.53\%.$$

Note :

Table 46 : Centrifuge application to various wastewaters

Application		*Solids*		*Cake (%solids)*	*Polymer added* (lb/ton)*
Effluent	Treatment	As fed	As discharged		

Table 47 : Solid-Bowl scroll centrifuge

Paper mill; Paper	*Primary; primary secondary*	*Coarse, fibrous, claylike*	*Relatively dry*	*28-40*	*None*
Municipal	Primary raw	Coarse, fibrous claylike	Relatively dry	30-40	1.5-2.5
Municipal	Primary digested, mixed digested	Coarse, fibrous, slimy	Slimy to dry; depends on primary-secondary ratio	20-30	3-6
Municipal	Primary raw, secondary	Coarse, fibrous, slimy	Slimy to dry; depends on primary-secondary rate	18-22	4-6
Refinery	-	Gritty, coarse	Dry to pudding	20-25	None
Paper mill municipal	-	Slimy, thickened	Thick pudding	18-22	10-20
Paper mill, water treatment	Lime sludge; water softening	Claylike	Dry	40-60 (depends on % hydroxide)	None
Steel mill	Pickle liquor, neutralized	Some floccy, some clay	Very thick pudding (can be shoveled)	20-30	1-2

Table 48 : Disc-Type centrifuge with nozzles.

Paper, municipal	*Waste activated*	*Slimy*	*Thickened (for further dewatering or digestion)*	*6-7*	*None (or <1)*
Refinery	Liquid-liquid solids	Oil-water emulsion; some fine claylike solids	Oil-water emulsion split; solids centrated	Oil (<1%) water; solid (7-10%)	None
Water treatment plant	Alum floc	Slimy, floccy	Thin, floccy	5-7	<1

Table 49 : Solid-Bowl Basket (Imperforate) centrifuge.

Municipal	*To improve recovery* +	*Floccy, slimy*	*Thick pudding*	*10-14*	*None*
Water treatment chemical waste	Alum floc, hydroxide sludges	Floccy	Very thick pudding	15-25	None (or <1)

*Recovery 85-90%

+Following solid-bowl scroll

Example 3.53

Dewatering Systems

Determine the total chemicals required, total dry solids processed cake characteristics, effluent characteristics, and energy requirements for (1) belt filter, and (2) centrifuge. Use the following data:

Sludge flow : 200,000 gal/d

Initial sludge concentration : 1.5% (15,000 mg/L)

[A] Belt filter

Hydraulic loading : 25 gal/min/belt-width (m)

Solids loading : 500 lb/h.meter of belt width

Cake solids content : 25%

Polymer required : 8 lb/tonne solids as 1% solution

Wash water : 10 gal/min. meter belt width

Total energy (direct and indirect) : 20 kWh/tonne-solids

Recovery : 85%

[B] Centrifuge

Cake solids content : 25%

Recovery : 98%

Polymer required : 7.5 lb/tonne of solids as 1% solution

Total energy (direct and indirect) : 60 kWh/tonne-solids

1. Belt filter

1.1 Required chemicals

- Amount of solids processed

  = (0.2 MGD)(8.34)(15,000 mg/L) = 25,020 lb/d

- Total chemicals requirements

$$= (25,020 \text{ lb/d})(8 \text{ lb/t})\left(\frac{1}{2000 \text{ lb}}\right)$$

$$= 100 \text{ lb/d}$$

$$= (100 \text{ lb/d})\left(\frac{99 \text{ lb}}{1 \text{ lb}}\right)$$

= 9908 lb water/d at 1% solution.

- Amount of total solids processed = (25,020 + 100) lb/d = 25,120 lb/d

1.2 Required belt size

- $$\text{Belt width} = \frac{(200,000 \text{ gal/d})}{(1440 \text{ min/d})(25 \text{ gal/min belt width (m)})}$$

  = 5.5 m [Based on hundraulic]

- $$\text{Belt width} = \frac{(25,120 \text{ lb/d})}{(24 \text{ h/d})(500 \text{ lb/h. meter of belt width})}$$

  = 2.0 meter belt width [Based on solids loading]

Final selection is based on hydraulic loading = 5.5 m (Critical).

1.3 Required cake characteristics

Dry solids = (25,120 lb/d)(0.85) = 21,352 lb dry solids/d

$$\text{Wet solids} = \frac{21,352 \text{ Ib dry solids/d}}{0.25} = 85,408 \text{ Ib wet sludge/d}$$

Density of sludge = 100 lb/ft^3 of sludge

$$\text{Sludge volume} = \frac{85,408}{100} = 854 \text{ ft}^3/\text{d of sludge generated}.$$

1.4 Required effluent characteristics

- Total influent = 200,000 × 8.34 = 1,668,000 lb/d

Influent solids	= 25,020 lb/d
Influent water	= 1,642,980 lb/d
Sludge water = $(21,352)\left(\frac{75}{25}\right)$	= 64,056 lb/d
Rejected influent water	= 1,578,924 lb/d
Water with channels	= 9908 lb/d

Backwash water :

= (5.5 m)(10 gal/min.m)(1440 min/d)(8.34) lb/d

= 660,528 lb/d

Total effluent water = (660,528 + 1578,924 + 9908) lb/d

= 2,250,260 lb/d.

1.5 Required effluent solids concentration

= (25,120 lb/d)(0.15) = 3768 lb/d [Not recovered]

Effluent solids concentration

$$= \frac{3768 \text{ lb/d}}{(3768 + 2,250,260)\text{lb/d}} \times 10^6 = 1672 \text{ mg/L}.$$

1.6 Required total effluent volume

$$= \frac{(3768 + 2,250,260)\text{lb/d}}{8.34} = 270,267 \text{ gal/d}$$

1.7 Energy requirements

$$= \left(\frac{20 \text{ kWh}}{\text{tonne of solids}}\right)\left(\frac{25,120 \text{ lb} - \text{solids/d}}{2000 \text{ lb/tonne}}\right)$$

= 251 kWh.

2. Centrifuge.

2.1 Required total chemicals

$$= (25,020 \text{ lb/d})(7.5 \text{ lb/t})\left(\frac{\text{t}}{2000 \text{ lb}}\right)$$

= 93.8 lb/d

$$\text{Water required} = (93.8 \text{ lb/d})\left(\frac{99}{1}\right) = 9289 \text{ lb water/d at 1\% solution}.$$

2.2 Required total solids processed

= (25,020 + 93.8) lb/d = 25,114 lb/d

2.3 Required size of centrifuge

$$= \frac{200,000 \text{ gal/d}}{1440 \text{ min/d}} = 139 \text{ gal/min}$$

Select any size: 150-175 gal/min

2.4 Required cake characteristics

(25,114 lb/d)(0.98) = 24,612 lb dry solids/d

$$\text{Wet sludge} = \frac{(24,612 \text{ lb/d})}{0.25} = 98,447 \text{ lb wet sludge/d}$$

Assume the density of sludge = 100 lb/ft^3 of sludge

$$\text{Volume of sludge generation} = \frac{98,447}{100} = 985 \text{ ft}^3/\text{d}$$

2.5 Required effluent characteristics

- Effluent water

  Influent water = (200,000)(8.34) = 1,668,000 lb/d

  $$\text{Sludge water} = (24,612)\left(\frac{75}{25}\right) = 73,836 \text{ lb/d}$$

  Rejected influent water = 1,594,164 lb/d

  Water with channels = 9289 lb/d

  Total effluent water = (9289 + 1,594,164) lb/d

  = 1,603,453 lb/d

2.6 Required effluent solids

= (25,114)(0.02) = 502.8 lb/d [Not recovered]

Effluent solids concentration

$$= \frac{502.8}{(502 + 1,603,453)} \times 10^6 = 313.2 \text{ mg/L}$$

Total effluent volume

$$= \frac{(502 + 1,603,453) \text{lb/d}}{8.34 \text{ lb/gal}} = 192,321 \text{ gal/d}$$

2.7 Energy requirements

$$= (60 \text{ kWh/t})\left(\frac{25,114 \text{ lb/d}}{2000 \text{ lb/t}}\right) = 753 \text{ kWh}$$

Note :

1. Solids recovery (R) is expressed as:

$$R = \frac{(\text{lb solids fed} - \text{lb solids in centrate})}{\text{lb solids fed}} \times 100$$

2. Shear number

   $G^x \, t = k$

   k is a fraction of polymer design. Exponent x is a fraction of the conditioned sludge characteristics, in compassing all factors relating to the sludge properties, and the conditioning process.

Table 50 : Shear rate of flocs.

Floc Type	*Shear Rate* (s^{-1})
Fragile flocs (biological)	10-30
Medium flocs (water treatment)	20-50
High strength flocs (chemical precipitation)	40-100

3. General dewatering device capabilities

Table 51 : General dewatering device capabilities.

Dewatering System	*% Water*
Final settling	1
Thickening	2
Basket centrifuge	5-20
Solid bowl centrifuge	15-30
Vacuum filter	10-30
Belt press	10-30
Filter press	30-50
Diaphragm filter press	35-55

4. Dewatering device recoveries

Table 52 : Dewatering device recoveries

System	Common Capture (%)
Gravity	88-95
Basket centrifuge	80-98
Solids bowl centrifuge	90-98
Belt filters	85-95
Vacuum filter	88-95
Filter press	> 98
Dry beds	> 99
Sludge lagoons	> 99

5. Comparison of device with disposal method

Table 53 : Comparison of device with disposal method

Device	*Incineration*	*Composting*	*Agricultural Land Application*	*Landfill*
Centrifuges				
Basket			×	×
Solid bowl	×	?	×	×
Belt filter	×	?	×	×
Vacuum filter	×	?	×	×
Filter press	×	×	×	×
Drying bed		×	×	×
Sludge lagoon			×	×

6. Direct dewatering energy requirements

Table 54 : Direct dewatering energy requirements

Dewatering Device	*Direct Costs*		
	Fuel kCal/mt (BTU/t)	*Electricity kWh/mt (kWh/t)*	*Total Equivalent Electricity (kWh/t)*
Basket centrifuge		99-132 (90-120)	99-132 (90-120)
High-speed solid bowl centrifuge		66-99 (60-90)	66-99 (60-90)
Low-speed solid bowl centrifuge		33-66 (30-60)	33-66 (30-60)
Belt filter press		11-28 (10-25)	11-28 (10-25)
Vacuum filter		44-66 (40-60)	44-66 (40-60)
Fixed-volume filter press		44-66 (40-60)	44-66 (40-60)
Diaphragm filter press		39-61 (35-55)	39-61 (35-55)
Drying beds	5,600 (20,000)	1-2 (1-2)	3-4 (3-4)
Sludge lagoons	24,000-41,000 (88,000-146,000)	1-2 (1-2)	10-18 (9-16)

Basis: (1) 50-50% digested mix of primary and WAS at 3% feed. (2) 11,080 kj/kWh (10,500 BTU/kWh) and electricity efficiency at 32.5% of generation

7. Indirect (conditioning) dewatering energy requirements

Table 55 : Indirect (conditioning) dewatering energy requirements

Dewatering Device	*Conditioning Chemicals*	*Dosage g/kg (lb/t)*	*Indirect Costs kWh/mt (kWh/t)*
Basket centrifuge	Polymer	3 (6)	0.7 (0.6)
Solid bowl centrifuge	Polymer	4 (8)	0.9 (0.8)
Belt filter press	Polymer	6 (12)	1.3 (1.2)
Vacuum filter	Lime	150 (300)	99 (90)
	Ferric chloride	40 (80)	44 (40)
Filter press	Lime	120 (240)	79 (72)
	Ferric chloride	50 (100)	55 (50)

Industrial facilities are not operated as a public service, so that although any skill can be assigned to a treatment task, such an assignment in the long run must be economically justified. Economically justified means not only the costs assigned to the treatment facility, but as it affects the entire production facility. This could become an extreme burden on small plants, where many functions involved in operating and maintaining total waste treatment may have to be performed by one or two general operators, and routine maintenance often deferred. In this scenario the sludge management facilities are often shortchanged. The process Engineer should keep this in mind and select an operating device consistent with the facility realities. This almost always means employing the simplest device that will work, and as maintenance free as possible. In any event, early inquires should be made as to what labour is available for the facilities.

8. Site Specific and Physical Considerations

After performance requirements have been defined and evaluated, and labour and energy costs considered, final dewatering device selection will depend on site specific consideration such as (1) plant size, (2) facility restraints, and (3) total environmental considerations.

9. Plant Size

Plant size (total manufacturing, waste treatment, or both) is an important consideration in evaluating applicable dewatering devices. Smaller facilities should first consider offsite, contracted sludge disposal, or uncomplicated on-site systems. Complex equipment is ineffective in small plants because of the lack of available labour manpower and specific skills that can be totally devoted to operating and maintaining the devices. EPA published guidelines based on U.S. experience are summarized in Table 56 relating dewatering devices with (municipal) waste treatment plant size. These guidelines are not absolute criteria but one of the factors to be considered.

10. Facility Restrictions

Dewatering facility location depends on its size required area, and its relative distance from the sludge generating facilities. The allocated processing area includes that for the dewatering device plus associated accessories, as well as the potential need for future expansions. On-site facility restrictions could influence all aspects of the dewatering system design, starting with the dewatering technology selected, and including all ancillary systems or structures. Some constraints that may affect the design include.

(1) Existing dewatering facilities and conditioning facilities, will strongly influence plant preference in any expansion or new proposed facilities. In such cases, upgraded technology can only be suggested on the basis of large economic advantages, especially if current applied technology is adequate, successfully applied, and the costs acceptable.

(2) The nature of the manufacturing facility may impose some restrictions. Concern over chemical handling and use would not be a problem with a chemical manufacturing facility, but may cause some concern at other type facilities, requiring selection of "safe" chemicals.

(3) At some facilities there may be an inherent tendency to consolidate all utility operations in a common treatment area, sharing available personnel. Compatibility with chemicals used in the waste treatment, water treatment, or manufacturing facility may influence the process selected.

(4) Where chemicals are not used in significant quantities, prepared bulk liquid chemicals may be better consideration than dry chemicals, with necessary preparation and dilution facilities eliminated. In fact, the predominant consideration may be the required chemical quantities, their costs, and the cost of the associated facilities, subordinating all other considerations.

(5) Frequently, it is advantages to house the dewatering equipment requiring a new building or selection of a suitable existing facility. Expansions or modification of.

Table 56 : Dewatering Compatibility with Plant Size

Dewatering Device	*Municipal Plant Size, cum/day (mgd)*		
	< 3785 (<1)	*3785-37,850 (1-10)*	*> 37,850 (>10)*
Basket Centrifuge	×	×	
Solid bowl Centrifuge	×	×	
Belt filter press	low pressure	×	×
Vacuum filter	×	×	
Filter press	×	×	
Drying bed	×	×	
Sludge lagoon	×	×	

11. Belt filter practical limits

Table 57 : Belt filter practical limits

Feed concentration	0.5-10%
Polymer addition	
Range	1-10 g/kg of dry sludge
Common	2-7.5
Cake dryness	
Range	10-35%
Common	20-25%
SS Recovery	
Typical	85-95%

12. Belt filter polymer effects

Table 58 : Belt filter polymer effects

Municipal Source	*Sludge Type*	*Feed (%)*	*Dry Feed Loading kg/hr/ meter**	*lb/hr/ meter**	*Dry Polymer, mg/kg dry sludge ppm (Weight)*	*Cake Solids (%)*
Treatment plant	WAS	0.5-4	40-230	100-500	1000-10,000	20-35
Aerobic digested	P + WAS	1-3	90-230	200-500	2000-8000	12-20
Oxygen plant	WAS	1-3	90-180	200-400	4000-10,000	15-23
Anaerobic digested	WAS	3-4	40-136	100-300	2000-10,000	12-22
Treatment plant	P + WAS	3-6	180-590	400-1300	1000-10,000	20-35
Treatment plant	P + TF	3-6	180-590	400-1300	2000-8000	20-40
Anaerobic digested	P + WAS	3-9	180-680	400-1500	2000-8000	18-44
Anaerobic digested	P	3-10	360-590	800-1300	1000-5000	25-36
Raw	P	3-10	360-680	800-1500	1000-5000	28-44
Thickened and aerobic digested	P + WAS	4-8	40-230	300-500	2000-8000	12-30
Thermal conditioned	P + WAS	4.8	290-910	600-2000	0	25-50

**Meter belt width*

13. Importance basket centrifuge operating data

Table 59 : Importance basket centrifuge operating data

Municipal Source	*Sludge Type Recovery (%)*	*Feed (%)*	*Dry Polymer, mg/kg*	*Dry sludge ppm (weight)*	*Cake Solids (%)*
Raw	WAS	0.5-1.5	0	8-10	85-90
Raw	WAS	0.5-1.5	500-1,500	12-14	90-95
Anaerobic digested	50% P + 50% WAS	1-2	0	12-14	75-80
Anaerobic digested	50% P + 50% WAS	1-2	750-1,500	10-12	85-90
Anaerobic digested	50% P + 50% WAS	1-2	2,000-3,000	8-10	93-95
Aerobic digested	WAS	1-3	0	8-11	80-95
Aerobic digested	WAS	1-3	500-1,500	12-14	90-95
Treatment plant	TF	2-3	0	9-10	90-95
Treatment plant	TF	2-3	750-1,500	10-12	95-97
Treatment plant	70% P + 30% TF	2-3	0	9-11	95-97
Treatment plant	70% P + 30% TF	2-3	750-1,500	7-9	94-97
Treatment plant	50% P + 50% WAF	2-3	500-1,500	12-14	93-95
Treatment plant	60% P + 40% RBC	2-3	0	20-24	85-90
Treatment plant	60% P + 40% RBC	2-3	2,000-3,000	17-20	98+
Raw	P	4-5	1,000-1,500	25-30	95-97

14. Solids bowl centrifuge operating data

Table 60 : Solids bowl centrifuge operating data

Municipal Source	*Sludge Type*	*Feed (%)*	*Dry Polymer, mg/kg dry sludge ppm (weight)*	*Cake Solids (%)*	*Recovery (%)*
Treatment plant	WAS	0.5-3	5000-7500	8-12	85-90
Aerobic digested	WAS	1-3	1500-3000	8-10	90-95
Anaerobic digested	P	2-5	3000-5000	28-35	95+
Anaerobic digested	P	2-5	3000-5000	28-35	98 +
Treatment plant	P + WAS	4-5	1500-3500	18-25	90-95
Raw	P	5-8	500-2500	25-36	90-95
Raw	P	5-8	0	28-36	70-90
Treatment plant	P + TF	7-10	0	35-40	60-70
Treatment plant	P + TF	7-10	1000-2000	30-35	98 +
Raw	P	9-12	500-1500	25-35	82-92
Raw	P	9-12	0	30-35	65-80
Thermal conditioned	WAS	9-12	0	35-40	75-85
High lime	30-50	10-12	0	90-95	
Treatment plant	P + WAS	13-15	500-2000	29-35	90-95

15. Basic equation for centrifuges

- Application of Stoke's Law for particle settling will a velocity:

  $$V = \frac{(\rho_S - \rho_L)d^2 g}{18\mu}$$

  Centrifugal force (G) = w^2 r/g

  $$V = \frac{(\rho_S - \rho_L)d^2 G}{18\mu}$$

  where ρ_s is the solids density, ρ_L is the liquid density, w is the rotational speed, r is the radius of rotation, d is the particle diameter, μ is the water viscosity, and g is the acceleration due to gravity

  Centrifugal force (F) = ma = m(rw^2)

  Therefore, the machine centrifugal force (F) = KDN^2

  where D is the diameter, and N is the bowl speed

- Centrifuge performance can be correlated to the product of the number of gravity forces times the detention time:
  - Clarifier operating at gravity force and are hour detention time operates at 60 g-minute level.

- Centrifuge operating at 3000 g and 0.5 minute detention time has a (3000× 0.5) = 1500 g-minute effectiveness.

- A machine's centrifugal force limits are part of a manufacturer's proprietary design dependent on bowl speed and interval diameters:

$$F = KN^2\left(\frac{(D_p + D_b)}{2}\right)$$

where F is the multiple of gravity force (expressed in G's), N is the bowl speed (rpm), D_b is the bowl diameter (in.), D_p is the inner pool diameter (in.), and K is the constant dependent on the unit of g and the diameter (= 0.0000142).

Commercial units have a bowl speeds which develop forces equivalent to 500 to 3000 G.

- Centrifuges manufacturer's have developed a comparison factor (Q/S) to theoretically compare machine performance based on the volumetric feed rate (Q), and a common capacity factor sigma (Σ), expressed in area units. Sigma is defined as the equivalent settler area, which theoretically produces results equivalent to the centrifuge:

$$\frac{Q}{\Sigma} = \frac{Q}{2\pi}L\left(\frac{w^2}{g}\right)\left(\frac{3}{4}r_b^2 + \frac{1}{4}r_p^2\right)$$

where Q is the flow rate (cm^3/s), L is the clarification length of the bowl (cm), w is the bowl speed (rad/s), g is the gravity constant (980 cm/s^2), r_b is the radius from axis of rotation to the bowl wall (cm), r_p is the radius from axis of rotation to pond bowl (cm), and S is the machine property specific to the centrifuge (sq cm).

Example 3.54

Dewatering of Sludge through a Filter Press

Calculate the density of cake solids, volume of dewatered cake, and filter press volume. Assume the following data:

Dry solids concentration	:	5%
Specific gravity	:	1.02
Dry solids density	:	150 lb/ft^3
Sludge flow rate	:	50 gal/min
Plot plant data reveals that a cake solids of 25% can be activated in	:	100 min

Solution

1. Density of cake

$$\frac{Sc}{62.4\ \text{lb/ft}^3} = \frac{150\ \text{lb/ft}^3}{150\ \text{lb/ft}^3 - 0.25\left(150\ \text{lb/ft}^3 - 62.4\ \text{lb/ft}^3\right)}$$

$$= \frac{150}{150 - 21.9} = 1.17$$

$$Sc = 1.17 \times 62.4 = 73.125\ \text{lb/ft}^3$$

2. Dewatered Cake Volume

Sludge flow = 50 gal/min × 1440 min/day × 8.34 × 1.02

= 6,12,500 lb/d = 0.60 × 10^6 lb/d (say)

Therefore, dry solids = 0.60 × 10^6 × 0.05 = 30,000 lb/d

$$\text{Volume of dewaterd cake} = \frac{30,000\ \text{lb/d}}{73.125\ \text{lb/ft}^3} = 410\ \text{ft}^3/\text{d}$$

3. Filter press volume

$$\text{No. of cycles per day} = \frac{24\ \text{hr}}{3\ \text{hr/cycle}} = 8\ \text{cycle/d}$$

$$\text{Volume of filter press} = \frac{410\ \text{ft}^3/\text{d}}{8\ \text{cycles/d}} = 51.2\ \text{ft}^3$$

Example 3.55

Belt Filter Press for Wastewater Sludge Dewatering

Determine the required belt width to dewater 50,000 gal/d of 2% thickened sludge. Assume the following data:

Total cake solids	: 35%
Solids capture	: 95%
Belt speed	: 9.5 ft/min
Throughput	: 250 lb per 0.5 m belt width per hour
Polymer dosage	: 4.5 lb/tonne day solids

Solution

1. Required filter belt width

Assume two-8 hour shifts with a total of 16 hour of belt press operation 5 days/week

(0.0500 mil gal/d)(20,000 mg/L)(8.34) = 8340 lb/d of solids

or $$\frac{(8340 \text{ lb/d})(7 \text{ d})}{(16 \text{ h/d})(5 \text{ d})} = 730 \text{ lb/hr}$$

Therefore,

$$\text{Belt width} = \frac{730 \text{ lb/hr}}{(250 \text{ lb/hr}/0.5 \text{ m belt})}$$

$$= \frac{730 \text{ lb/hr}}{500 \text{ lb/hr.m}} = 1.46 \text{ m.}$$

2. Required pressure and belt tension

Upper belt pressure	=	5 bar (72.5 lb/in.^{2})
Lower belt pressure	=	5 bar (72.5 lb/in.^{2})
Belt tensions	=	45 bar (653 lb/in.^{2}).

Note :

1. The press belt can be pressed on the filter belt by means of a pressure roller system whose rollers can be individually adjusted horizontally and vertically.
2. The sludge to be dewatered is fed on the upper face of the filter belt and is continuously dewatered between the filter and press belts.
3. The pressure rollers of the pressure belt are adjusted in such a way that the belts and the sludge between them describe an S-shaped curve. This configuration induces parallel displacement of the belts relative to each other due to difference in radius, producing shear in the cake. After dewatering in the shear zone, the sludge is removed by a scraper.
4. Available data indicate that over a range of 0.5 to 12% dissolved solids, a filter is relatively insensitive to concentration, but is very sensitive to rate of flux per unit area.
5. On the average belt, wash water flow approximately equals the sludge application rate.
6. Three zones of dewatering in a belt filter press :
   - Draining zone
   - Press zone
   - Shear zone
7. Press belt, filter belt, pressure rollers, support rollers, and drive roller are the components of a belt filter press.

Example 3.56

A Belt Filter Press

Determine the hydraulic feed rate, the solids loading rate, the polymer dosage, and the solids recovery for a belt filter press. Use the following data:

Effective belt width	:	1.5 m
Anaerobically digested sludge feed rate	:	100 gal/min
Polymer dosage	:	5.5 gal/min with 0.20% (by weight of polymer)
Washwater usage	:	45 gal/min
Total solids in the feed sludge	:	4.5%
SS in the wastewater (filtrate, polymer feed and wash water)	:	2000 mg/L
Total solids in the cake	:	38%

Solution

1. Required hydraulic loading

$$= \frac{100 \text{ gal/min}}{1.5 \text{ m}} = 66.67 \text{ gal/m.min}$$

2. Required solids loading

$$= (\ 66.67 \text{ gal/m.min})(0.035)(8.34)(60 \text{ min/hr})$$

$$= 1167.6 \text{ lb/m.hr}$$

3. Required polymer dosage

$$= \frac{(5.5 \text{ gal/min})(60 \text{ min/hr})(0.002)\,(8.34)(2000 \text{ lb/tonne})}{(1.5 \text{ m})(1167.6 \text{ lb/m.hr})}$$

$$= 6.3 \text{ lb/tonne}$$

4. Required volumetric flow of sludge cake (V)

$$V = \frac{W}{\left(\frac{s}{100}\right)8.34} = \frac{W}{\left(\frac{100-p}{100}\right)8.34}$$

where V is volume of wet sludge (gal), W is the weight of dry solids (lb), s is the solids content (%), and p is the water content (%)

$$V = \frac{(1000 \text{ gal/min})(0.035)(8.34)}{(0.38)(8.34)(1.05)}$$

$$= 8.8 \text{ gal/min}$$

Flow of filtrate = [Sludge feed] – [Cake volume]

Flow of filtrate = 100 – 8.8 = 92.2 gal/min

Wastewater flow = Filtrate + Polymer feed + Washwater

= 92.2 + 5.5 + 45

= 142.7 gal/min.

5. Required solids concentration in wastewater

$$= \frac{2000 \text{ mg/L}}{10{,}000 \text{ mg/L percent}} = 0.2\%$$

Solids in wastewater :

$$= \frac{(142.7 \text{ gal/min})(60 \text{ min/hr})(0.002)(8.34)}{1.5 \text{ m}}$$

= 95 Ib/m.hr.

6. Required solids recovery

$$= \frac{1167.6 \text{ lb/hr} - 95 \text{ lb/m.hr}}{1167.6 \text{ lb/m.hr}}$$

= 92%.

Note :

1. Table 61, Typical operating parameters for belt filter press dewatering of polymer flocculated wastewater sludges

Table 61

Sludge Type	*Feed Solids (%)*	*Loading*		*Cake Solids(%)*	*Polymer Dosage (lb/ton)*
		Hydraulic[a]	*Solids[b]*		
• Anaerobically digested primary only	4-6	40-50	1000-1600	25-35	3 - 6
• Anaerobically digested primary plus waste activated	2-5	40-56	500-1000	15-26	6-12
• Aerobically digested without primary	1 to 3	30-45	200-500	11-22	8-14
• Raw primary and waste activated	3 to 6	40-50	800-1200	16-25	4-10
Thickened waste activated	3 to 5	40-50	800-1200	14-20	6-8
Extended aeration waste activated	1 to 5	30-45	200-500	12-22	8-14
Heat-treated primary with waste activated	4 to 10	35-50	1000-1800	30-40	1-2

[a] : *gal/m.min*

[b] : *lb/m.hr*

Example 3.57

Vacuum Filters : Air Suction Rate

Determine the horse power required for vacuum pump on a rotary vacuum filter. A rotary vacuum-drum filter is required to handle a slurry containing 20 lb of water per 1 lb of solid material. Assume the following data:

Ratio of α/β	: 0.6
19 lb of filtrate (not including washwater) is obtained for each 21 lb of slurry	
Temperature of surroundings and of slurry	: 70°F
Pressure drop to be maintained	: 7.5 psi
Surrounding pressure	: 1 atm
Fraction of the drum area submerged (ψ_f)	: 0.25
Fraction of the drum area available for air suction (ψ_a)	: 0.10
Slurry rate	: 10,000 lb/h

Solution

1. Required mathematical expressions

 - Rate at which air is drawn through the dewatering section of vacuum filter :

$$\frac{dV_a}{dt} = \frac{A_D \psi_a \Delta P}{\left(R_F^{'} + R_K\right)\mu_a}$$

$$R_F + R'_K = C(l + l'_f)$$

$$(l + l'_f) = \frac{W\left(V_R + A_D \psi_a V_F^{'}\right)}{\rho_c A}$$

Therefore, $$\frac{dV_a}{dt} = \frac{A_D^2 \psi_a \Delta P}{\beta W\left(V_R + A\psi_a V_f^{'}\right)\mu_a}$$

Integrating this equation between t = 0 to t, and $V_a = 0$ to $V_a = V_{aR}$, and the expression for noncompressible cake is:

$$v_{aR} = \frac{A_D^2 \psi_a \Delta P}{\beta W\left(V_R + A\psi_a V_f^{'}\right)\mu_a N_R}$$

If the cake is compressible, a rough correction for variation in b with change in DP can be estimated as:

$$\beta = \beta'\,(\Delta P)^{s'}$$

Neglecting the filter media resistance, the equation 1 can be simplified to:

$$\text{Volume of air resolution} = V_{aR} = \frac{A_D^2 \psi_a \Delta P}{\beta W\, V_R \mu_a N_R}$$

$$\text{Volume of air per unit time} = V_{aR}\, N_R$$

$$= \frac{A_D^2 \psi_a \Delta P}{\beta W\, V_R \mu_a}$$

$$\text{Volume of air per unit time} = \frac{A_D \psi_a}{\beta\, \mu_a}\left[\frac{\alpha\mu\, N_R\, \Delta P}{2W\psi_f}\right]^{0.5}$$

$$\frac{\text{Volume of air/time}}{\text{Weight of dry cake/time}} = \frac{\psi_a}{\psi_f}\,\frac{\mu}{\mu_a}\,\frac{\alpha}{2\beta\, W}$$

where V_a = Air volume at temperature and pressure of surroundings drawn through cake in time (t)

Ψ_a = Fraction of total surface available for suction

μ_a = Viscosity of air at temperature and pressure of surroundings

C' = Proportionality constant

l = Cake thickness

R'_K = Cake resistance

R'_F = Filter-medium resistance

l'_f = Fictitious cake thickness

V'_f = Fictitious volume of filtrate per unit of air-suction area necessary to lay down a cake of thickness l'_f

β = C'/ρ_c (Specific resistance)

β' and s' = Constants

A_D = Total available area

ψ_f = Fraction of area submerged in slurry

μ = Viscosity of filtrate

ΔP = Pressure drop across filter

α = C/ρ_c (Specific cake resistance)

C = Constant

W = lb of dry cake solids per unit volume of filtrate

ρ_c = Cake density (lb dry cake solids per unit volume of wet filter cake)

N_R = Revolution per unit time

2. Required air suction rate

$$\frac{\text{Volume of air/time}}{\text{Weight of dry cake/time}} = \frac{\psi_a}{\psi_f}\frac{\mu}{\mu_a}\frac{\alpha}{2\beta W}$$

$\psi_a = 0.1$

$\psi_f = 0.25$

Temp. = 70°F

$\mu = 0.982 \text{ CP} = 0.982 \times 2.42 \text{ lb/ft.h (water)}$

$\mu_a = 0.018 \text{ CP} = 0.018 \times 2.42 \text{ lb/ft.h (air)}$

$$\frac{\alpha}{\beta} = 0.6$$

Water density at 70°F = 62.3 lb/ft^3

lb-filtrate/lb dry cake solids = 19

$$W = \frac{1}{19/62.3} = 3.28\frac{\text{lb dry cake}}{\text{ft}^3 - \text{filtrate}}$$

$$\text{Weight of dry cake/hour} = (10{,}000)\left(\frac{1}{2}\right)$$

$$= 476.2 \text{ lb/h}$$

$$\text{Volume of air/hour} = \frac{(476.2)(0.1)(0.982 \times 2.42)(0.6)}{(0.3)(0.018 \times 2.42)(2 \times 3.28)}$$

$$= 792 \text{ ft}^3\text{/h at } 70° \text{ F and 1 atm}$$

[No air leakages a/b = 0.6]

3. Required Horsepower

For single-stage isoentropic compression (PV^k = Constant), the horsepower is given as:

$$h_P = \frac{3.03 \times 10^{-5} k}{k-1} P_1 \, q_{f_1}\left[(P_2/P_1)^{\frac{k-1}{k}} - 1\right]$$

where k = Ratio of heat capacity of gas at constant pressure to heat capacity of gas at constant volume (1.4)

P_1 = Vacuum-pump intake pressure [= (14.7 – 5) 144 psf]

P_2 = Vacuum delivery pressure [= (14.7) 144 psf]

q_{f_1} = Cubic feet of gas per minute at vacuum-pump intake conditions

$$= \frac{(792)(14.7)(144)}{(60)(14.7-5)(144)} = 20 \text{ ft}^3/\text{min at } 70^\circ \text{f and } 9.7 \text{ psi}$$

$$h_P = \frac{(3.03\times10^{-5})(1.4)(14.7-5)(144)(20)}{0.5(1.4-1)}\left[\left(\frac{14.7}{9.7}\right)^{\frac{1.4-1}{1.4}} - 1\right]$$

(Vacuum pump and motor efficiency = 50%)

= 0.74 (» 1 hp motor will be OK).

Assumptions are:

- Filter media resistance is negligible
- Any effects by air leakage are taken into account in the value given for a and b
- k = 1.4 for air
- b volume is based on the temperature and pressure of the air surrounding the filter
- Filter removes all of the solids from the slurry.

Note :

1. Pressure and temperature at exit conditions

$$p_2 = p_1\left[\frac{V_1}{V_2}\right]^k = p_1\left[\frac{T_2}{T_1}\right]^{k/k-1}$$

$$T_2 = T_1\left[\frac{V_1}{V_2}\right]^{k-1} = T_1\left[\frac{p_2}{p_1}\right]^{\frac{k-1}{k}}$$

2. Multi-stage compressor horsepower

$$h_p = \frac{3.03\times10^{-5}kN_s}{(k-1)} p_1 q_f\left[\left(\frac{p_2}{p_1}\right)^{\frac{k-1}{kN_s}} - 1\right]$$

where p_1 = Intake pressure (lb_f/ft^2)

V_1 = Specific volume of gas at intake conditions (ft^3/lbm)

p_2 = Final delivery pressure (lb_f/ft^2)

q_f = Cubic feet of gas per minute at intake conditions

T_1 = Absolute temperature of gas at intake (°K)

T_2 = Absolute temperatures of gas at final delivery conditions (°K)

N_s = Number of stages of compression

V_2 = Specific volume of gas at final delivery conditions (ft^3/lbm).

Example 3.58

Vacuum Filter : Continuous System

A rotating drum vacuum filter with negligible filter media resistance delivers 400 ft^3 of filtrate per hour when a given sludge mixture is filtered under known conditions. Determine the amount of filtrate if the pressure drop is doubled (all other conditions remaining constant and the sludge is non compressible).

Solution

1. Required mathematical expression

- Rate of filtration is given as:

$$\frac{dV}{dt} = \frac{A_D \psi_f \Delta P}{(R_K + R_F)\mu}$$

$$(R_K + R_f) = C(l + l_f)$$

Thickness of cake is (l)

$$l_{\text{leaving filtering zone}} = \frac{W\,V_R}{\rho_c\,A_D}$$

Average cake thickness during the cake deposition period can be assumed to be one-half of the thickness at the entrance and exit of the filtering zone:

$$l_{avg} = \frac{W\,V_R}{2\rho_c\,A_D}$$

$$\text{Therefore, } l + l_f = \frac{W\left(V_R/2 + A_D \psi V_F\right)}{\rho_c\,A_D}$$

$$\alpha = \frac{C}{\rho_c}$$

$$\frac{dV}{dt} = \frac{2A^2{}_D \psi_f \Delta P}{\alpha\,W\left(V_R + 2A_D \psi_f V_F\right)\mu}$$

Integrating this equation between the limits V = 0 and V = V_R and t = 0 and t = $1/N_R$ gives:

$$V = V_R^2 + 2A_D \psi_f V_F V_R = \frac{2A_D^2 \psi_f (\Delta P)}{\alpha\,W\,\mu\,N_R}$$

[For non-compressible cake]

If $\alpha = \alpha' (\Delta P)^s$, then

$$V_R^2 + 2A_D\psi_f V_F V_R = \frac{2A_D^2\psi_f (\Delta P)^{1-s}}{\alpha' W \mu N_R}$$

Volume of filtrate per revolution (V_R)

$$= A_D\left[\frac{2\psi_f\,\Delta P}{\alpha W \mu N_R}\right]^{0.5}$$

Volume of filtrate per unit time ($V_R N_R$)

$$= A_D\left[\frac{2\psi_f N_R\,\Delta P}{\alpha W \mu}\right]^{0.5}$$

Weight of dry cake per unit time ($V_R N_R W$)

$$= A_D\left[\frac{2\psi_f N_R W\,\Delta P}{\alpha \mu}\right]^{0.5}$$

where V = Volume of filtrate delivered in time (t)

A = Area of the filtering surface (= $A_D\ y_f$)

A_D = Total available area

ψ_f = Fraction of area submerged in slurry

ΔP = Pressure drop across the filter

R_K and R_F = Resistance of the cake and filter media

μ = Viscosity of cake

C = Proportionality constant

l = Thickness of cake

l_f = Fictitious cake thickness

W = lb dry-cake solids per unit volume of filtrate

V_f = Fictitious volume of filtrate per unit of filtering area necessary to lay down a cake thickness l_f

α = Specific cake resistance (= C/ρ_c)

C = Constant

ρ_c = Cake density (lb of dry-cake solids per unit volume of wet filter cake)

V_R = Volume of filtrate delivered per revolution

s = Compressibility coefficient

N_R = Number of revolution per unit

2. Required effect of doubling the pressure drop over the cake

   Volume of filtrate per unit time ($V_R N_R$) is:

$$V_1 = 400\ \text{ft}^3/\text{h} = A_D \left[\frac{2\psi\, N_R}{\alpha\, W\, \mu}\right]^{0.5} (\Delta P)^{0.5} \quad (1)$$

   Volume of filtrate per unit time with doubling the pressure (= $2\Delta P_1$) is:

$$V_2 = A_D \left[\frac{2\psi_f\, N_R}{\alpha\, W\, \mu}\right]^{0.5} (2\Delta P_1)^{0.5} \quad (2)$$

   Dividing equation (2) by equation (1) yields:

$$V_2 = 400(2)^{0.5}$$

$$= 400(1.41) = 566\ \text{ft}^3 \text{ of filtrate/h.}$$

Note :

1. Filtration zones in a rotary vacuum filter
   - Pick-up zone
   - Washing zone
   - Drying zone
   - Blow zone
   - Dead zone (Cake discharger is between blow zone and dead zone).

Example 3.59

Vacuum Filtration

Calculate the area required for dewatering of combined primary and secondary sludges from an effluent treatment plant. Use the following equation to determine the loading rate

$$L = 35.7 \left[\frac{P(1-s)}{\mu R}\right]^{0.5} \frac{C^m}{t^n}$$

where L is the loading rate (lb/ft^2.hr), P is the vacuum (psi), R is the r $\times 10^{-7}$ sec^2/g, C is the solids deposited per unit volume filtration (g/µL), µ is the filtrate viscosity (centipoises), and t is form time (min).

Assume the following data:

- m = 0.20 and n = 0.60
- 30 percent solids obtained at a 4 min dry time
- Compressibility coefficient (s) : 0.9

- Specific resistance (r) : $1.8 \times 10^7\ sec^2/g$
- Vacuum filtration unit operates 20 hours per day, 7 days a week
- Vacuum : 20 inches $\left(\frac{14.7}{30} \times 20\right) = 9.8$ psi
- Submergence of drum : 25%
- Initial solids concentrate : 5%
- Waste flow : 50 gal/min.

Solution

1. Cycle time

$$\frac{4 \text{ min}}{0.75} = 5.33 \text{ min}$$

2. Solids deposited per unit volume filtrate

$$C = \frac{1}{\left(\frac{C_i}{100 - C_i}\right) - \left(\frac{C_f}{100 - C_f}\right)} = \frac{1}{\left(\frac{95}{5}\right) - \left(\frac{70}{30}\right)}$$

$$= 0.06 \text{ g/cm}^3$$

3. Loading rate (L)

$$L = 35.7\left[\frac{(9.8)^{0.1}}{1 \times 1.8}\right]^{0.5} \frac{(0.06)^{0.2}}{(5.33)^{0.6}} = 6.23 \text{ lb/ft}^2\text{.hr}$$

4. Solids rate applied to the vacuum filter

$$= \frac{50 \times 1440 \times 50{,}000 \times 8.34 \times 10^{-6}}{20} = 1501 \text{ lb/hr}$$

$$\text{Required area} = \frac{1501}{6.23} = 240.96 \text{ ft}^2$$

Example 3.60

Vacuum Filtration

A petrochemical plant has a sludge consisting of lime from a neutralising process and several oil waste streams. The combined waste has a solids concentration of 50 g/L and an average flow of 45,000 gpd calculate the vacuum filter area based on 5 day per week operation.

Assume the following data:

Dynamic viscosity	: 1.0 Cp
m	: 0.6
n	: 0.6
Compressibility coefficient (s)	: 0.8
R_o	: 3.5 s^2/g
Operating pressure (P)	: 10 psig
Cake solids	: 30%
Cycle time	: 5 min.

Solution

1. Initial solids concentration

   50 g/L is approximately equal to 5.0%, and therefore moisture constant is 95%

2. Weight of solids deposited per unit volume of filtrate

$$C = \frac{1}{\left(\frac{C_i}{100 - C_i}\right) - \left(\frac{C_f}{100 - C_f}\right)}$$

   where C_i is the initial moisture constant (%), and C_f is the final moisture constant

$$C_i = 95\% \text{ and } C_f = 70\%$$

$$C = \frac{1}{\left(\frac{95}{100 - 95}\right) - \left(\frac{70}{100 - 70}\right)}$$

$$= \left[\frac{95}{5} - \frac{70}{30}\right]^{-1}$$

$$= 0.06 \text{ g/mL}$$

3. The filter loading

$$L_f = 35.7\left(\frac{CP^{1-s}}{\mu R_{ot_f}}\right)^{0.5},$$

   Neglecting the initial resistance of the filter media

   This equation is modified for prediction of filtration rates of various type of sludges

$$L_f = 35.7\left(\frac{P^{1-s}}{\mu\, R_o}\right)^{0.5} \frac{C^m}{t_f^n}$$

where R_o is the $r_o \times 10^{-7}\ s^2/g$, P is the vacuum (psig), C is the solids deposited per unit volume of filtrates (g/mL), m is the filtrate viscosity (Cp), t_f is the cycle time (min), and L_f is the filter loading ($lb/ft^2.hr$).

$$L_f = 35.7\frac{\left(10^{0.2}\right)^{0.5}}{1\times(3.5)^{0.5}} \times \frac{(0.06)^{0.6}}{(5)^{0.6}}$$

4. Weight of sludge solids

$$= 50{,}000\ mg/L \times 0.045\ \text{mil-gal} \times 8.34$$

$$= 18{,}765\ lb/d$$

5. Required filter area

$$= \frac{\text{Load applied (W)}}{\text{Loading rate}\left(L_f\right)}$$

$$= \frac{18{,}765\ lb/d}{1.69\ lb/ft^2.hr \times 24\ h/d} = 463\ ft^2$$

Note :

1. n varies between 0.4 to 1.0 (experimental determination).
2. m varies between 0.25 to 1.0 (experimental determination).
3. Final cake moisture is related to the cake thickness, the drying time, the pressure drop, the liquid viscosity, and the air rate through the cake.
4. In a porus cake, the change in cake moisture with increased drying time or increased vacuum is small, because the high air rates through the cake cause a rapid initial drying to equilibrium.
5. Non-porus cake required larger drying time and high vacuum to attain maximum cake constant.
6. A rotary drum passes through a slurry tank in which solids are retained on the drum surface under vacuum. The drum submergence can vary from 10 to 60%. As drum passes through the slurry, a cake is built up and water removed by filtration through the deposited solids and the filter media. The time the drum remains submerged in the slurry in the form time (t_f). As the drum emerges from the slurry tank, the deposited cake is further dried by the liquid transfer to air drawn through the cake by the applied vacuum. This period of the drum's cycle is called the dry time (t_d). This conventionally-converted to cycle time for final specification.

$$L_c = L_f\left(\%\frac{\text{Submergence}}{100}\right)\times 0.8$$

The factor 0.8 compensates for the area of the filter drum where the cake is removed and the media washed. The total cycle time on a filter may vary from 1 to 6 minutes. Submergence of drum from 10 to 60% results in a maximum spread of form (t_f) time of 0.1 to 3.5 minutes and spread of dry time (t_d) of 2.5 to 4.5 minutes.

7. In general, the filter yield from highly compressible cakes is relatively unaffected by increases in form vacuum varying from 12 to 17 inches of Hg.
8. The greater the value of compressibility coefficient(s), the move compressible is the sludge. When S = 0, the specific resistance is independent of pressure and the sludge is incompressible.
9. Raw sewage sludge is easier to filter than digested sewage sludge, and primary sewage sludges are easier to filter than secondary sludges.
10. Filterability is influenced by particles size, shape, density, and electrical charge on the particle. The large the particles size, higher is the filter rate, and lower the cake moisture.
11. Coagulants must be added to enhance filterability and reduce the specific resistance.
12. The cationic polymers effect charge neutralisation and the anionic effects polymer particles bridging and agglomeration of the particles. Excessive coagulant dosages result in a charge reversal and an increase in specific resistance.

Example 3.61

Vacuum Filtration

A Buchner funnel test on a conditioned sludge of 2% solids content a $\bar{r}_p$ value of 10^{11} m/kg at a pressure (difference) of 90 kN/m^2 at 18°C. Scrappings from the newly formed cake gave a value of 15% solids in the wet cake on the filter which had a diameter of 2m, the sludge holding level being 0.8 m below the drum axis. Assume the vacuum to be equivalent to 90 kN/m^2, calculate the yield for a cycle time of 200s.

Solution

The fractions of filter cycle (F_f) under sludge surface is

$$F_f = \frac{\cos^{-1}\left(\frac{h}{r}\right)}{180} = \frac{\cos^{-1}(0.8)}{180} = 0.205$$

where h is the perpendicular distance from the top surface of sludge contained in a container, and r is the radius of the vacuum filter drums

If cycle time (θ_R) = 200 s, Pressure difference (P) = 90 kN/m^2, specific resistance to dewatering (apparent) corrected for the operating pressure $(\bar{r}_p)$ = 10^{11} m/kg, and filtrate viscosity (h) = 1.14 Ns/m^2, then mass of dry solid sludge per unit of liquid in the sludge (C_1) is calculated as

$$C_1 = 10^3 \times \frac{W_s}{(1-W_s)}\left(kg/m^3\right)$$

(Assuming water has a density of 10^3 kg/m^3)

$$C_1 = \frac{0.02}{0.98} \times 10^3 = 20.5\ kg/m^3$$

- The cake correction factor (F_c) can be obtained using:

$$F_c = \frac{1-W_s}{1-\left(W_s/\overline{W}_c\right)}$$

where W_s and $\overline{W}_c$ are the mass of dry SS per unit mass of sludge and average mass of dry suspended solids per unit mass of compressible cake, respectively

$$F_c = \frac{1-0.20}{1-\left(\dfrac{0.20}{0.15}\right)} = 1.13$$

- The yield (Y) from the vacuum filter is calculated using :

$$Y = \left[\frac{2PC_1F_f}{\eta\bar{r}_p\theta_R}\right]^{1/2} F_e$$

where $F_e = \dfrac{\text{Mass of liquid/unit mass of sludge}}{\text{Mass of filterate obtained/unit mass of sludge}}$,

C_1 is the mass of dry solids/unit volume of liquid in the sludge, θ_R is the time per revolution or cycle time, $\bar{r}_p$ is the apparent value of specific resistance from Buchner test (using solids concentration of the sludge) at pressure P-i.e., The pressure of vacuum filtration, Y is the yield (mass dry SS produced in unit time per unit surface area), and F_f is the fraction of filter area used for a cake formations.

$$Y = \left[\frac{2 \times 90 \times 10^3 \times 20.4 \times 0.205}{1.14 \times 10^{-3} \times 10^{11} \times 200}\right]^{1/2} \times 1.13$$

$= 6.49 \times 10^{-3}$ kg/m^2.s or 5661 kg/m^2.d

Example 3.62

Vacuum Filtration

Determine the filter area. Assume the following data.

Dry solids to be removed by the filter	: 1000 lb/h
Operating vacuum	: 30 in. Hg
Percent submergence	: 50%
Wash time (t_w)	: 20% of total cycle time (t_c)
Scale-up factor (F)	: 0.8
Solids content of slurry to the filtered	: 10%, (C_i = 90%)
% of solids in the filter cake	: 30%, (C_f = 70%)
Polymer coagulant added prior to filtration	: 10 mg/L
Dynamic viscosity	: 1 cp
Drying time (t_d)	: 0.6 min.
Specific resistance (r)	: 3.416×10^{11} cm/g = 3.416×10^{4} practical units.

Solution

1. Required design equations
   - Filtration parameter (C)

     It is defined as mass of solids (g) deposited on the filter (solids in filter cake) per unit volume of filtrate (cm^3). Select 100 g of incoming slurry for which C_i is the % weight of water. The solids deposited on the filter are, therefore, $(100 - C_i)$ g. The grams of filtrate corresponds to the difference between the grams of water in the slurry (C_i) and the grams of water retained on the cake. If C_f is the moisture percent in the filter cake, the grams of water retained in it, corresponds to the difference between total mass of cake, *i.e.* $(100 - C_i)/ [100 - C_f)/100]$ g, and the grams of solids in the cake $(100 - C_i)$ g. In addition, the grams of filtrate (taken as water) equals numerically the cm^3 of filtrate, the parameter C is :

$$C = \frac{100 - C_i}{C_i - \left\{\dfrac{(100 - C_i)}{\left[(100 - C_f)/100\right]} - (100 - C_i)\right\}}$$

$$= \frac{1}{C_i/(100 - C_i) - C_f/(100 - C_f)}$$

- Filtration design equation

$$\frac{t_f}{V} = \left(\frac{\mu\, r\, C}{2P\, A^2}\right) V; \;\; [R_m = 0]$$

In terms of filter loading (L_f = lb of deposited cake/ft^2.h):

CV = weight of cake, filter loading (L_f) based on form time (form loading):

$$L_f = \frac{CV}{A\, t_f} = \frac{\text{lb of deposited cake}}{\text{ft}^2\text{.h}}$$

or $$\frac{V^2}{A^2} = \frac{(2P\, t_f)}{(\mu\, r\, C)}$$

If $$t_f = \frac{t_f^{\,2}}{t_f}; C = \frac{C^2}{C}, \text{ then :}$$

$$\left(\frac{CV}{At_f}\right)^2 = \frac{2PC}{\mu\, r\, t_f}$$

$$L_f = \left[\frac{(2PC)}{(\mu\, r\, t_f)}\right]^{0.5}$$

$$L_f = 1118\left[\frac{PC}{\mu\, r\, t_f}\right]^{0.5}$$

where P = psi

C = g/cm^3

μ = cp

r = cm/g $\times 10^{-7}$ = practical units

t_f = min.

L_f = lb/ft^2.h

Total cycle time (t_c) = Cake form time (t_f) + Cake drying time (t_d) + Cake wash time (t_w).

Loading based on cycle time or cycle loading (L_c) is:

$$L_c = L_f\left(\frac{t_f}{t_c}\right)$$

$$\text{Filter area} = \frac{\text{lb/h of solids to be removed}}{L_c\left(\text{lb/ft}^2\text{.h}\right)}, \text{ft}^2$$

$$\text{Design filter area} = \frac{\text{Filter area}}{F}$$

where F = Scale factor (0.7 to 0.9)

- Percent submergence = $\left(\frac{t_f}{t_c}\right)\Big/100$

  Usually drum submergences between 10% and 60%, i.e., 0.6 $t_c > t_f >$ 0.1 t_c

  Therefore, form time is usually between 10% and 60% of the total time.

- Wash time (t_w) is usually 20% of the total cycle time,

  $t_w = 0.2t_c$

  In general, if f_w is the fraction of total cycle time taken as wash time,

  $t_w = f_w\ t_c\ (f_w = 0.2)$

  Therefore, $t_c = \dfrac{t_d}{\left[(1-f_w)-(\%\ \text{submergence})/100\right]}$

2. Required filtration parameter (C)

$$C = \frac{1}{\left[\dfrac{C_i}{(100-C_i)} - \dfrac{C_f}{(100-C_f)}\right]}$$

$$= \frac{1}{\left(\dfrac{90}{100-90} - \dfrac{70}{100-70}\right)}$$

$$= \frac{1}{9.0-2.33} = 0.15\ \text{g/cm}^3$$

3. Required total cycle time (t_c), form time (t_f) and wash time (t_w)

   $t_w = 0.2t_c$, $f_w = 0.2$

   $t_d = 0.6$ min

$$t_c = \frac{t_d}{\left[(1-f_w)-(\%\ \text{submergence})/100\right]}$$

$$t_c = \frac{0.6}{\left[(1-0.2)-50/100\right]} = 2\ \text{min}$$

$$t_f = \frac{t_c(\%\ \text{submergence})}{100} = \frac{2(50)}{100} = 1\ \text{min}$$

$$t_w = 0.2 \times 2 = 0.4\ \text{min}$$

$$t_c = t_f + t_d + t_w$$

$$= 1 + 0.6 + 0.4 = 2\ \text{min (OK)}$$

4. Required filter loading, cycle loading, and area

$$\text{Filter loading}(L_f) = 1118\left(\frac{PC}{\mu\, r\, t_f}\right)^{0.5};$$

$$P = \frac{30\ \text{in. Hg}}{29.92\ \text{in. Hg}} \times 14.7\ \text{psi} = 14.74\ \text{psi}$$

$$= 1118\left[\frac{14.74(0.15)}{(1.0)(2.416\times 10^4)(1.0)}\right]^{0.5}$$

$$= 8.9\ \text{lb/ft}^2.\text{h}$$

$$\text{Cycle loading}(L_c) = L_c\left(\frac{t_f}{t_c}\right)$$

$$= 8.9\left(\frac{1}{2}\right) = 4.45\ \text{lb/ft}^2.\text{h}$$

$$\text{Filter area}(A) = \frac{\text{lb/h of solids to be removed}}{F \times L_c}$$

$$= \frac{1000}{0.8(4.45)} = 281\ \text{ft}^2$$

$$\text{Cake yield} = \frac{1000\ \text{lb/h}}{281\ \text{ft}^2} = 3.56\ \text{lb/ft}^2.\text{h}$$

Note :

1. Ease of filterability of wastewater sludges

   It decreases with the degree of treatment, *i.e.*,
   r(raw sludge) < r(primary sludge) < r(secondary sludge)

   Filterability is influenced by particle size, shape and density, and electrical charge on the particle. The larger the particle size, the higher is the filtration rate (lower specific resistance), and thus the final cake moisture is lower. Addition of polymers helps in agglomeration of particles, and thereby increasing filtration rate.

2. Vacuum application
   - A vacuum of 10 in. Hg [P (vacuum) = 0 – 10 in. Hg], then DP = – (– 10) = 10 in. Hg
   - A vacuum of 10 in. Hg [P = –10 in Hg gauge] = (29.92 – 10)

$$= 19.92 \text{ in Hg absolute pressure} = \frac{19.92 \text{ in. Hg}}{29.92 \text{ in. Hg}} \times 14.7$$

$$= 9.79 \text{ psi}$$

3. Units of specific resistance of cake (r)
   - Applied vacuum (in. Hg)

     (in. Hg)(33864) = dyne/cm^2
   - Applied vacuum in psi

     (psi)(68948) = dyne/cm^2
   - Filtration parameter (C) in g/mL or g/cm^3
   - $S = \frac{(t/V)}{V} = \frac{\text{see/mL}}{\text{mL}} = \left(\text{see/cm}^3\right)/\text{cm}^3 = \text{s/cm}^6$
   - Dynamic viscosity (m) = g/cm.s = Poise
   - Specific resistance $(r) = \frac{\left(2P\,A^2\right)}{(\mu\,C)S}$

$$= \frac{\left(\text{Dyne/cm}^2\right)\left(\text{cm}^4\right)}{(\text{g/cm.s})\left(\text{g/cm}^3\right)\left(\text{s/cm}^6\right)}$$

$$= \frac{\text{g}\left(\text{cm/s}^2\right)/\text{cm}^2\left(\text{cm}^4\right)(\text{s})}{\text{g}(\text{cm/s})\left(\text{g/cm}^3\right)\left(\text{cm}^6\right)} = \text{cm/g}.$$

Example 3.63

Vacuum Filtration

Determine the value of the specific resistance for the activated sludge. Assume the following data:

Vacuum values (Dp) : 15, 30, 60 cm Hg

Filter diameter : 4.7 cm (Lab model)

Sludge concentration : 25 g/L

Volume of filtrate (V_m) collected at various vacuum conditions under

Table 62 : Laboratory conditions

Time (min)	*10*	*20*	*40*	*80*	*120*	*160*	*200*	*240*	*280*	*320*
Vacuum [Dp, cm Hg], Filtered collected [V_m (cm^3)]										
15	11.7	16.9	24.3	34.7	42.7	49.3	55.3	60.7	65.6	70.2
30	15.4	22.2	31.9	45.6	56.1	64.9	72.7	79.8	86.8	96.3
60	20.0	28.8	41.3	59.1	72.7	84.1	94.2	103.4	111.8	119.6

Solution

1. Required specific resistance of sludge at various Dp's

   The basic equation is expressed as:

$$\frac{1}{V_w} = \frac{\mu\, m\, r\, V_w}{2A^2 \Delta p} + \frac{\mu\, R_c}{A\, \Delta p}$$

where A = Cake area

V_w = Volume of water filtered (filtrate collected)

m = Mass of solids deposited per unit of filtrate

r = Specific resistance of the filter cake

R_c = Initial resistance including the resistance of the filter cloth and initial layer of cake

m = Liquid viscosity

Plot t/V_w versus V_w at three vacuum conditions:

The three slopes $\left[\dfrac{\mu\,(m)\, r\, V_w}{2A^2 \Delta p}\right]$ are:

Slope (15) = 3.73

Slope (30) = 2.20

Slope (60) = 1.30

$$r = \frac{2(\text{Slope})_p\, \Delta p\, A^2}{\mu(m)},\ m = 25\ \text{g/L}\left[= 0.025\ \text{g/cm}^3\right]$$

$$r(15) = \frac{2(3.73)(15)(13.5)(981)\left[\frac{\pi}{4}\right]^2 (4.7)^4}{0.01(0.025)}$$

$= 1.79 \times 10^{12}$ cm/g

$$r(30) = \frac{2(2.20)(30)(13.5)(981)\left[\frac{\pi}{4}\right]^2 (4.7)^4}{0.01(0.025)}$$

$$= 2.13 \times 10^{12} \text{ cm/g}$$

$$r(60) = \frac{2(1.30)(60)(13.5)(981)\left[\frac{\pi}{4}\right]^2 (4.7)^4}{0.01(0.025)}$$

$$= 2.50 \times 10^{12} \text{ cm/g}$$

Initial resistance is estimated as follows:

$$R_{cp} = \frac{A\Delta p}{\mu}\left[\frac{t}{V_w} - \frac{\mu(m)(r)(V_w)}{2A^2\Delta p}\right]$$

At, t = 0, all the three plots coverage at V_w = -3.0 cm^3

$$R_{cp} = \frac{A\Delta p}{\mu}\left[\frac{\mu(m)}{2A^2}\left\{\frac{2.Slope.\Delta p\ A^2}{\mu(m)\Delta p}\right\}V_m\right]$$

$$= -\frac{A\Delta p}{\mu}.\frac{\mu(m)(Slope)\Delta p\ A^2}{(2A^2)\mu(m)\Delta p}.V_m$$

$$= -\frac{A(\Delta p)(Slope)V_m}{(\mu)}$$

$$R_C(15) = \frac{-\frac{\pi}{4}(4.7)^2(15\times13.5\times981)(-3)(3.73)}{0.01}$$

$$= 3.86 \times 10^9 \text{ cm}^{-1}$$

$$R_C(30) = \frac{\frac{\pi}{4}(4.7)^2(30\times13.5\times981)(3)(2.20)}{0.01}$$

$$R_C(30) = \frac{\frac{\pi}{4}(4.7)^2(30\times13.5\times981)(3)(2.20)}{0.01}$$

$$= 4.55 \times 10^9 \text{ cm}^{-1}$$

$$R_C(60) = \frac{\frac{\pi}{4}(4.7)^2(60\times13.5\times981)(3)(1.30)}{0.01}$$

$$= 5.38 \times 10^9 \text{ cm}^{-1}$$

The schematic diagram is given as follows:

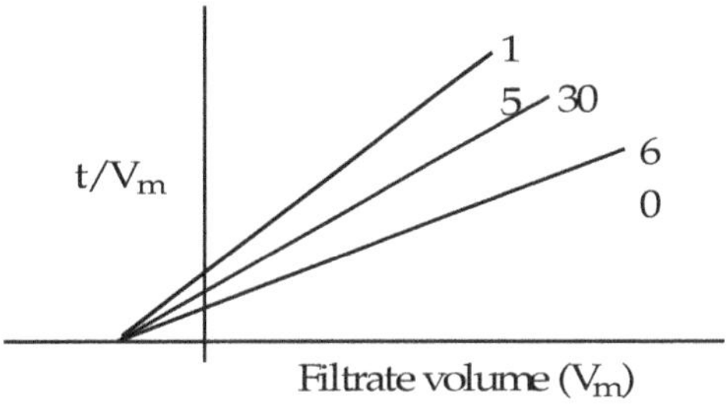

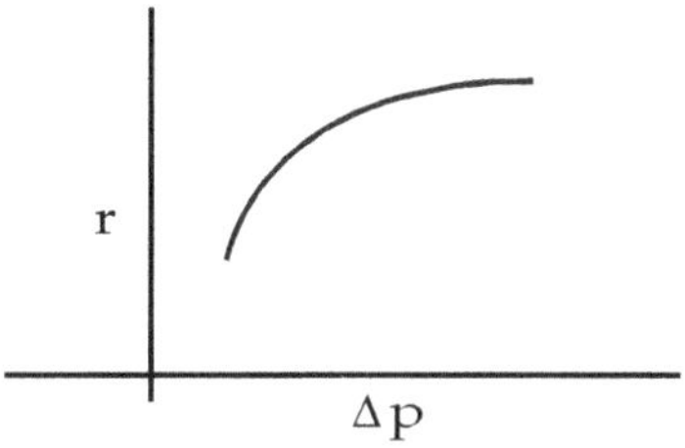

Similarly, filter resistance can be calculated, as all values are known. These points are not sufficient to describe the relationship between r and Dp. Once these values are known, the expression can be used in the design of continuously operating units (it is necessary to remember that vacuum is applied for only a fraction of the cycle time).

Example 3.64

Vacuum Filtration

[A] Calculate the weight of cake produced and volume of filtrate per 1000 gal of 5.0% sludge applied and also the chemical consumption (polymer) per tonne of dry solids.

Solution

1. Polymer dose (Assume filter yield of 7.0 lb/ft^3.hr with cake density of 29% solids)

   At the filter yield, it has been found that polymer dosage is 0.6%

   Polymer dosage $= 0.006 \times 2000 = 12$ lb/ton dry solids

2. Dry solids per 1000 gal of sludge

$$= 0.05 \times 1000 \times 8.34 = 417 \text{ lb}$$

3. $$\frac{\text{Weight of cake produced}}{\text{1000 gal of sludge filtered}} = \frac{417 \text{ lb}}{0.29} = 1438 \text{ lb}$$

4. Cake volume

$$= \frac{\text{Weight}}{\text{Sp. gravity} \times 8.34 \times S} = \frac{1438}{1.05 \times 8.34} = 164 \text{ gal}$$

5. Filtrate volume

$$= (1000 - 164) = 836 \text{ gal}$$

[B] Calculate the specific resistance for a sludge. Assume the following:

Filter diameter	: 7.5 cm

Pressure applied an a Buchner funnel : 10 psi (vacuum)

For 40 mL of filtrate, the dried cake solids : 4.0 g

The slopes of t/V versus V is 0.72 s/mL = b

$$P = 10 \text{ psi} = 703 \text{ dynes/cm}^2 = 6.90 \times 10^4 \text{ N/m}^2$$

$$A = 3.14(3.75 \text{ cm})^2 = 44.2 \text{ cm}^2 = 0.00442 \text{ m}^2$$

$$b = \frac{0.72 \text{ seconds}}{\text{mL}^2}\left(\frac{100 \text{ cm}}{\text{m}}\right)^6 = 72 \times 10^{10} \text{ s/m}^6$$

$$\mu = 1.308 \text{ Cp at } 50^\circ \text{F} = 0.0013 \text{ N-s/m}^2$$

$$\text{Weight} = \frac{4 \text{ g}}{50 \text{ mL}} = 0.08 \text{ g/mL} = 80 \text{ kg/m}^3$$

$$r = \frac{2PA^2 b}{\mu W}$$

$$= \frac{2 \times 6.9 \times 10^4 \times (0.00442)^2 \times 72 \times 10^{10}}{0.0013 \times 80} = 1.89 \times 10^{13} \text{ m/kg}.$$

Note :

1. The theoretical filtration equation is $\frac{dV}{dt} = \frac{PA^2}{\mu(r W V + R_f A)}$ for constant pressure through out filtration, integration of this equation yields:

$$\frac{t}{V} = \frac{\mu r WV}{2PA^2} + \frac{\mu R_f}{PA}$$

$$\frac{t}{V} = bV + a \quad \left[\text{plot of } \frac{t}{V} \text{ versus V gives slope} = b\right]$$

$$b = \frac{\mu r W}{2PA^2}$$

or $$r = \frac{2PA^2 b}{\mu W}$$

where V is the filtrate volume (m^3 or mL), t is the time (sec), P is the pressure difference (Newtons/m^2 or dynes/cm^2), A is the filter area (m^2 or cm^2), μ is the viscosity of filtrate (N-sec/m^2 or poise), r is the specific resistance of sludge cake (m/kg or sec^2/g), W is the mass of the dry solids per unit volume of filtrate (kg/m^3 or g/cm^2), and R_f is the resistance of filter medium (sec^2/cm^2).

2. This equation assumes (a) laminar flow (b) uniform solids deposition during filtration and a constant increase in filtrate flow resistance as cake increases in thickness.

Example 3.65

Vacuum Filtration of Wastewater Sludges

An ETP produces 0.04 mil gal/d of sludge containing 6% solids what size vacuum filters are required, assuming a yield of 5.5 lb/ft^2.hr and 10 hours of operation per day. Chemicals required for sludge conditioning are 9% lime and 2.5% $FeCl_3$. The Cake is 25% solids (organics + chemical additions).

Solution

1. Dry sludge solids

$$= 40{,}000 \times 8.34 \times 0.06 = 20{,}016 \text{ lb/d}$$

2. Filter area

$$= \frac{20{,}016 \text{ lb/d}}{5.5 \text{ lb/ft}^2.\text{hr} \times 12 \text{ hr/d}} = 303 \text{ ft}^2$$

3. Chemical dosage

3.1 Lime

$$= 0.09 \times (2000) = 180 \text{ lb/ton dry solids (1 ton = 2000 lb)}$$

3.2 $FeCl_3$

$$= 0.025 \times (2000) = 50 \text{ lb/ton dry solids}$$

4. Total cake solids

$$= \text{Sludge solids} + \text{Chemical additions}$$

$$= 20{,}016 + (0.09 + 0.02) \times 20{,}016$$

$$= 20{,}016(1 + 0.11) = 22{,}218 \text{ lb/d}$$

5. Weight of wet filter cake

$$= \frac{22{,}218 \text{ lb/d}}{(0.25 \times 2000 \text{ lb/ton})} = 44.43 \text{ ton/d}.$$

Table 63 : Typical values from vacuum filtration of chemically conditioned wastewater sludges

Sludge Type	*Filter Yield*	*Cake Solids*	*Chemical Dosage (%)*		
	(lb/ft^2.hr)	*(%)*	*$FeCl_3$*	*Lime*	*Polyelectrolyte*
Raw primary	6-10	25-40	1-4	6-8	1-4
Primary + TF humus	4-8	20-35	2-4	6-8	3-6
Primary + Waste activated	3-5	15-25	3-6	5-10	5-10
Anaerobically digested	5-10	20-35	3-6	5-10	4-10

Polyelectrolyte : Actual dosages range from 0.2 to 2 times these values 2% conditioner means 2 lb of coagulant per 100 lb of dry solids in the filter cake.

2. Filter yield

$$\text{Yield}\left(\text{lb/ft}^2.\text{hr}\right) = \frac{\text{Dry sludge solids (lb)} \times \text{cycles/hour}}{\text{Filter leaf area}\left(\text{ft}^2\right)}$$

Example 3.66

Vacuum Filtration for Thickener Sludge

Calculate the surface area required for a vacuum filter for thickening of sludge from a wastewater treatment facility-thickener. Assume the following data:

Vacuum operating cycle	:	20 hours per day
Solids content in the thickened sludge	:	3.0%
Feed sludge rate to vacuum filter (dry basis)	:	50,000 lb/d
Cycle time	:	10 minutes

Solution

The cake yield can be calculated using the following equation:

$$L = 3.85\frac{C^{0.767}}{T^{0.656}}$$

where L is the cake yield (lb/ft^2-hr), C is the percent solids in feed (%), and T is the cycle time (minutes/revolution)

$$L = 3.56\frac{(3)^{0.767}}{(10)^{0.656}}$$

Surface area for 20 hours/day basis and assume a safety factor of 0.60 for design purposes.

$$A = \frac{50,000}{(20\ \text{hour/d})(0.6)X} = \gamma\ \text{ft}^2.$$

Example 3.67

Rotary Vacuum Filtration

A STP produces 20,000 lb of dry solids per day at a concentration of 5 percent solids. Polymer is added as a dewatering conditioner. It has been observed that using a cycle time of 4 min, 40% submergence, and 25 in. vacuum, all producing an 85 percent moisture cake. Calculate the filter area for 100 hours per week.

Solution

1. Filter loading equation (L)

$$L = 35.7\left(\frac{x\,CP}{\mu\,R\,t}\right)^{0.5}$$

(Specific resistance) $r = 0.40 \times 10^7\ P^{0.9}$

$= 0.40(25)^{0.9} = 7.25$

$R = r \times 10^7\ sec^2/g$

$= 7.25 \times 10^{-7} \times 10^7\ sec^2/g$

$= 7.25\ sec^2/g.$

2. Solids deposited per unit filtrate volume (C)

$$C = \frac{1}{\left(\frac{C_i}{100 - C_i}\right) - \left(\frac{C_f}{100 - C_f}\right)}$$

where C_i is the % solids by weight in the influent, and C_f is the % solids by weight in the filter cake

$$C = \frac{1}{\left(\frac{95}{5} - \frac{85}{15}\right)} = \frac{1}{13.33} = 0.075$$

3. Filter loading

$$L = 35.7\left[\frac{(0.40)(0.075)(12.25)}{1.0 \times 7.25 \times 4}\right]^{0.5} = 4.02\ lb/ft^2 - hr$$

where t is the cycle time (min), x is the fraction of cycle time for cake formation, μ is the filtrate viscosity (centipoise), C is the solids deposited/unit of filtrate (g/mL), P is the pressure (lb/m^2), and R is the $r \times 10^7\ sec^2/g$.

4. Required filter area (A)

$$= \frac{20{,}000\ lb/day \times 7\ day/week}{100\ hr/week} = 1400\ lb/hr$$

$$\text{Required filter area}\,(A) = \frac{1400}{4.02} = 348\ ft^2$$

Note :

The fundamental filtration equation is:

$$\frac{t}{V} = \frac{\mu\, r\, C}{2P\,A^2}V + \frac{\mu\,R_m}{PA}.\,V; \qquad [R_m \text{ is the filter media resistance}]$$

or $$\frac{V^2}{A^2} = \frac{2Pt_f}{\mu r C}; \quad t_f = x\,t$$

If W = dry weight of cake = CV

$$\frac{C^2V^2}{A^2} = \frac{2P \times t\,C}{\mu r C} = \frac{W^2}{A^2}$$

$$\frac{W}{A} = \left(\frac{2P \times t\,C}{\mu r}\right)^{0.5}$$

$$C = \frac{1}{\left(\frac{C_i}{100 - C_i}\right) - \left(\frac{C_f}{100 - C_f}\right)}$$

Filter loading

$$L = \frac{W}{A} \cdot \frac{60}{t}$$

or $$L = 35.7\left(\frac{P \times C}{\mu R t}\right)^{0.5}$$

and $R = r \times 10^7\ sec^2/g$.

Example 3.68

Rotary Vacuum Filter

Use the data obtained from the Buchner funnel apparatus for the determination of specific resistance in the laboratory at 60°F. This data are used to design a rotary vacuum filter operating with a vacuum of 20 in Hg, a total cycle time of 1.5 min, submergence of 24%, and a sludge cake concentration 50 kg/m^3 of filtrate. Determine the yield of the vacuum filter. Assume the following data :

Table 69 : Buchner funnel data for design of rotary vacuum filter

Time (t, min.)	1	2	3	4	5
Filtrate volume (V, mL)	5.2	9.2	12.4	16.0	18.4
t/v (s/mL)	11.5	13.0	14.5	15.0	16.3

- Buchner funnel area (A) : 45 cm^2.

Solution

1. Required value of the slope for t/V versus V plot and a

$$\frac{t}{V} = \left[\frac{\mu C \alpha}{2\Delta p A^2}\right] V + \frac{\mu R_m}{\Delta p A} \qquad (1)$$

Since liquid is contained in the sludge, the mass of cake/volume of filtrate is given by the material balance [C is not the same as initial solids concentration (C_o)] :

$$C = \frac{C_c (C_f - C_o)}{(C_o - C_c)} \qquad (2)$$

where C_c = Solids concentration of sludge cake

C_f = Solids concentration of the filtrate

Total cycle time for a continuous vacuum filter is given by:

t_c = t (cake forming time) + t_d (dewatering time) + t_d (cake discharge time)

The cake formation time (t) is only a fraction (f) of the total cycle time (t_c):

$$t = f\ t_c \qquad (3)$$

Rearranging equation (1) and substituting the value of t gives :

$$\frac{CV}{A t_c} = G = \frac{\left[\left(\frac{2C \alpha \Delta p f}{\mu t_c}\right) + \left(\frac{R_m}{t_c}\right)^2\right]^{0.5} - \frac{R_m}{t_c}}{\alpha}$$

Mass of cake formed in a cycle/unit area is called filter loading (G). If the resistance of the filter medium is negligible confered to the sludge cake, the filter loading is :

$$G = \frac{C V}{A t_c} = \left[\frac{2C \Delta p f}{\alpha \mu t_c}\right]^{0.5}$$

$\alpha = \alpha_o (\Delta p)^s$ [Variation of specific resistance with applied vacuum]

where s = compressibility coefficient

- Plot t/V versus V

$$\text{Slope} = \frac{\mu C \alpha}{2\Delta p A^2} = 0.374 \text{ s/cm}^6$$

Δp = 20 in. Hg = 67.7×10^4 dynes/cm^2

$$\alpha = \frac{(0.374)(2)(\Delta p) A^2}{\mu C}$$

$$= \frac{(0.374)(2)(67.7 \times 10^4)(45)^2}{(0.011)(0.05)} = 1.9 \times 10^{12} \text{ cm/g}.$$

2. Required vacuum filter yield (loading)

$$G = \left[\frac{2C\Delta p\, f}{\alpha \mu\, t_c}\right]^{0.5}$$

$$= \left[\frac{2(0.05)(67.7 \times 10^4)(0.20)}{1.9 \times 10^{12}(0.011)(1.5 \times 60)}\right]^{0.5}$$

$= 0.848 \times 10^{-4}$ g/cm^2.s.

Note :

1. Buchner funnel apparatus determines the value of specific resistance and compressibility coefficient of the sludges.
2. Filter leaf test is used to design a vacuum filtration system.

   A filter leaf consists of a round permeable disc (about 10 cm dia.), over which the filter medium is stretched. The disc is connected to a vacuum source through a graduated cylinder in which the filtrate is collected. The filter leaf is immersed in the sludge for the expected time of submergence in a rotating filter. The filter leaf is then removed from the sludge for the time period the filter would be in the drying phase on a rotary filter. Sludge cake is then scrapped off the filter leaf, dried and weighed.

   The filter yield (G) is :

$$G = \frac{W}{A\, t_c}$$

where W = Weight of dried cake

A = Area of filter leaf

t_c = Total cycle time.

Example 3.69

Rotary Vacuum Filtration for Wastewater Sludges

Determine the required area for a rotary filter to process sludge containing 10 tons/d of dry solids. Assume the following data :

Experimental data :

Cake form time	: 1.5 min
Dry time	: 2.5 min

Filter area	:	0.1 ft^2
Weight of dry solids deposited	:	0.06 lb
Vacuum (pressure)	:	10 psi
Final moisture content	:	20%
Filter operation	:	10 hrs/d.

Solution

1. Required cycle time

$$t_c = t_f + t_d + t_w$$

where t_f = Cake formation time

t_d = Dry time

t_w = Cake removal time (typically = 0.2 t_c)

$$t_c = 1.5 + 2.5 + 0.2\ t_c$$

$$t_c = \frac{4.0}{0.8} = 5 \text{ min}$$

2. Required filter loading (L)

$$L = \frac{W}{At_c}$$

where W = Filter load (lb/ft^2.d)

A = Filter leaf area

t_c = Cycle time

$$L = \frac{0.06 \text{ lb}}{\left(0.1 \text{ ft}^2\right)\left(\dfrac{5 \text{ min}}{60 \text{ min/hr}}\right)} = 7.2 \text{ lb/ft}^2\text{.hr}$$

3. Required rate of solids removal (R) and the filter area (A)

$$R = \frac{(10 \text{ tonnes/d})(2000 \text{ lb/tonne})}{10 \text{ hrs/d}} = 2000 \text{ lb/hr}$$

$$A = \frac{2000 \text{ lb/hr}}{7.2 \text{ lb/ft}^2\text{.hr}} = 278 \text{ ft}^2$$

Example 3.70

Air Suction Rate in Rotary Vacuum Filters [Estimation of Power Requirements of Vacuum Pumps]

1. Prediction of air suction rate

The rate at which air sucked through the dewatering of a rotary vacuum filter can be expressed in a form similar to filtration equation :

$$\frac{dV}{dt} = \frac{A\psi\Delta P}{(R_k + R_f)\mu} \quad (1)$$

where V is the volume of air at temperature and pressure of surroundings sucked through the cake in time-t, ψ is the fraction of total surface available for air suction, and ¼ is the viscosity of air at temperature and pressure of surroundings.

The resistance of the cake (R_k) and filter media (R_f) are proportional to the cake thickness (l), and fictitious cake thickness (l_f)

$$R_k + R_f = C(l + l_f) \quad (2)$$

where C is the constant.

If the cake is non-compressible, l must be equal to the thickness of the cake leaving the filtering zone.

$$l + l_f = \frac{W(V_R + Q\psi V_f)}{\rho_c A} \quad (3)$$

where V_f is the fictitious volume of filtrate per unit of air-suction area necessary to lay down a cake thickness l_f

$$\frac{dV}{dt} = \frac{A^2\psi\Delta P}{\beta W(V_R + A\psi\, V_f)\mu} \quad (4)$$

$^2 = C/\rho_c$ (specific air-suction cake resistance)

Integrating between the limits corresponding to V = 0 and V = V_R (where V_R is the volume of air per revolution) :

$$V_R = \frac{A^2\psi\Delta P}{\rho_w (V_R + A\psi V_f)\mu\, N} \quad (5)$$

$$\beta = \beta_o (\Delta P)^s$$

Neglecting the filter medium resistance :

- Volume of air per revolution (V_R) $= \dfrac{A^2\psi\Delta P}{\beta\, W\mu\, N}$ (6)
- Volume of air per unit time (V_R N) $= \dfrac{A^2\psi\Delta P}{\beta\, W\mu}$ (7)
- Volume of air per unit time $= \dfrac{A\psi}{\beta\mu}\left(\dfrac{\alpha\,\mu\, W\,\Delta P}{Z\, W\,\psi}\right)^{0.5}$ (8)

$$\bullet \quad \frac{\text{Volume of air per unit time}}{\text{Weight of dry cake per unit time}} = \left(\frac{\psi_a}{\psi_f}\right)\left(\frac{\mu}{\mu_a}\right)\left(\frac{r}{2\,\beta\,W}\right) \tag{9}$$

If the constants for a given filter (ψ_f, ¼, r, W) are known and the assumption of no air leakage is adequate, the total amount of suction air can be estimated. This value combined with a knowledge of air temperature and the pressure at the intake and delivery side of the vacuum pump can be used to estimate the power requirements of the vacuum pump.

Table 65 : Estimation of house power motor required for vacuum pump on a rotary vacuum filter.

Slurry contains	:	20 lb of water/lb of solid material
Ratio of specific resistance (F/²)	:	0.6
Filtrate	:	19 lb of filtrate (not including washwater) is obtained for each 21 lb of slurry
Temperature of surrounding air and that of slurry	:	70°F
Pressure drop to be maintained by the vacuum pump	:	5 psi
Fraction of the drum submerged in the slurry	:	0.3
Fraction of the drum area available for air suction	:	0.1
Slurry to be handled	:	50,000 lb/hr
Any effects caused by air leakage are taken into account in the values for (r/p) Cp/Cv	:	1.4
Vacuum pump and motor have an overall efficiency based on isentropic compression	:	50%

The value of F/2 is based on the temperature and pressure of the air surrounding the filter.

Filter removes all of the solid from slurry.

2.1 Volume of air per unit time Eq. (9) :

$$\frac{\text{Volume of air per unit time}}{\text{Weight of dry cake per unit time}} = \left(\frac{\psi_a}{\psi_f}\right)\left(\frac{\mu}{\mu_a}\right)\left(\frac{r}{2\,\beta\,W}\right)$$

$$\psi_a = 0.1$$

$$\psi_f = 0.3$$

μ for water = 0.982 Cp (at 70° F) = 0.982 × 2.42 lb/ft – hr

μ_a (for air) at 70° F = 0.0184 Cp = 0.018 × 2.42 lb/ft – hr

$$\frac{\rho}{\beta} = 0.6$$

Density of water at 70°F = 62.3 lb/ft^3

Pounds filtrate per lb of dry cake = 19 lb

$$W = \frac{1}{19/62.3} = 3.28 \text{ lb dry-cake solids/ft}^3 \text{ filtrate}$$

Weight of dry cake/hr = (50,000 × 1/21) = 2380 lb/hr

$$\text{Volume of air/hr} = \frac{(2380)(0.1)(0.982 \times 2.42)(0.6)}{(0.3)(0.018 \times 2.42)(2 \times 3.28)}$$

= 3960 ft^3 at 70° F and 1 atm.

2.2 Horse power equipments

Theoretical horse power for isentropic single-stage compression :

$$= \frac{3.03 \times 10^{-5} k}{(k-1)} \left(P_1 \, q_{fm_1}\right) \left[\left(\frac{P_2}{P_1}\right)^{\frac{k-1}{k}} - 1\right]$$

where k is the ratio of heat capacities at constant pressure and volume (Cp/Cv) = 1.4

P_1 is the vacuum intake pressure = (14.7 - 5) × 144 psf

P_2 is the vacuum delivery pressure = 14.7 × 144 psf

q_{fm_1} is the ft^3 of air per minute at vacuum intake

$$\text{Condition} = \frac{(3960)(14.7)(144)}{(60)(14.7-5)(144)} = 100 \text{ ft}^3/\text{min at } 70^\circ\text{F and 9.7 psi}$$

Required horse power of motor for vacuum pump (hp)

$$h_p = \frac{3.03 \times 10^{-5} \times 1.4(14.5-5)(144) \times 100}{(0.5)(1.4-1)} \left[\left(\frac{14.7}{9.7}\right)^{\frac{1.4-1}{1.4}} - 1\right]$$

= 3.7 hp (say 4 hp).

Example 3.71

Continuous Filters

1. Derive the general design equations for continuous rotary or rotary-disk filters.

   It is convenient to develop the design equation in terms of the total area available for filtering service, even through only a fraction of this area is in direct use at any time. A is the total area available and the fraction of this area immersed in the slurry is represented as ψ. The effective area of the filtering surface becomes A È and the basic filtration equation becomes:

$$\frac{dV}{dt} = \frac{A\psi\,\Delta P}{(R_K + R_f)\mu} \tag{1}$$

Cake resistance (R_k) varies directly with thickness (l) of the cake, and the proportionality can be expressed as

$$R_k = C\,l \text{ at time t} \tag{2}$$

The fictitious resistance of the medium (R_f) various directly with the fictitious cake thickness (l_f) :

$$R_f = C\,l_f \tag{3}$$

Total resistance is :

$$R_k + R_f = C(l + l_f) \tag{4}$$

With a continuous filter, the cake thickness at any given location on the submerged filtering surface does not vary with time.

The thickness, however, does vary with location as the cake builds up on the filtering medium during the passage through the slurry. The thickness of the cake leaving the filtering zone is a function of the slurry concentration, cake density, and volume of the filtrate delivered per revolution. This thickness can be expressed by the following equation:

$$l(\text{leaving filtereing zone}) = \frac{W\,V_R}{\rho_c\,A} \tag{5}$$

where V_R is the volume of filtrate delivered per revolution, ρ_c is the cake density (lb-dry solids per unit volume of wet filter cake leaving filter zone).

An average cake thickness during cake deposition period can be assumed to be one half of the thickness at the entrance of the filtering zone. Since no appreciable amount of cake should be present on the filter when it enters the filtering zone :

$$l_{avg} = \frac{W\,V_R}{2\rho_c\,A} \tag{6}$$

$$(l + l_f) = \left(l_{avg} + l_f\right) = \frac{W\left(V_{R/2} + A\psi\,V_f\right)}{\rho_c\,A} \tag{7}$$

$$\text{Therefore, } \frac{dV}{dt} = \frac{2A^2\,\psi\Delta P}{\alpha\,W\left(V_R + 2A\,\psi\,V_f\right)\mu}, \qquad r = \frac{C}{\rho_c} \tag{8}$$

Integrating between limits of V = 0 and V = V_R and t = 0 and t = 1/N (N is the No. of revolution per unit time) :

$$V_R^2 + 2A\,\psi\,V_f\,V_R = \frac{2A^2\,\psi\Delta P}{r\,W\,\mu\,N} \tag{9}$$

or $$V_R^2 + 2A\,\psi\,V_f\,V_R = \frac{2A^2\,\psi(\Delta P)^{1-s}}{r_o\,W\,\mu N} \tag{10}$$

If V_f fictitious volume of filtrate per unit of filtering area) based is small as the resistance of the filtering medium is negligible and the filter cake is non-compressible :

- Volume of filtrate per revolution (V_R) = $A\sqrt{\dfrac{2\psi\Delta P}{r\,W\,\mu\,N}}$ (11)

- Volume of filtrate per unit time (V_R N) = $A\sqrt{\dfrac{2\psi\,N\Delta P}{r\,W\,\mu}}$ (12)

- Weight of dry cake per unit volume (V_R NW) = $A\sqrt{\dfrac{2\psi\,N\,W\,\Delta P}{r\,\mu}}$ (13)

2. Effect of pressure difference on capacity of a rotary vacuum filter

   A rotary vacuum filter with negligible filter medium resistance delivers 225 ft^3 of filtrate per hour when a given sludge is filtered under known conditions. How many ft^3 of filtrate will be delivered per hour if the pressure drop (DP) over the cake is doubled, all other conditions remaining the same? Assume the sludge filter cake is non-compressible.

   Using equation (12).

$$(V_R\,N)_2 = 100 = A\sqrt{\frac{2\psi\,N\,\Delta P}{r\,W\,\mu}}$$

$$\frac{(V_R\,N)_2}{(V_R\,N)_1} = \frac{A\sqrt{2\,N/r\,W}\,(\Delta P)_2^{0.5}}{A\sqrt{2\psi}\,N/rW\,(\Delta P)_1^{0.5}}$$

$$(V_R\,N)_2 = (V_R\,N)_1\left[\frac{(\Delta P)_2}{(\Delta P)_1}\right]^{0.5} = 100\left(\frac{2\,\Delta P_1}{\Delta P_1}\right)^{0.5}$$

$$= 1.41 \times 100 = 141 \text{ ft}^3/\text{h}.$$

Example 3.72

Rotary Dryers

1. Derive an expression for dryer length and the retention time in a counter-current rotary air dryers. The following assumptions are considered for mathematical analysis of rotary dryers.

   - No heat transfer takes place between the dryer and the surroundings
   - All required heat is supplied by the hot air

- No evaporation and drying takes place during the preliminary, and final heatup periods
- Only free moisture is present (unbound)
- All drying takes place at the wet bulb temperature.

A schematic diagram of a rotary counter current air dryer is shown in Figure 28.

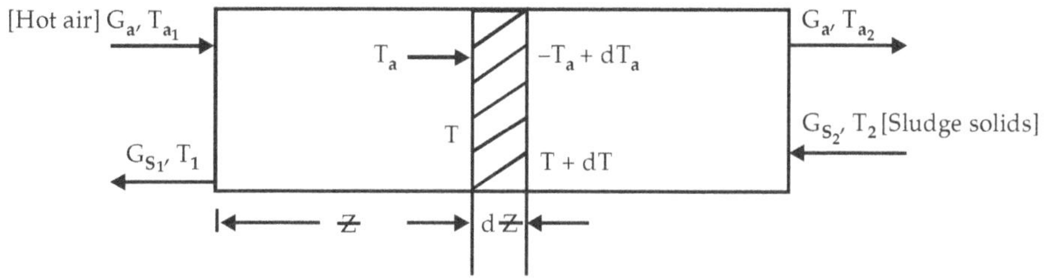

Figure 28 : Schematic of rotary dryer.

Air cools through its course in the dryer by supplying the necessary heat of evaporation and by heating the solids.

A steady state heat balance across a section (dZ) for the sludge solids and air is expressed as

$$dQ = -G_s A_c \, C_s dT = -G_a \, A_c f \, d \, T_a \tag{1}$$

where G_s = Mass flux of solids through the dryer

G_a = Mass flux of air through the dryer

A_c = Dryer cross-sectional area

C_s = Specific heat of the solids

f = Humid heat of the air [Amount of BTU required to raise the temperature of and lb of dry air and its vapour content by °F]

T = Solids temperature

T_a = Air temperature

$$dQ = U_s A_s A_c \left(T_a - T\right) dZ \tag{2}$$

where A_s = Surface area of the solids particles/volume

U = Overall heat transfer coefficient

Combining equations (1) and (2) :

$$-G_c A_c f dT_a = U A_s A_c \left(T_a - T\right) dZ \tag{3}$$

Integrating equation (3) between T_{af} and T_{ai} and Z_f and Z_i, assuming constant f and UA_s :

$$-\int_{T_{af}}^{T_{ai}} \frac{dT_a}{(T_a - T)} = \int_{Z_f}^{Z_i} \frac{UA_s}{G_a f} dZ = \frac{UA_s (Z_i - Z_f)}{G_a f} \tag{4}$$

In evaporating zone, the temperature of the solids is the wet bulb temperature (T_i) which is constant :

$$-\int_{T_{af}}^{T_{ai}} \frac{dT_a}{(T_a - T_i)} = \ln\left[\frac{T_{af} - T_i}{T_{ai} - T_i}\right] \tag{5}$$

Constant temperature zone equation becomes :

$$\frac{UA_s}{G_a f}(Z_i - Z_f) = \ln\left[\frac{T_{af} - T_i}{T_{ai} - T_i}\right] \tag{6}$$

The heatup zone (f and C_s are constant) :

$$dT = \frac{G_a f}{G_s C_s} dT_a \tag{7}$$

[T is a Linear function of T_a between T_2 and T_i, Figure 29]

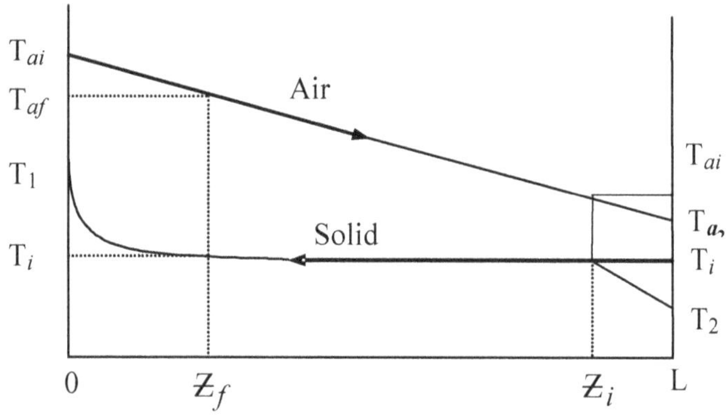

Figure 29 : Plot of Temperature Vs. Length of dryer.

Integrating,

$$-\int_{T_{ai}}^{T_{a2}} \frac{dT_a}{(T_a - T)} = \frac{T_{ai} - T_{a_2}}{(T_a - T)_m} \tag{8}$$

where $(T_a - T)_m$ = Logarithmic average temperature

$$= \frac{(T_{ai} - T_i) - (T_{a_2} - T_2)}{\ln\left[\left(\frac{T_{ai} - T_i}{T_{a_2} - T_2}\right)\right]} \tag{9}$$

Therefore, heatup zone is expressed as :

$$\frac{UA_s = (L - \not{Z}_i)}{G_a\, f} = \frac{(T_{ai} - T_{a_2})}{(T_a - T)_m} \tag{10}$$

A similar expression holds for the final section of the dryer. In this section, the air temperature will vary between the inlet temperature T_{a_1} and the temperature T_{af} that corresponds to the end of the constant temperature section, Figure 29.

$$\frac{UA_s \not{Z}_f}{G_a\, f} = \frac{(T_{a_1} - T_{af})}{(T_a - T)_m} \tag{11}$$

where $(T_a - T)_m$ refers to a new set of temperature for the two streams, Figure 29.

The total dryer length will be the sum of the lengths for all sections above.

The equations apply only for the initial heatup period and for constant temperature drying, until the unbound water has evaporated [not for bound water].

A steady-state momentum balance on the particles in the dryer yields :

$$\rho_s\, V_s \frac{dV}{dt} + \rho_s g\, V_s \sin\theta - F_d = 0 \tag{12}$$

where ρ_s = Density of particle

V_s = Volume of particles

V = Velocity of particles along the dryer

t = Time of fall of the particles

F_d = Drag force

θ = Angle of the dryer with the horizontal

Drag force (F_d) is expressed as

$$F_d = C_o V_s U_a^n \tag{13}$$

where U_a = Air velocity

C_o and n = Constant

Integrating with initial condition, t = 0 V = 0 gives particles velocity :

$$V = \frac{t}{\rho_s}\left(C_o U_a^n - \rho_s\, g \sin\theta\right) \tag{14}$$

Average length ($\overline{Z}$) traveled by a particle per revolution can be determined in a similar fashion :

$$\overline{Z} = 1.85\, D_d \left(\frac{C_o U_a^n}{\rho_s\, g\, \text{Cos}\,\theta} - \tan\theta \right) \tag{15}$$

where : D_d = Dryer diameter

Retention time for a dryer of length (L), rotating at a rate of ω (rpm) is expressed as :

$$t = \frac{L}{1.85\, D_d\, (\omega) \left[\frac{C_o U_a^n}{\rho_s\, g\, \text{Cos}\,\theta} - \tan\theta \right]} \tag{16}$$

This expression is similar to the equation of Salman and Mitchell :

$$t = \frac{L}{C'D_d\,(\omega)(\theta - U_a)}, \quad C' = \text{Constant} \tag{17}$$

2. A rotary dryer is considered for drying of 10,000 lb/h of a partly dewatered municipal sludge. The sludge enters the dryer at 70°F with 25% solids and exits when the unbound moisture is evaporated, which corresponds to 33% solids. Air of wet bulb temperature of 75°F is heated to 260°F before entering the dryer counter currently to the sludge. Determine the required dryer length. Assume the specific heat of the solids (C_s) is 0.8 BTU/lb°F.

2.1 Required temperature T_{a_2}

It is assumed that the solids are raised fast to the wet bulb temperatures and remain at this temperature during drying. No heating of the solids takes place above the wet bulb temperature

T_{a_1} = 260°F and T_1 = 75°C

$$\ln\left[\frac{T_{a_1} - T_i}{T_{a_2} - T_i}\right] = 1.5$$

Solving for T_{a_2} ; T_{a_2} = 116.3°F

2.2 Required energy and mass balances

- Input mass :

Solids (m_s) = 0.25 × 10,000 = 2500 lb/h

Water (m_w) = 0.75 × 10,000 = 7500 lb/h

- Outlet mass :

$$\text{Total (m)} = \frac{2500}{0.33} = 7576 \text{ lb/h}$$

$$\text{Water } (m_w) = 0.67 \times 7576 = 5076 \text{ lb/h}$$

Water removed (W)

$$W = (7500 - 5076) = 2424 \text{ lb/h}$$

- Heat required to evaporate water (Q_V)

$$Q_V = WDH_V = (2424 \text{ lb/h})(1051 \text{ BTU/lb})$$

[ΔH_V assumed as 1051 BTU/lb]

$$Q_V = 2.547 \times 10^6 \text{ BTU/h}$$

Amount of air required

$$Q_v = -G_a A_c \left(T_{a_2} - T_{a_1}\right) \quad [1]$$

[Humidity factor (f) = 0.265 BTU/lb.°F at 75°F]

$$\text{Therefore, } G_a A_c = \frac{2.547 \times 10^6}{0.265(116.3 - 260)} = 6.69 \times 10^4 \frac{\text{lb dry air}}{\text{hour}}$$

2.3 Required dryer diameter

A 10% excess air is considered to account for heat losses

Therefore, $G_a A_c = 7.36 \times 10^4$ lb day air/h

[Amount of water in this air is neglected]

$$A_c = \frac{7.36 \times 10^4}{1000}$$

[Assuming, a superficial mass velocity (G_a) of 1000 lb/ft².h]

Therefore, $D_d = 9.68$ ft (≈ 10 ft and $A_c = 78.5$ ft²)

$$\text{Therefore, } G_a = \frac{7.36 \times 10^4}{78.5} = 937.6 \text{ lb/ft}^2\text{.h}$$

2.4 Required heat transfer coefficient using Friedman and Marshall equation

$$UA_s = 15\frac{G_a^{0.16}}{D_d} = \frac{15(937.6)^{0.16}}{12} = 3.74\frac{\text{BTU}}{\text{h.°F}}$$

2.5 Required dryer length (L)

Using equation (6), the length for the constant temperature zone (L_C) is given as :

$$L_C = \frac{G_a f}{UA_s} \text{In}\left[\frac{T_{af} - T_i}{T_{a_2} - T_i}\right] \quad (6)$$

$$= \frac{(937.6)(0.265)}{(3.74)} \text{In}\left[\frac{T_{af} - T_i}{116.3 - 75}\right]$$

$$= 99.6 \text{ ft}$$

[$T_{a2} = T_{ai}$ and $T_{af} = T_{a_1}$; pre-heating zone has been neglected]

$$\text{The ratio of } \frac{L}{D_d} = \frac{99.6\text{ft}}{10.0\text{ft}} = 9.96 \text{ (OK)}$$

Example 3.73

Gravity Dewatering

1. Derive an expression for the filtration time as a function of height of sludge due to gravity dewatering.

   Dewatering of sludge on sand beds and lagoons starts with gravity drainage of water (Cake filtration and flow through porous media under constant pressure, and the flow is laminar). Hagen-Poisseuille cannot describe the rate of drainage fully (resistance to flow depends on the shape and the size of the particles, the fraction of voids, the length of the tortuous path of the liquid through the cake and the resistance of the supporting bed). The driving force is the overall head loss (pressure drop) across the sludge and the bed.

   Kozney's equation for flow through packed beds (excluding support resistance) is given by the following relationship :

$$\frac{1}{A}\frac{dv}{dt} = u = \left[\frac{e^3}{C(1-e)^2\left(\frac{S}{V}\right)^2}\right]\frac{\Delta P_s}{\mu L} \qquad (1)$$

where $\frac{1}{A}\frac{dv}{dt}$ = Linear velocity of the liquid (m)

A = Bed cross-sectional area normal to the flow

e = Bed porosity (volume of voids/sludge cake volume)

C = Constant (= 5 for in compressible bed)

S = Surface area of a particle

V = Volume of a particle

ΔP_s = Pressure drop across the sludge cake

L = Sludge cake thickness

μ = Liquid viscosity

If C = 2, equation (1) can be reduced to the Hagen-Poisseuille equation.

Specific resistance of cake (α) is expressed as :

$$\alpha = \frac{C(1-e)\left(\frac{A}{V}\right)^2}{\rho_s\left(e^3\right)} \tag{2}$$

where ρ_s is the density of the solids

The weight of solid in cake is expressed as :

$$W_s = m_s\, g = \rho_s\, g\, (1 - e)\, L \tag{3}$$

$$= \rho_s\, g\, VX_s$$

where X_s is the weight fraction of initial solids.

Using equations (2) and (3), the Kozeny's equation (1) is re-expressed (pressure drop across the sludge cake) as :

$$\Delta P_s = \frac{(\alpha)(\mu)(m_s)(\mu)}{A} = \frac{\alpha\,\mu\,\rho\,X_s\,V_u}{A}$$

Pressure drop across the supporting bed (DP_b) can be determined as:

$$\Delta P_b = \mu\ u\ R_b \tag{5}$$

where R_b = Supporting bed resistance (L^{-1} old used supporting media have higher resistance than new ones).

Normally, the value of R_b ranges from 0.05 to 0.1 of the specific cake resistance (α)

$$R_b = 0.1\ R_b \tag{6}$$

Total pressure drop across the cake and the supporting media is given as

$$\Delta P = \Delta P_s + \Delta P_b \tag{7}$$

Kozeny's equation (1) can be re-expressed as

$$\frac{dt}{dv} = \frac{\mu}{A(\Delta P)}\left[\left(\frac{\alpha\,\rho\,X_s}{A}\right)V + R_b\right] \tag{8}$$

Integrating for the limits t = 0, V = 0 gives the filtration time (t) :

$$t = \frac{\mu}{\Delta P}\left[\frac{\alpha\,\rho x_s}{2}\left(\frac{V}{A}\right)^2 + R_b\left(\frac{V}{A}\right)\right] \tag{9}$$

[Effluent volume-time relationship is represented by a parabola and considering α = constant]

Rearranging equation (9) allows the determination of α and R_b by determining volume of filtrate collected (V) over a time (t) at a specified pressure gradient (ΔP)

$$\frac{t}{v} = \left[\frac{\rho\,\alpha\,\mu\,X_s}{2A^2\Delta P}\right]V + \frac{\mu\,R_b}{A\,\Delta P} \qquad (10)$$

$$\text{Slope} = \frac{\rho\,\alpha\,\mu\,X_s}{2A^2\Delta P}\left[\alpha \text{ can be determined}\right]$$

$$\text{Intercept} = \frac{\mu R_b}{A\Delta P}\left[R_b \text{ can be determined}\right]$$

Compressibility of sludge cake can be taken into account through an empirical relationship :

$$\alpha = \alpha_o\left(\frac{h_p}{h_o}\right)^s \qquad (11)$$

where h_p = Pressure head

S = Compressibility coefficient of sludge

α_o = Specific resistance at a reference pressure head (h_o)

[α = 0 (incompressible sludge), s = 0.2 to 0.8 (compressible sludge)

Rate of drainage can be expressed in terms of height of the sludge bed as:

$$\frac{1}{A}\frac{dV}{dt} = -\frac{dL}{dt} \text{ and} \qquad (12)$$

$$\frac{V}{A} = L_o - L \qquad (13)$$

$$hp = \frac{\Delta p}{\rho g} = h_s + h_b \qquad (14)$$

where L = Height of sludge at any time t > 0

L_o = Initial height of the sludge

Total pressure drop (ΔP) in terms of pressure head (hp) is expressed as:

Pressure head due to sludge (h_s) = L, [height of sludge at any time (t)]

Rewriting the basic equations in terms of the sludge :

$$-\frac{dL}{dt} = \frac{\rho g}{\mu}\left[\frac{h_b + L}{\alpha\,\rho X_s\left(L_o - L\right) + R_b}\right] \qquad (15)$$

and R_b = K a [equation (6)], a = a_o $(h/h_o)^s$ [Equation (11)], equation (15) can be re-written as :

$$-\frac{dL}{dt} = \frac{\rho g\, h_o^s}{\mu\,\alpha_o}\left[\frac{\left(h_b + L\right)^{(1-s)}}{\rho\, x_s\left(L_o - L\right) + K}\right] \qquad (16)$$

[Sludge height decreases with times]

Integrating for the limits : t = 0, L = L_o

t = t, L = L

$$t = \frac{\mu\alpha_o (h_b + L_o)^s}{\rho\, g\, h_o^s}\left\{\left(\frac{\rho X_s L_o + K + \rho X_s h_b}{s}\right)\times\left[1-\left(\frac{h_o + L}{h_b + L_o}\right)^s\right]\right.$$

$$\left. - \frac{\rho X_s (h_b + L_o)}{(s+1)}\left[1-\left(\frac{h_b + L}{h_b + L_o}\right)^{s-1}\right]\right\} \qquad (17)$$

If h_b [supporting medium pressure head] is small compared to the total pressure head (h_b << L (and L_o) :

$$t = \frac{\mu\alpha_0 X_s L_0^{s+1}}{g^s (s+1) h^s}\left\{(s+1)\left[1+\frac{K}{s\,\rho\, X_s L_o}\right]\left[1-\left(\frac{L}{L_o}\right)^s\right]-s\left[r-\left(\frac{L}{L_o}\right)^{s+1}\right]\right\} \qquad (18)$$

Supporting medium resistance is usually important during initial stages of drainage. For long times or negligible medium resistance, the term containing K can be neglected and equation (18) becomes :

$$t = \frac{\mu\alpha_0 X_s L_0^{s+1}}{gs(s+1)h_0^s}\left\{(s+1)\left[1-\left(\frac{L}{L_o}\right)^s\right]-s\left[1-\left(\frac{L}{L_o}\right)^{s+1}\right]\right\} \qquad (19)$$

[Similar to Nebiker relationship (JWPCF, Res. Suppl (pt 2), (19) p-R 255, Aug 1969) for negligible media resistance].

Ionic coagulants may change the specific resistance of the sludge. Addition of magnesium ions results in precipitation of $Mg(OH)_2$ in the form of gelatinous (highly hydrated floc) that increases the sludge specific resistance.

2. Empirical relationships

- Sludge bed area (A) requirement to dewater 1 m^3 of sludge :

  A = 0.8 − 0.02 S_o

  where S_o = Initial percent solids content

  A = Sludge bed area (m^2)

- Rate of solids dewatered per unit area per day (Y)

  $$Y = 0.33 S_0^{1.16} \quad \left[kg/m^2.d\right]$$

- Dewatering time (t) is related to the initial depth (h_0) of the sludge application

  $$t = S_o h_o^{1.6} \quad \text{[Specific resistance ignored]}$$

- Sludge application rate (Y) is related to the specific resistance (α)

$$Y = \left(kg/m^2.yr\right) = \frac{10^7}{\alpha^{0.5}}, \quad \left[\alpha \text{ in } s^2/g\right]$$

3. Determine the specific resistance of the cake and of the supporting media (sand). Assume the following data :

Initial sludge concentration	:	3.7% solids
Column diameter	:	4 in.
Initial height of sludge (L_o)	:	21 in.
Sludge density (r_s)	:	62.2 lb m/ft^3
Water viscosity	:	6.18×10^{-4} lb m/ft s
Gravity constant (gc)	:	4.15×10^8 lb m.ft/lb$_f$.h^2

Table 66 : Sludge drainage an data sand with time at pressure drop of 0.53 atm

Time (d)	0	3	5	8	10	15	20	25
Effluent volume (in.^{3})	0	14.5	18.8	24.5	28.4	34.5	39.5	45.8

3.1 Required specific resistance of sludge and supporting media (sand) resistance Plot (t/V) versus effluent volume (V)

$$\text{Slope} = \frac{\alpha\, \mu\, \rho\, X_s}{2A^2 \Delta P} = 0.0112\left(d/in.^6\right)$$

$$= 8.03 \times 10^5 \ (h/ft^6)$$

$$= \frac{(\alpha)(2.2\ lb/ft.h)\left(62.4\ lb/ft^3\right)(0.037)}{2\left(0.0872\ ft^2\right)\left(112\ lb_f/ft^2\right)\left(4.15 \times 10^8\ lbm\ ft/lb_f.h^2\right)}$$

or $\alpha = 1.113 \times 10^{15}$ ft/lb$_m$

$$\text{Intercept} = \frac{\mu\, R_b}{A \Delta P} = 0.044\left(d/in.^3\right)$$

$$= \frac{(2.2\ lb\ m/ft.hr)(R_b)}{\left(0.0872\ ft^2\right)\left(1122\ lb_f/ft^2\right)\left(4.15 \times 10^8\ \frac{lb_m ft}{lb_f.h^2}\right)}$$

$R_b = 3.3 \times 10^3$ ft^{-1}.

3.2 Determine the compressibility coefficient (s) of the sludge and the reference resistance (α_o) of the sludge at 1 ft of water pressure head. Assume the following data :

Table 67 : Variation of specific resistance of sludge with pressure

Specific resistance (a, ft/lb)	2.25×10^{14}	3.5×10^{14}	4.2×10^{14}	5.0×10^{14}
Pressure head (ft of water)	2.29	4.92	6.56	9.84

Refer equation (11)

$$\alpha = \alpha_o \left(hp/h_o\right)^s$$

$$\log \alpha = \log\left[\alpha_o/h_o^s\right] + s \log hp$$

Plot α versus pressure head on a long-log graph paper

Slope = s = 0.535

Therefore, $\frac{\alpha_o}{h_o^s} = 1.48 \times 10^{14}$

$\alpha_o = 1.48 \times 10^{14}$ ft/lb, [h_o =1 ft of water head

Compressibility coefficient (α) is :

$\alpha = 1.48 \times 10^{14} (hp)^{0.535}$ ft/lb.

Example 3.74

Pressure Filtration

A sludge of initial solids concentration 3% required 5 h to process to cake containing 35% solids in a particular press at a pressure of 500 kN/m^2. The specific resistance at 500 kN/m^2 for the conditioned sludge was 10^{11} m/kg. Calculate pressing time for a final cake solids of 25%.

A theoretical description of the effect of specific resistance on the pressing time (T) is given by

$$T = \frac{\bar{r}_P \left(\overline{W}_c - W_s\right)^2}{PW_s\left(1 - W_s\right)}$$

It was found that pressing time determined by above expression was on an average of more than 2.3 times less than those found in practice.

An approximation to above expression is

$$T = \frac{\left(\overline{W}_c - W_s\right)^2}{W_s\left(1 - W_s\right)}$$

The pressing time (T) for a final solids content of 25% would be given by:

$$T_{25} = T_{35} \frac{\left(\overline{W}_{c,25} - W_s\right)^2}{W_s\left(1 - W_s\right)} \times \frac{W_s\left(1 - W_s\right)}{\left(\overline{W}_{c,25} - W_s\right)^2}$$

where the suffixes 25 and 35 refer to cakes of 25 and 35% solids concentration

$$T_{25} = \frac{5.0(0.25-0.03)^2}{(0.35-0.03)^2} = 2.4 \text{ h}$$

For a final cake of 25% and a worse starting specific resistance of 10^{12} m/kg at 500 kN/m^2, the pressing time would be

$$T = 2.4\frac{10^{12}}{10^{11}} = 2.4 \text{ h}$$

For a final cake solids of 35% with $\bar{r}_{500} = 10^{11}$ for an initial cake solids of 8% the pressing time would be

$$T = 5.0 \times \frac{0.03(1-0.03)(0.35-0.08)^2}{0.08(1-0.08)(0.35-0.03)^2} = 1.41 \text{ h}$$

If two conditioned sludges of similar moisture contents (MC) and specific resistances at 49 kN/m^2, but with compressibility coefficients of 0.6 and 1.0 are pressed to make final cake solids at 800 kN/m^2, the pressing times will be in proportion to $\bar{r}_P$ values at 800 kN/m^2 :

$$\frac{T_{0.6}}{T_{1.0}} = \frac{\left(\frac{800}{49}\right)^{0.6}}{\left(\frac{800}{49}\right)^{1.0}} = 0.33$$

where $T_{0.6}$ and $T_{1.0}$ refer to pressing times with S (coefficient of compressibility) values of 0.6 and 1.0, respectively. Thus, for the high compressibility cake, the pressing time is almost three times as long as despite the similarity of the low-pressure specific resistance.

Note :

- In practice many other considerations apply apart from the theoretical parameters, e.g., the extent to which the filter cloth is blinded by fines is critical, high-pressure washing at regular intervals being necessary.
- The liquor from a filter press is often highly polluting, depending upon the prior treatment of the sludge, and the filtrate is usually returned to works inlet.
- The pressure is usually applied in a horizontal bed press with several times of individual plates, the pressing time being measured in several hours, to achieve final MC of 60% for primary or mixed sludges.

Example 3.75

Dewatering of Thickened Sludge using High Pressure Filtration

Estimate the pressure filter sizing utilizing high pressure filtration to dewater the thickened excess sludge at 43300 gal/d, 2.5% solids, or nominally 9000 lb/d. Assume the following data :

Sludge solids	:	9000 lb/d
Conditioning solids	:	0.2% of sludge solids
Cake solids	:	40%
Cake density	:	70 lb/ft^3
Cake thickness	:	1.2 in.
Filter chamber volume	:	2.0 ft^3
Cycle time	:	100 min filtration + 20 min turn around = 120 min (2 hour/cycle)
Filter pressure	:	240 psia
Operating cycle	:	12 h/d and 7 day/weak (and using chemical conditioning)
Design requirements	:	40% solids (cake) + 60% moisture

Solution

1. Required total solids, filter cake and filter cake volume
   - Total solids (dry weight basis)

     = (9000 lb/d) + 0.2(9000 lb/d) = 10800 lb/d
   - Filter cake weight

$$= \frac{10800 \text{ lb/d}}{0.40} = 27000 \text{ lb/d of cake at 40\% solids}$$

   - Cake volume

$$= \frac{27000 \text{ lb/d}}{70 \text{ lb/ft}^3} = 386 \text{ ft}^3/\text{d}.$$

2. Required number of chambers, number of filter cycles per day and number of filter chambers per day
   - Number of filter chambers per day :

$$= \frac{386 \text{ ft}^3/\text{d}}{2.0 \text{ ft}^3/\text{chamber}} = 193 \text{ chambers/d}$$

- Number of filter cycles per day :

$$= \frac{12 \text{ h/d}}{2 \text{ h/cycle}} = 6 \text{ cycles/d}$$

- Number of filter chambers per cycle :

$$= \frac{193 \text{ chambers/d}}{6 \text{ cycles/d}} = 32 \text{ chamber/cycle.}$$

Example 3.76

Plate and Frame Pressure Filter for Dewatering of Wastewater Sludges

Determine the size of a plate and frame pressure filter to dewater sludge. Use the following data :

Maximum loading	:	50,000 lb/d (dry TSS)
Average loading	:	26,000 lb/d (dry TSS)
Minimum sludge concentration	:	2%
Maximum sludge concentration	:	3%
Total cycle time	:	4 hrs (includes sufficient time for cloth washing and cake removal)
Average cake solids	:	45%
Minimum cake solids	:	35%
Cake density	:	75 lb/ft^3
Chemicals for conditioning	:	120 lb $FeCl_3$/ton dry solids + 250 lb-lime/ton dry solids (Total = 370 lb/ton).

Solution

1. Required volume of sludge to be treated

$$\text{Maximum volume} = \frac{50,000 \text{ lb dry SS/d}}{0.02 \text{ lb dry SS/lb sludge} \times 8.34 \text{ lb/gal}}$$

$$\approx 300,000 \text{ gal/d}$$

$$\text{Average volume} = \frac{26,000}{(0.03)(8.34)} = 104,000 \text{ lb/d.}$$

2. Required dewatered volume

$$\text{Maximum} = \frac{(50,000 \text{ lb/d}) + (370 \text{ lb/tonne} \times 5 \times 10^{-4} \text{ lb/ton} \times 50,000 \text{ lb/d})}{(0.35 \text{ lb TSS/lb cake})(75 \text{ lb/ft}^3)}$$

$$= \frac{(50{,}000 + 9250)}{26.25}$$

$$= 2257 \text{ ft}^3/\text{d}$$

$$\text{Minimum} = \frac{(26000) + (370 \times 5 \times 10^{-4} \times 26{,}000)}{(0.4)(75)}$$

$$= 1027 \text{ ft}^3/\text{d}$$

3. Required number of filter cycles per day

$$\text{Minimum number of cycles per day} = \frac{(2 \text{ shifts/d})(8 \text{ hr/cycle})}{(4.0 \text{ h/cycle})}$$

$$= 4 \text{ cycles/d}$$

$$\text{Minimum number of cycles per day} = \frac{(1 \times 8)}{4}$$

$$= 2 \text{ cycles/d.}$$

4. Required dewatered volume of sludge or pressure filter volume per cycle

$$\text{Maximum filter volume per cycle} = \frac{2257 \text{ ft}^3/\text{d}}{4 \text{ cycles / d}}$$

$$= 564 \text{ ft}^3/\text{cycle}$$

$$\text{Minimum filter volume per cycle} = \frac{1027 \text{ ft}^3/\text{d}}{2 \text{ cycles / d}}$$

$$= 514 \text{ ft}^3/\text{cycle.}$$

5. Required filter size

   Assume the volume per chamber of the press = 3 ft^3

$$\text{Therefore, the minimum No. of chambers} = \frac{564 \text{ ft}^3}{3 \text{ ft}^3}$$

$$= 188.$$

Note :

Hydraulic pressures have also been applied to further dewater filter cake sludges for incineration. Board mill sludge has been dewatered to 40% solids from 30% solids at a pressure of 300 lb/in.^2 and a pressing time of 5 min.

Example 3.77

Estimation of Filtering Area Required for a Plate and Frame Filtration Operation

1. Basic design equation (Constant pressure filtration)

$$\frac{dV}{dt} = \frac{A^2(\Delta P)^{1-s}}{rW(V + A\,V_f)\mu} \tag{1}$$

where V is the volume of filtrate delivered in time t, A is the area of the filtering surface, ΔP is the pressure drop across the filter, r is the specific cake resistance, W is designated as the lbs of dry-cake solids per unit volume, V_f is the fictitious volume of filtrate per unit volume of the filtering area necessary to lay down a cake of thickness l_f, s is the compressibility coefficient of the cake, and m is the viscosity of the filtrate.

Integrating this equation between t = 0 to t = t and V = 0 to V = V :

[A, ΔP, s, r, W, V_f and h are constant]

$$V^2 + 2A\,V_f\,V = \left[\frac{2\,A^2(\Delta P)^{1-s}}{r\,W\,\mu}\right]t \tag{2}$$

Generallly V_f is small, therefore,

$$V^2 = \left[\frac{2\,A^2(\Delta P)^{1-s}}{r\,W\,\mu}\right]t \tag{3}$$

2. Determine the total filtering surface required. Assume the following data:

(a) General

Initial solids content of the slurry : $\frac{5\text{ lb} - \text{dry solids}}{\text{ft}^3\text{ of solid free liquid}}$

Viscosity of the liquid (μ) : 1 Cp

Filter delivery (solid free liquid) : 400 ft^3 of solid-free filtrate in 2 hours period

Pressure difference over the filter unit is constant : 25 psi.

Table 68 : Experimental data obtained in a plate-frame filter press with a total filtering area of 8 ft^2

Total Volume (V) of Filtrate (ft^3)	*Time (hr.) from the Start of Filtration (t) at Constant Pressure Difference (ΔP) of : 20, 30 and 40 psi*		
	20	30	40
5	0.34	0.25	0.21
8	0.85	0.64	0.52
10	1.32	1.00	0.81
12	1.90	1.43	1.17

The slurry (with filter aid) was identical to that which was to be used in the large filter. The filtrate obtained was free of solids, and a negligible amount of liquid was retained in the cake.

2.1 **Approximate Solution :** An approximate solution could be obtained by interpolating for values of V at ΔP = 25 psi and then using two of there values to set up equation (2) in the form of two equations involving only the two unknowns V_f and $[(DP)^{1-s}/rW\mu]$. By simultaneous solution, the values of V_f and $[(DP)^{1-s}/rWm]$ could be obtained. The final required area could then be obtained directly from equation (2).

2.2 **Graphical Method :** The solution in item 2.1 puts too much reliance on the precision of the individual experimental measurements, a more involved procedure using all experimental data is recommended.

Rearranging equation (2) :

$$\frac{t\,\Delta P}{V/A} = \left[\frac{r\,W\,\mu\,(\Delta P)^S}{2}\frac{V}{A}\right] + \left[r\,W\,V_f\,\mu\,(\Delta P)^S\right] \tag{4}$$

$$\text{At constant } \Delta P \text{ a plot of } \left[\frac{t\,\Delta P}{V/A}\right] \text{Vs} \left[\frac{V}{A}\right]$$

should give a straight line with slope $\left[\frac{r\,W\,\mu\,\Delta P^S}{2}\right]$ and an intercept at $\frac{V}{A}$ = 0 of $\left[r\,W\,\mu\,V_f\,\Delta P^S\right]$. Any time the same variable appears in both the ordinate and abscissa of a straight line plot, an analysis for possible mis-interpretation should be made. In this case, the value of t and DP change sufficiently to make a plot of this type acceptable.

The following slopes and intercepts are obtained with the use of experimental data $\left(\text{Plot of} \left[\frac{t\,\Delta P}{V/A}\right] \text{Vs} \left[\frac{V}{S}\right]\right)$

Table 69

ΔP (p&t)	*Slope* = $\frac{\left[r\,W\mu(\Delta P)^s\right]}{2}$ *(lb-hr/ft⁴)*	*Intercept* = $\left[r\,W\,\mu\,V_f\,\Delta P^s\right]$ *(lb-hr/ft³)*
20×144	2380	70
30×144	2680	80
40×144	2920	90

Values of r, s, and V_f could be obtained by simultaneous solution with any three of the appropriate values presented in the preceding list. A letter procedure to evaluate the design constants is obtained by using the following procedure :

$$\log(\text{slope}) = s\,\Delta P + \log\left[\frac{r\,W\,\mu}{2}\right]$$

$$\log(\text{int ercept}) = s\Delta P + \log\left[r\,W\,\mu\,V_f\right]$$

A log-log plot of slopes obtained from the plots of $\left(\frac{t\,\Delta P}{V/A}\right) Vs \left(\frac{V}{A}\right)$ versus DP gives a straight line with a slope of s and an intercept at log (ΔP) = 0 of log $\left(\frac{r\,W\,\mu}{2}\right)$. In this way, s and r can be evaluated, and the consistency of the data checked. With the use of this plot :

$$s = 0.3$$

$$\frac{r\,W\,\mu}{2} = 220$$

Similarly, a log-log plot of $\left(\frac{t\Delta P}{V/A}\right) Vs \left(\frac{V}{A}\right)$, intercepts Vs ΔP should give a straight line with a slope of s and an intercept at log (ΔP) = 0 of log ($r\,W\,\mu\,V_f$) from which V_f could be evaluation,

Because the value of V_f is relatively small, the intercepts read are not precise. Therefore,

$$W = 5\ \text{lb/ft}^3$$

$$\mu = 2.242\ \text{lb/ft.hr}$$

$$r_f = \frac{(220)(2)}{(5)(2.42)} = 36\ \text{hr}^2/\text{lb}$$

$$V_f = \frac{80}{r_f\,W\,\mu\,(\Delta P)^s} = \frac{80}{(36)(5)(2.42)(30\times 144)^{0.3}}$$

$$= 0.015\ \text{ft}^3/\text{ft}^2$$

Substituting, all there values in equation (2) :

$$V^2 + 0.03\,AV = \frac{2\,A^2\,(\Delta P)^{1-0.3}}{(36)\,W\,\mu}\,t$$

Given $V = 400\ \text{ft}^3$

$\Delta P = 25\times 144$ psf

$$W = 5\ lb/ft^3$$

$$\mu = 2.42\ lb/ft.hr$$

$$t = 2\ hr$$

Therefore,

$$(400)^2 + (0.03)(400)\,A = \frac{2A^2\,(25\times144)^{0.7}\,(2)}{(36)(5)(2.42)}$$

$$A = 240\ ft^2.$$

Note :

(r, s, V_f, A, ΔP, V, t, W and μ) are the constants and may be used to scale-up to a similar filter with perhaps 100 times the area of the experimental unit. To reduce scale-up errors, the constants should be obtained experimentally with the same slurry mixture, same filter and approximately the same pressure drop as are to be used in the final designed filter. Under these conditions r and s will apply adequately to the larger unit. V_f is usually small enough for changes in its value due to scale-up to have little effect on the final results.

Example 3.78

Cycle Time for Maximum-Amount of Production from a Plate and Frame Filter Press

The relation between the volume of filtrate and the time in operation is expressed as :

$$V^2 = 2.25 \times 10^4\ (t + 0.11)$$

where V is the ft^3 of filtrate delivered in filtering time t (hr)

The cake formed in each cycle must be washed with an amount of water equal to 1/16 times the filtrate delivered per cycle. The washing rate remains constant and is equal to 1/4 of the filtrate delivery rate at the end of filtration. The time required per cycle for dismantling, dumping, and re-assembling is 6 hours. Determine the total cycle time necessary to permit the maximum output of filtrate during each 24 hours.

Solution

1. Total time per cycle

   Let t = hours of filtering time per cycle

   Filtrate delivered per cycle (V)

$$V = [2.25 \times 10^4\ (t + 0.11)]^{0.5}$$

$$= 150\ (t + 0.11)^{0.5}\ ft^3/hr$$

Rate of filtrate delivery at the end of cycle is :

$$\text{Washing rate} \times 4 = \frac{dV}{dt} = \frac{150}{2}(t+0.11)^{-0.5}\ ft^3/hr$$

$$\text{Time for washing} = \frac{\text{Volume of wash water}}{\text{Washing rate}}$$

$$= \frac{(4)(2)(150)(t+0.11)^{0.5}}{16(150)(t+0.11)^{-0.5}} = \frac{(t+0.11)}{2},\ hr$$

$$\text{Total time per cycle} = \left[t + \frac{(t+0.11)}{2} + 6\right]\ hr$$

$$= [1.5t + 6.06]\ hr$$

$$\text{Cycle per 24 hr} = \frac{24}{[1.5t+6.06]}$$

Filtrate delivered/24 hr is :

$$V(\text{Cycle per 24 hr}) = 150(t+0.11)^{0.5}\frac{24}{(1.5t+6.06)}$$

2. Maximum output of filtrate per 24 hr

$$\frac{d}{dt}\left[ft^3 - \text{filtrate per 24 hr}\right] = 0$$

$$\frac{dV}{dt} = \frac{d}{dt}\left[\frac{150(t+0.11)^{0.5}(24)}{(1.5\,t+6.01)}\right] = 0$$

Optimum time (t) = 3.8 hours

Therefore, total time necessary to permit maximum output of filtrate* = (1.5) + (3.8) + 6.06 = 11.8 hour.

Example 3.79

Standard Rate Anaerobic Digestion of Sludge

1. Required capacity for a standard-rate digester

During the digestion period, sludge becomes more concentrated as a supernatant separates from solids. For greater economy of digestion capacity, the supernatant is withdraw frequently during the process. The reduction in sludge volume with time appears to follow a parabolic relation. The capacity requirement for a standard rate digester can be expressed as :

$$V = \left[v_t + \frac{1}{3}(v_o - v_t)\right]t$$

where v_o and v_t are the volumes occupied by a daily accretion to the sludge initially and at the end of the retention period. The volume of the daily accretion to the sludge can be computed with :

$$v = v_s + v_w = \frac{W}{s} + \frac{W(1-x)/x}{X.\rho}$$

The specific gravity of solids can be estimated from :

$$S = \frac{1}{\frac{p}{S_v} + \frac{(1-p)}{S_f}}$$

where v_s and v_w are the volumes occupied by the solids and water, respectively (ft^3), W is the weight of daily accretion to the sludge (lb/d), S is the specific gravity of solids, x is the weight fraction of solids in the sludge, ρ is the density of water (lb/ft^3), S_v and S_f are the specific gravities of the volatile and fixed solids, respectively, and p is the % of solids that are volatile (expressed as a decimal).

A detention period of 30 to 50 days is normal at 85 to 95°F for the digestion of sludges from domestic sewage.

2. Determine the digester capacity on the basis of 1200 lb dry solids introduced to the process daily. Assume the following data:

Initial sludge solids and water	:	5% (solids) and 95% (water)
Volatile solids content (initial)	:	80% (volatile) (S_v = 1)
Fixed solids content (initial)	:	20% (fixed) (S_f = 2.5)
Solids content of the digested sludge	:	10%

Reduction of volatile solids for 30 days digestion period : 50%.

3. Specific gravities

$$\text{Undigested solids (S)} = \frac{1}{\frac{0.8}{1} + \frac{0.2}{2.5}}$$

$$\text{Digested solids (S)} = \frac{1}{\left(\frac{0.4}{0.6 \times 1}\right) + \left(\frac{0.2}{0.6 \times 2.5}\right)}$$

$$= 1.25.$$

3.1 Initial volume (v_o) and final volume (v_t) of sludges

$$V_o = \frac{1200}{(1.14)(62.4)} + \frac{1200(1-0.05)/0.05}{62.4}$$

$$= 16.87 + 365.38$$

$= 382.25\ \text{ft}^3$

$$v_t = \frac{720}{(1.25)(62.4)} + \frac{720(1-0.10)/0.10}{62.4}$$

$= 9.23 + 103.85$

$= 113.08\ \text{ft}^3$.

3.2 Digested volume (Standard rate digestion, V)

$$V = \left[v_t + \frac{1}{3}(v_o - v_t)\right] t$$

$$= \left[113.08 + \frac{1}{3}(382.25 - 113.08)\right] \times 30$$

$= 6084\ \text{ft}^3$.

Example 3.80

Anaerobic Sludge Digestion [Standard Design]

Determine the sizes of the two stage digesters required to treat the sludge from a community. Assume the following data :

Population	:	40,000 persons
Raw waste solids (per capita)	:	0.10 kg/capita.d
Primary settling removal	:	55%
Secondary sludge	:	0.05 kg/capita.d
Primary sludge solids content	:	5% + 95% moisture
Thickening required	:	4% of TS
Specific gravity of sludge	:	1.02
Digested sludge solids content	:	8% + 92% moisture
Raw sludge volatile solids content	:	75%.

Solution

1. Required weight of volatile solids to be added to the digester
   - kg of primary sludge :

     = (40,000 persons)(0.10 kg/capita.d)(0.55)

     = 2200 kg SS/d
   - kg of waste sludge from secondary biological unit:

     = (40,000 persons)(0.05 kg/capita.d)

     = 2000 kg SS/d

- Total volatile solids added to the digester:

  = (2200 + 2000)(0.75)

  = 3150 kg VSS/d.

2. Required volumes of primary and secondary waste sludges

- Primary sludge

$$= \frac{2200 \text{ kg/d}}{(0.05)(1.02)\left(1000 \text{ kg/m}^3\right)}$$

= 43.1 m^3/d

- Secondary waste sludge

$$= \frac{2000 \text{ kg/d}}{(0.04)(1.02)\left(1000 \text{ kg/m}^3\right)}$$

= 49 m^3/d.

- Total feed to the digester :

  = 43.1 m^3/d (Primary) + 49.0 m^3/d (Secondary)

  = 92.1 m^3/d.

3. Required volume of the first stage digester (V_1)

$$V_1 = \frac{3150 \text{ kg VSS/d}}{\text{Loading rate}}$$

$$= \frac{3150 \text{ kg VSS/d}}{1.28 \text{ kg VSS/m}^3\text{.d}} = 2461 \text{ m}^3$$

[Volume of the second digester also should be 2461 m^3].

4. Required HRT in the primary digester and the quantity of digested sludge produced

- $\text{HRT}(\theta) = \theta_c = \frac{V}{Q} = \frac{2461 \text{ m}^3}{92.1 \text{ m}^3/\text{d}} = 26.7 > 10 \text{ d} \left(\text{min. OK}\right)$

- Assume θ = 27 days and VSS reduction = 54%

Solids mass balance :

$$\begin{bmatrix}\text{Mass of total solids} \\ \text{in digested sludge}\end{bmatrix} = \left[\text{Mass of fixed solids}\right] + \left[\text{Mass of VSS remaining}\right]$$

= (2200 + 2000)(0.25)+(2200+2000)(0.75)[1 – 0.54]

= 1050 kg Fixed solids/d+1449 kg VSS remaining/d

= 2499 kg/d of digested solids on dry basis

Volume of total sludge produced :

$$= \frac{2499 \text{ kg/d}}{(0.08)(1.02)\left(1000 \text{ kg/m}^3\right)}$$

$$= 30.6 \text{ m}^3\text{/d.}$$

5. Required heat

$$\text{Heat lost} = \left(4190 \text{ W/100 m}^3\right)\left(\frac{2461}{100}\text{m}^3\right),$$

[Assumed for 2461 m^3 of Sludge]

$$= 1.03 \times 10^5 \text{ W}$$

Heat required (H) = $WC(T_2 - T_1)$

where H = Amount of heat required (Joules)

C = Mean specific heat of raw sludge

W = Mass of sludge entering the tank (kg/hr)

T_2 = Temp. of sludge in the tank (°C) [35 °C]

T_1 = Temp. of raw sludge entering tank (°C) [10 °C]

$$H = \left[\frac{92.1 \text{ m}^3\text{/d} \times 1.02 \times 1000 \text{ kg/m}^3}{24 \text{ h/d}}\right]\left(4200 \text{ Joules/kg}^\circ\text{C}\right)(35-10)$$

$$= 1.14 \times 10^5 \text{ W}$$

Total heat requirements = Heat lost + Heat to be added

$$= [1.03 \times 10^5 + 1.14 \times 10^5] = 2.17 \times 10^5 \text{ W.}$$

Note :

1. Heat losses from digester can be estimated using the following equation:

$$Q = UA\ (T_2 - T_1)$$

where Q = Heat loss from tank (W, BTU/h)

U = Heat transfer coefficient ($W/m^2.°C$ or $BTU/ft^2.h$ °F)

A = Surface area of tank element ($m^2 - ft^2$)

T_1 = Temperature outside the tank (°C – °F)

T_2 = Temperature inside the tank (°C – °F)

The temperature (T_2) would be the normal operating temperature of the digester. The average ambient temperature, air temperature for the coldest two week expected should be used for the temperature outside the tank (T_1).

An easier method of estimating digester heat losses that does not require the consideration of heat losses through each element of the digester has been used.

Table 70, Approximately the following data can be used :

Table 70

Digester Condition	Heat losses [W/100 m^3 (BTU/1000 ft^3.h)]
Mixed + Insulated	[1260 – 4190] (1200 – 4000)
Mixed + Uninsulated	[1570 – 5230] (1500 – 5000)
Unmixed + Insulated	[840 – 2720] (800 – 2600)
Unmixed + Uninsulated	[1260 – 4190] (1200 – 4000)

Table 71 : Digester heat transfer coefficients

	U	
Digester element	*W/m².°C*	*BTU/ft².hr.°F*
6 in. concrete roof	2.84	0.50
Floating cover with built-up insulated roof	1.36	0.24
12 in. concrete walls with airspace insulation	1.99	0.35
12 in. concrete walls wet earth covered	1.42	0.25
12 in. concrete walls dry earth covered	1.02	0.18
Floor	0.68	0.12

Example 3.81

Anaerobic Sludge Digestion

Determine the capacity of the two-stage digesters required to treat the sludge from a community. The wastewater flow is about 40,000 m^3/d. Assume the following data :

Raw wastewater SS	:	275 mg/L
Volatile content of raw SS	:	75%
Raw wastewater BOD_5	:	250 mg/L
Primary effluent SS	:	125 mg/L
Primary effluent BOD_5	:	165 mg/L
Secondary excess biological solids	:	0.28 kg VSS/kg BOD_5
Thickened primary raw sludge concentration	:	6% (94% moisture)
Thickened secondary waste sludge concentration	:	4% (96% moisture)

Digested sludge solids concentration	:	8% + 92% moisture
All sludges specific gravity	:	1.02
Primary digester temperature	:	35°C.

Solution

1. Required VSS added to the digester

$$\text{VSS in primary sludge} = \frac{(275-125)\,\text{mg/L} \times 40{,}000\ \text{m}^3/\text{d} \times 0.75}{1000}$$

$$= 4500\ \text{kg VSS/d}$$

$$\text{Organic}\left(\text{BOD}_5\right)\text{ load to the aeration basin} = \frac{165\ \text{mg/L} \times 40{,}000\ \text{m}^3/\text{d}}{1000}$$

$$= 6600\ \text{kg/d}$$

Excess VSS produced from = (6600 kg/d)(0.28 kg VSS/kg BOD)
biological system = 1848 kg VSS/d

Total impact to the digester = 4500 + 1848
= 6348 kg VSS/d

2. Required volume of sludges

Volume of primary sludge

$$= \frac{4500\ \text{kg/d}}{(0.06)(0.75)(1.02)\left(1000\ \text{kg/m}^3\right)}$$

$= 98\ \text{m}^3/\text{d}$

Volume of excess wasted sludge from biological unit:

$$= \frac{1848\ \text{kg/d}}{(0.04)(0.75)(1.02)\left(1000\ \text{kg/m}^3\right)},$$

[Waste sludge volatile content = 75%]

$= 60.4\ \text{m}^3/\text{d}$

Total volume of sludge input to the digester:

$= 98\ \text{m}^3/\text{d}$ [primary] + $60.4\ \text{m}^3/\text{d}$ [Excess sludge]

$= 158.4\ \text{m}^3/\text{d}$

Percent total solids fed to the digester:

$$= \frac{(6)(98.0)+4(60.4)}{98+60.4} = 5.24\% \text{ Total solid}.$$

3. Required capacity (V) of the first digester

Assume T = 35° C, θ_c = 10 days, VS loading = 4.01 kg/m^3.d

$$\text{Digester capacity}(V) = \frac{6348 \text{ kg VSS/d}}{4.01 \text{ kg/m}^3\text{.d}}$$

$$= 1583 \text{ m}^3$$

$$\text{Residence time}(\theta = \theta_c) = \frac{V}{Q} = \frac{\text{Digester volume}}{\text{sludge feed rate}}$$

$$= \frac{1583 \text{ m}^3}{158.4 \text{ m}^3/\text{d}} = 9.99 \text{ d} (\sim 10 \text{ days, OK}).$$

4. Required digested sludge and gas production

At, θ = 10 days, volatile solids content of 75%, and volatile solids reduction of about 55%

- Sludge mass balance :

$$\begin{bmatrix}\text{Mass of total solids} \\ \text{in the digested sludge}\end{bmatrix} = \begin{bmatrix}\text{Mass of} \\ \text{fixed solids}\end{bmatrix} + \begin{bmatrix}\text{Mass of volatile solids} \\ \text{remaining in the digester}\end{bmatrix}$$

$$\text{Total solids in raw sludge} = \frac{6348 \text{ kg VSS/d}}{0.75} = 8464 \text{ kg/d}$$

[Mass of total solids in the digested sludge] = (8464)(0.25) + (8464)(0.75(1 - 0.55)

= 2116 kg fixed solids/d + 2857 kg VSS/d

= 4973 kg TS/d (dry basis)

$$\text{Volume of digested sludge} = \frac{4973 \text{ kg TS/d}}{(0.08)(1.02)(1000 \text{ kg/m}^3)} = 61 \text{ m}^3$$

- Gas production rate :

Assume gas production = 0.90 m^3/kg VSS destroyed

Total VSS destroyed = (8464 kg TS/d)(0.75)(0.55)

= 3491 kg VSS/d

Gas produced = (0.90 m^3/kg VSS destroyed) (3491 kg VSS/d)

= 3142 m^3/d

Energy produced = $(3142\ m^3/d)\ (5850\ kg\ Cal/m^3)$

= 1.84×10^7 kg Cal/m^3.

Note :

Table 72 : Suggested mean cell residence time for use in the design of complete mix digester

Operating Temp. (°C)	*q_c (min) (day)*	*θ_c (day)[Suggested for design]*
18	11	28
24	8	20
30	6	14
35	4	10
40	4	10

Table 73 : Effect of concentration and HRT on volatile solid loading factor

Sludge Conc. (%)	*HRT (d); kg VSS/m³.d*			
	10	*12*	*15*	*20*
4	3.06	2.55	2.04	1.53
5	3.83	3.19	2.55	1.91
6	4.59	3.83	3.06	2.30
7	5.36	4.46	3.57	2.68
8	6.12	5.10	4.08	3.06
9	6.89	5.74	4.59	3.44
10	7.65	6.38	5.10	3.83

The loading factors can be used to size high rate digesters. If mixed organics and chemical sludges are to be digested, the volume of the digester must be increased over that calculated using the above loading factors to accommodate the greater volume of fixed solids reaching the digester. This adjustment can be made by multiplying the table values by the ratio of the actual percentage of volatile solids in the sludge to be digested to the 75% VSS sludge used as the basis for calculating tabular values.

3. GLUMRB standards for anaerobic reactor design and sizing

   - *CSTRs :* Completely mixed systems shall provide intimate and effective mixing to prevent stratification and to ensure homogenity of the digester content. The system may be loaded at a rate upto 80 lbVSS/1000 ft^3.d (1.28 kg VSS/m^3.d) in the active digestion unit when grit removal facilities are not provided, the reduction of digester volume caused by grit accumulation should be considered.

   - *Moderately mixed system:* For digestion where mixing is accomplished only by circulating sludge through an external heat exchanger, the system may be loaded upto 40 lbVSS/1000 ft^3.d (0.64 kg VSS/m^3.d) in the active digestion unit.

- Standard digesters: Loading ratio: 0.03 to 0.1 lb VSS/ft^3.d (0.48 to 1.6 kg VSS/m^3.d).
- High rate digester: Loading rates: 0.10 to 0.40 VSS/ft^3.d (1.6 to 6.4 kg VSS/m^3.d).

Example 3.82

Anaerobic Digestion

Determine the amount of digested sludge per day alongwith biogas generation and digester volume. Assume the following data :

Water content of raw sludge	:	97.5%
Volatile solids content	:	75%
VSS content destruction in the digester	:	60%
Digester detention time	:	30 days
Raw sludge temperature	:	21°C
Digested sludge solids	:	5%
Population	:	32,000
Per capita per day generation	:	0.2 lb/capita.d.

Solution

1. Sludge production from PST

$$0.2 \times 32{,}000 = 6{,}400 \text{ lb/d}$$

(*i*) Consider 60% removal of SS in PST

Sludge production = 6,400 × 0.6 = 3840 lb/d.

(*ii*) Sludge generated from the aeration basin with of Y = 0.55 lb sludge from the $BOD_{removed}$

Assume,

Applied BOD to aeration	:	3700 lb/d
Effluent BOD	:	800 lb/d
Therefore, BOD removal	:	2900 lb/d
Activated sludge generated	:	2900 lb/d $\times$ 0.55 = 1595 lb/d

Total Sludge = (Primary + Secondary) sludge

= 3840 + 1595 = 5435 lb/d.

2. Sludge volume generated

(*i*) Volume of total sludge to be handled in the digester

$$= \frac{5435}{0.025} = 217{,}000 \text{ lb/d}$$

$$= \frac{217{,}000}{\text{Density of sludge} = (1)}$$

$$= \frac{217{,}000}{8.34} = 3480 \text{ ft}^3/\text{d}.$$

(*ii*) Volatile fraction in solids and fed to the digester

$= 3480 \text{ ft}^3/\text{d} \times 0.75$

$= 4100$ lb/d.

(*iii*) Volatile solids destroyed in the digester

$= 4100 \times 0.6 = 2460$ lb/d.

(*iv*) Non volatile fraction in the effluent of the digester

$= (4100 - 2460) = 1640$ lb/d.

(*v*) Non volatile fraction fed to the digester is equal to the non-volatile fraction in the incoming sludge

$= (5435 - 4100)$ lb/d $= 1335$ lb/d.

(*vi*) Digester volume output

$$= \frac{(\text{Volatile solids in the effluent}) + (\text{Non-volatile component})}{\%\text{ solids in the digested solids}}$$

$$= \frac{(1640 + 1335)\text{lb/d}}{0.05} = 59{,}500 \text{ lb/d}$$

$$= \frac{59{,}500 \text{ lb/d}}{8.34 \text{ lb/gal}} = 7100 \text{ gal/d}.$$

(*vii*) Digester volume

Rate of sludge input to digester : $3480 \text{ ft}^3/\text{d}$

Volume of digester $= V \times \theta = 3480 \times 30$

$= 1{,}04{,}400 \text{ ft}^3$.

3. Biogas generation

Standard gas production rate from digester sludge $= 8.5 \text{ ft}^3/\text{lb VSS}$

Therefore, total gas generation rate $= 8.5 \times 4100 = 34{,}100 \text{ ft}^3/\text{d}$

Methane generation rate $= 34{,}100 \times 0.60$

$= 20{,}460 \text{ ft}^3/\text{d}$.

Example 3.83

Anaerobic Sludge Digester

1. Basic equation for total digestion capacity (V)

$$V = \left[\frac{V_1 + V_2}{2} T_1\right] + V_2 T_2$$

where V_1 is the volume of daily raw sludge applied (gal/d), V_2 is the volume of accumulated digested sludge (gal/d), T_1 is the required time for digestion (days-30 days at 30-35°C), and T_2 is period of digested sludge storage (days).

2. Determine the volume of the digester (single-stage-floating cover digester). Assume the following :

Solids contributed per person per day	:	0.2 lb/person.d
Raw/Solids concentration	:	3.5%
Digested solids concentration	:	7.0%
Reduction of solids	:	45%
Digestion period	:	25 days
Storage for digested sludge	:	60 days

2.1 Required daily volume of raw sludge

$$V_1 = \frac{(0.2 \text{ lb/person.d})}{\left(\frac{3.5}{100}\right)(8.34)} = 0.69 \text{ gal/person.d}$$

2.2 Required daily digested sludge accumulation

$$V_2 = \frac{(0.55)(0.2 \text{ lb/person.d})}{\left(\frac{7.0}{100}\right)(8.34)} = 0.19 \text{ gal/person.d.}$$

2.3 Required total digester volume (V)

$$V = \left(\frac{V_1 + V_2}{2}\right)(T_1) + (V_2)(T_2)$$

$$= \left(\frac{0.69 + 0.19}{2}\right)(25) + (0.19)(60)$$

$$= (11) + (11.4) = 22.4 \text{ gal/person.}$$

3. An ASD has two floating-cover digesters each with a total capacity of 20,000 ft^3 with 15,000 ft^3 below the landing brackets. The remaining 5,000 ft^3 in each tank is the volume between lowered and fully raised cover position (storage volume). The digesters can be operated in both modes (parallel and or in series). The one digester is equipped with gas mixing to serve as a first-stage high rate process. Daily raw sludge production is 5000 gal containing 2000 lb of dry solids (60% volatile). The digested sludge contains 7% solids, and the process converts 50% of the VS to gas.

3.1 Required storage period for digested sludge in parallel operation

$$\text{VS applied to each tank} = \frac{(0.6)(2000)}{2} = 600 \text{ lb VS/d}$$

$$\text{Loading with covers lowered} = \frac{600 \text{ lb VS/d}}{15{,}000 \text{ ft}^3}$$

$$= 0.03 \text{ Ib VS/ft}^3\text{.d}$$

Required supernatant/d (R) = [Raw sludge volume (applied)/d – V_1] – [Digested sludge volume accumulated/d – V_2]

$$V_1 = 5000 \text{ gal/d}$$

$$V_2 = \frac{\text{Inert solids} + \text{Remaining VS}}{(\text{Solids content}/100)(8.34)}$$

$$= \frac{(800+400)100}{(7)(8.34)} = 2055 \text{ gal/d}$$

Therefore, $R = V_1 - V_2$

$$= 5000 - 2055 = 2945 \text{ gal/d}$$

$$\text{Storage period} = \frac{2 \times 5000 \text{ ft}^3 \times 7.48 \text{ gal/ft}^3}{2055 \text{ gal/d}} = 36.4 \text{ days}$$

$$= 36.4 \text{ days.}$$

3.2 Required solids content in the digested sludge leaving the first-stage digester (operation in series)

lbs VS/d applied to the final-stage digester = (0.6)(2000)

= 1200 lb VS/d

$$\text{Loading with cover lowered} = \frac{1200 \text{ lb VS/d}}{15{,}000 \text{ ft}^3}$$

$$= 0.08 \text{ lb VS/ft}^3\text{.d}$$

Required detention period with cover lowered

$$= \frac{(15{,}000)(7.48)}{(5000)} = 23.44 \text{ days}$$

$$\text{Loading with cover raised} = \frac{1200 \text{ lb VS/d}}{20{,}000 \text{ ft}^3} = 0.06 \text{ lb VS/ft}^3\text{.d}$$

$$\text{Required detention period with cover raised} = \frac{(20{,}000)(7.48)}{(5000)}$$

$$= 29.92 \text{ days}$$

Required solids content (S) in digested sludge leaving first-stage digester is :

$$S = \frac{(0.4)(2000) + (0.5)(0.4 \times 2000)}{(5000)\left(\frac{8.34}{100}\right)}$$

$$= \frac{1200 \times 100}{(5000)(8.34)} = 2.9\%$$

[Similar exercise can be done for the second-stage digester].

Note :

1. **Rectification of anaerobic digester (Temperature effect).** Initially, acid forming and methane forming populations are in balance and, because they are operating near optimum temperature, the majority of the VS are being converted to gaseous end products. When the heater fails, and temperature of the digesting sludge drops, a new equilibrium of population is established at reduced gasification efficiency. Therefore, raw organic matter accumulates in the digesting sludge. These dormant VS suddenly become available to the bacteria when the temperature is raised back to 95°F. The sharp increase in gas production is related primarily to CO_2 produced when the first-stage bacteria convert volatile solids to organic acids. The acid-splitting methane formers attempt to respond to the increased organic acid supply. However, their population have been reduced by the lack of food during low-temperature operation. Further more, the methane bacteria, being more sensitive, are inhibited by the acid conditions and, hence, total gas yield drops. The solution is to drop the temperature of the digesting sludge back to 75°F, and to increase it slowly, perhaps one degree per week, to allow the methane formers to adjust gradually to increasing production of organic acids.

4. Determine the volume of anaerobic digesters: A standard rate digester and a high rate two-stage digester. Assume the following :

Thickened sludge	: 100 m^3/d
Organic content	: 70%
Digestion period (standard rate anaerobic digester)	: 30 days
Digestion period (high rate first digester)	: 10 days
High rate second stage digester's dewatering period	: 3 days
Percent Reduction of organics	: 60%
Amount of thickened sludge	: 3000 kg/d
Digested sludge solids content	: 5%

Solution

1. Required volume of a single stage standard anaerobic digester :
   - Organic fraction in the thickened sludge :
     = 0.7 (3000 kg/d)
     = 2100 kg/d
   - Organic fraction remaining :
     = 0.4 (2100 kg/d)
     = 840 kg/d
   - Inorganic fraction remaining :
     = 0.3 (3000 kg/d)
     = 900 kg/d
   - Total undigested sludge (Mass remaining after digestion) :
     = (840 + 900) kg/d
     = 1740 kg/d
   - Digested sludge accumulation rate (V_2) :

$$= \frac{\text{Mass of dry sludge}}{\left(100\ \text{kg/m}^3\right)(S)}$$

$$= \frac{1740\ \text{kg/d}}{\left(100\ \text{kg/m}^3\right)(0.05)}$$

$$= 34.8\ \text{m}^3/\text{d}.$$

   - Anaerobic digestion volume :

$$V = \left(\frac{V_1 + V_2}{2}\right)t_1 + V_2 t_2$$

where V = Digester volume (m^3)

V_1 = Raw sludge loading rate (m^3/d)

V_2 = Digester sludge accumulation rate (m^3/d)

t_1 = Digestion period (d)

t_2 = Digested sludge storage period (d)

$$V = \left(\frac{100 + 34.8}{2}\right)(30\ \text{d}) + \left(34.8\ \text{m}^3/\text{d}\right)(90\ \text{d})$$

$$= 2022\ \text{m}^3 + 3132\text{m}^3$$

$$= 5154\ \text{m}^3$$

[t_2 = 90 days of accumulated sludge storage].

2. Required volume of two-stage high-rate anaerobic digester :
   - First-stage digester volume (V)

     $= V_1\ t_1$; [Digestion only]

     $= (100\ m^3/d)\ (10d) = 1000\ m^3$.
   - Second-stage digester volume (V)

$$V = \left(\frac{V_1 + V_2}{2}\right) t_1 + V_2\ t_2$$

$$= \left(\frac{100 + 34.8}{2}\right)(3d) + 34.8\ m^3/d\ (90\ d)$$

$$= (202 + 3132) m^3 = 3334\ m^3.$$

   - Total volume of two-stage digester

     $= (1000 + 3334\ m^3)$

     $= 4334\ m^3$.
   - Comparison in volume is :

     Standard digester volume = $5143\ m^3$

     High-rate two-stage digester volume = $\underline{4334\ m^3}$

     Difference in volume = $709\ m^3$.

Example 3.84

Bicarbonate Alkalinity in an Anaerobic Digester

Calculate the bicarbonate alkalinity in an anaerobic digester. Assume the following data :

pH level to be maintained in the digester	:	7
CO_2 in the gas	:	30% (by volume)

Solution

1. The equilibrium relationship between CO_2 at bicarbonate is given by :

$$H_2CO_3 = HCO_3^- + H^+$$

H_2CO_3 refers to $[CO_2 + H_2CO_3]$

CO_2 concentration is assumed equation H_2CO_3 concentration

$$K_1 = \frac{[H^+][HCO_3^-]}{[H_2CO_3]}$$

$$pH = pK_1 + \log\frac{[HCO_3^-]}{[CO_2]}$$

pK_1 = 6.35

2. 100% CO_2 at 1 atmosphere (closed system), the concentration of CO_2 is 1320 mg/L

   Therefore,

   30% CO_2 = 1320 × 0.3 = 396 mg/L

$$\text{mol/L of } CO_2 = \frac{396\times10^{-3}\ \text{mol/L}}{44\ \text{g/mol}} = 9\times10^{-3}\ \text{mol/L}$$

$$pH = pK_1 + \log\frac{[HCO_3^-]}{[H_2CO_3]}$$

$$7.0 = 6.35 + \log\frac{[HCO_3^-]}{[0.009]}$$

   Bicarbonate alkalinity = mol/L × 61 g/mol × 10^{-3}
   = mg/L as $CaCO_3$

Example 3.85

Bicarbonate Alkalinity in Anaerobic Digester

Calculate the bicarbonate alkalinity and how much bicarbonate alkalinity must be added to achieve an alkalinity of 4500 mg/L. Assume the following data:

Total alkalinity in the anaerobic digester	: 6000 mg/L as $CaCO_3$
Volatile acid in the anaerobic digester	: 2500 mg/L (as acetic acid)
Bicarbonate alkalinity to be achieved	: 4500 mg/L as $CaCO_3$
Present bicarbonate alkalinity	: 2500 mg/L as $CaCO_3$
Digester retention time	: 20 days.

Solution

1. Bicarbonate alkalinity

   Bicarbonate alkalinity = Total alkalinity – (0.71)(Volatile acid)
   [0.71 factor conversion of volatile acid].
   = 6000 – 0.71 × 2500
   = 4225 mg/L as $CaCO_3$

2. Additional amount of bicarbonate alkalinity

$$D_d = D_{max}(1 - e^{-l})$$

where D_d is the daily dose of alkalinity necessary to reach a required level (mg/L as $CaCO_3$), D_{max} is the required increase in alkalinity concentration (mg/L as $CaCO_3$), and l is the reciprocal of average digester detention time (d^{-1})

$$D_d = (4500 - 2500)(1 - e^{0.05}), \quad \left[1 = \frac{1}{20} = .05d^{-1}\right]$$

$$= 100 \text{ mg/L as } CaCO_3.$$

Example 3.86

Biogas from Sludge Digestion

Determine the volume of gas produced (its calorific value), the power of gas engine and generator, the volume of excess gas, and the power requirements of an activated sludge plant. Assume the following data :

Sewage sludge to be processed	:	20,000 lb/d
Sludge solids content	:	5%
Volatile content of sludge solids	:	70%
Methane content	:	70%
Destruction of organic solids	:	70%

Solution

1. Gas production

Volatile solids (organic) destroyed

$$= 20 \times 10^3 \times 0.05 \times 0.70 \times 0.7.0$$

$$= 490 \text{ lb/d}$$

$$1 \text{ ft}^3 \text{ of } CH_4 \text{ weighs} = \frac{16}{359} = 0.0446 \text{ lb}$$

$$1 \text{ ft}^3 \text{ of } CO_2 \text{ weighs} = \frac{44}{359} = 0.1225 \text{ lb}$$

[1 lb mole of gas occupies 359 ft^3 at STP and 1 ft^3 = 28.3 L and 1 lb=454 g]

Assume that weight of the gas = 1.25 lb/lb of VS destroyed.

Therefore, the total gas yield

$$= \frac{1.25 \times 490}{[0.70 \times 0.0446 + 0.30 \times 0.1225]}$$

$$= \frac{1.25 \times 490}{0.06797} = 9011 \text{ ft}^3/\text{d}$$

Heat available $= 9011 \times 958 \times 0.70$

$= 6.04 \times 10^6$ BTU/d.

2. **Gas Engine and Generator Power** : Assume the gas-engine has a water-heating efficiency of 50%, and volume of gas available for plant purposes other than digester heating or required to be burnt in a waste-gas burner (10.0×10^5 BTU/d)

$$\text{Required heat input} = \frac{10 \times 10^5}{0.5} = 2 \times 10^6 \text{ BTU/d}$$

Generator and motor efficiency = 75%

Engine efficiency = 25%

1 hp-hr = 2545 BTU

1 hp-hr = 0.7457 kW-hr

$$\text{Heat consumed} = \frac{2545}{0.25} = 10^4 \text{ BTU/brake hp-hr}$$

$$\text{Gas consumed} = \frac{10^4}{963 \times 0.70} = 14.8 \text{ ft}^3/\text{brake-hp-hr}$$

[Net fuel value of methane = 963 BTU/ft^3 at STP]

$$\text{Therefore, engine power} = \frac{2 \times 10^6}{24 \times 10^4} = 8.33 \text{ hp}$$

Electric power $= 8.33 \times 0.75 \times 0.7457$

$= 4.66$ kW

If the entire available gas is utilized for the engine, the engine and generator, the increase in their utility will rise to

$$= \frac{6.04 \times 10^6}{2.0 \times 10^6} = 3.02 \text{ folds.}$$

3. Excess volume of gas

$$= (6.04 \times 10^6 - 2 \times 10^6)\ \frac{\text{BTU}}{\text{d}} \text{ not used by engine}$$

$$\text{Therefore, volume of excess gas} = \frac{4.04 \times 10^6}{963 \times 0.7} = 59.93 \text{ ft}^3/\text{d}.$$

4. Activated sludge plant requirements

Assume : 550 hp-hr or 410 kW-hr/10^6 ft^3 of air at 7 psig

Energy requirement for air compressor to deliver

10^6 ft^3 of air at 7 psig = 550 hp-hr (410 kW-hr)

$$\text{Amount of gas needed} = 550 \text{ hp} - \text{hr} \times 14.8 \frac{\text{ft}^3 \text{of gas}}{\text{brake hp} - \text{hr}}$$

$$= 8159 \text{ ft}^3/\text{mil} - \text{ft}^3 \text{ of air}$$

or $$\frac{8159}{0.75} = 10.9 \times 10^3 \text{ ft}^3 \text{ gas for electric drive.}$$

Note :

1. Methane oxidation

$$CH_4 + 2O_2 = CO_2 + 2H_2O$$

Air contains 21% of oxygen by volume, therefore, $\frac{2}{0.21} = 9.5 \text{ ft}^3$ of an area required to burn 1.0 ft^3 of CH_4. If digester gas contains 70% methane, the requirement of oxygen = 9.5 × 0.7 = 6.65 ft^3 of air for 1 ft^3 of digester gas

2. Explosive mixture of CH_4 and O_2

The range is about 5.6-13.5% of methane by volume and the flame speed is miximal at 9.6%. Above 13.5% the mixture burns quickly after ignition.

3. Operating, protecting and regulating devices of gas collection and distributing system
   - Condensate traps and drains for water vapour.
   - Flame traps (preventing flash backs from gas burners and engines).
   - Pressure regulating valves.
   - Waste-gas burners.

4. Gas may be stored under a head of 3 to 6 inches of water in the open. The economical pressure for cylindrical and spherical pressure tanks is about 40 psig. Bottled gas is placed under a pressure of about 5000 psig (before being compressed, the gas may be passed through scrubbers to remove unwanted constituents : CO_2, H_2S and water vapour)

5. Efficiencies
   - Hot water boilers are neither as efficient (60%) nor as trouble free (SO_2^- corrosion) as steam boilers with heat exchangers for hot-water heating (80%).
   - Water-jacketed gas engineers equipped with exhaust gas boilers have a water efficiency of about 50% and a direct power efficiency of 22 to 27%, depending on the engine load (0.5 to full load, respectively).
   - Conversion of gas-engine power to electric-power and use of electric motor-drive equipment entail a loss of about 25% of the engine power. For equal performance in automotive engines, about 160 ft^3 of sludge gas (70% methane : 110,000 BTU) may be substituted for a gallon of gasoline (122,000 BTU).

6. Activated Sludge Plant : 10^6 ft^3 of air per day will treat about 1 mgd of sewage. Gas production from primary + secondary sludge is about 1.25 ft^3/capita : the available gas supply of 1.25 × 104 ft^3/d (1 mgd from 10^4 persons) is sufficient to provide the necessary engine (where wanted electrical) power and to keep the sludge-digestion tanks at optimal mesophillic temperatures.
7. 1 ft^3/capita of gas yield is sufficient from primary treatment units.

Example 3.87

Heat Requirements of a Digester

Determine the heat requirement of the digester when the temperatures of the incoming sludge, digesting volume, earth, and air are at 60, 100, 45 and 25°F respectively, the required area of stationery heating coils through which the water is circulated at an incoming temperature of 150°F and with a temperature drop of 10°F, as also the water requirement for heating through the coils. Assume the following data :

Digester diameter (Roof rises to 1 ft to a quality central insulated gas dome)	: 30 ft (Concrete with good insulation)
Coefficient of heat flow	: 0.25 BTU /ft^2.hr
Coefficient of heat flow through the circulating coil	: 15 BTU/ft^2.hr
Side walls	: 20 ft high and bottom slopes 5 ft to a central sludge withdrawal pipe
Daily sludge addition	: 20,000 lb

Solution

1. Required heat

Area of digester exposed to earth $= \pi D\left(H + \frac{1}{2}\sqrt{r^2 + (\text{Slope})^2}\right)$

$$= \pi \times 30\left(20 + \frac{1}{2}\sqrt{(15)^2 + (5)^2}\right), \quad [r = D/2]$$

$$= 3373 \text{ ft}^2$$

Heat requirements for = 20,000(100 – 60)

Incoming flow of sludge (Q) = 800,000 BTU/day

Heat lost = 0.25 × 24[3373(100 – 45) + 708(100 – 25)]

= 14,31,690 BTU/day

Total = 14,31,690 + 800,000 = 22,31,690 BTU/d

2. Required area for heating coils

   Arrange temperature of heating water = (135 – 5) = 130°F

$$\text{Required area of coils} = \frac{\text{Total heat required}}{[C\Delta T]}$$

$$= \frac{22,31,690}{24 \times 15(130 - 100)} = 206 \text{ ft}^2$$

3. Quantity of recirculating water

$$\text{Temperature drop} : 10^\circ \text{F} = 10 \text{ BTU/lb}$$

$$\text{Weight of water } = \frac{22,31,690}{10} = 223,169 \text{ lb/d}$$

$$\text{Volume of water } = \frac{223,169}{8.34} = 26758 \text{ gal/d}$$

   Heat lost from the heat source to digester must be added.

Note :

1. Heat used in evaporation of water into the sludge gases is small.
2. Coefficient of heat transfer (C) depends on the motion of ambient fluids (sludge inside and air or water outside of the tank), the thickness of specific portion of the digester, and also the opportunities for radiation
3. The value of C = 0.10, 0.15, 0.30 for exposure to dry earth, air, and wet earth respective (BTU/ft^2.hr).
4. For equal exposure of all parts of a digester, heat is conserved best if the tank geometry is spherical as structural economy will allow. The largest ratio of sludge volume to tank surface is thereby obtained.
5. For unequal exposures, the portions through which the greatest losses per unit area occur should be kept minimum
6. Heat requirements to the digester is met by :
   - Incoming sludge is heated outside the tank in counter current heat exchangers.
   - Hot water is circulated through fixed or moving coils inside the tank.
   - Gas in burnt under water or in a heater submerged in the sludge.
   - Steam is introduced into the sludge.

Example 3.88

Heat Requirement to Raise the Temperature of Incoming Sludge to Anaerobic Digester

Calculate the heat required to raise the temperature of incoming sludge to 35°C if the ambient water temperature is 7°C and Summer temperature is 19°C. Overall heat transfer coefficient for the digester is approximately 0.15 BTU/ft².hr °F.

Solution

Anaerobic digester heat requirement = [Heat required to raise temperature to digester 35°C] + [Heat losses from the anaerobic digester]

1. Heat required to raise the temperature of incoming sludge from 21°C to 35°C. Assume the mass inflow to the anaerobic digester is 100,000 lb/d and specific heat is 1.0 BTU/lb/°F.

$$1,00,000 \frac{\text{lb}}{\text{d}} \times 1.0 \frac{\text{BTU}}{\text{lb.}^\circ\text{F}} \times (95.0 - 65.8) = 3.0 \times 10^6 \frac{\text{BTU}}{\text{d}}$$

2. Heat losses from the anaerobic digester surfaces

   Surface area of the digester = Roof area + Floor area + Side area

   It is assumed approximate 20,000 ft² of the anaerobic digester losses will be more during Winter in comparison to Summer

   $$Q = U\ A\ \Delta T$$

   where Q is the heat loss rate from the digester (BTU/h), U is the overall heat transfer coefficient $\left(\frac{\text{BTU}}{\text{ft}^2.\text{hr.}^\circ\text{F}}\right)$, and DT is the temperature (°F)

   $$Q = 0.14 \times 20,000\ (95 - 44.6) = 141,120\ \text{BTU/hr}$$
   $$= 3.4 \times 10^6\ \text{BTU/d.}$$

3. Total heat required

   $$= (3.0 \times 10^6 + 3.4 \times 10^6)\ \text{BTU/d} = 6.4 \times 10^6\ \text{BTU/d}$$

4. Heat available from burning of methane

   10 ft³ is produced per lb of volatile solids to the anaerobic digester. Assume volatile fraction is about 75 % in the sludge.

   Total methane produced (60% in the biogas)

   $$= 100,000\ \text{lb/d} \times 0.75 \times 10\ \text{ft}^3/\text{lb} \times 0.60$$
   $$= 430,000\ \text{ft}^3/\text{d (at standard conditions)}$$

   Heat available through burning of methane

   $$= 430,000\ \text{ft}^3/\text{d} \times 960\ \text{BTU/ft}^3$$
   $$= 413 \times 10^6\ \text{BTU/d}$$

Assume, combustion efficiency of 50% :

Therefore, net heat available = 41.3×10^6 BTU/d $\times$ 0.5

= 20.65×10^6 BTU/d

Net heat available far exceeds the heat requirements and, therefore, no external heat source is required.

Example 3.89

Anaerobic Digestion of Sludge

Determine the effluent COD, the VSS production, the digester volume, biogas production, the mixing levels and the heating requirements for a completely mixed single stage anaerobic sludge digester. Assume the following data :

Waste sludge flow	:	8000 gal/d
Sludge concentration	:	2% (20,000 mg/L)
Volatile content of sludge	:	80% (16,000 mg/L)
Reduction of VSS	:	60%
Ultimate BOD of solids	:	1.42
Reactor temperature	:	35°C (95°F)
Ambient air/minimum waste temperature	:	5°C (40°F)
Digester depth	:	20 ft
Decay coefficient (k_d)	:	0.03 d^{-1} (35°C)
Yield coefficient (Y)	:	0.04
SRT	:	10 days

Solution

1. Required influent and effluent COD
   - Applied feed COD :

     = (0.008000 MGD)(8.34)(16,000 mg/L)(1.42)

     = 1516 lb S-COD/d [Influent VSS = 16,000 mg/L)
   - Effluent S-COD

     = (0.008000 MGD)(8.34)(16,000)(0.4)(1.42)

     = 606 lb S-COD/d [Effluent VSS = (16,000)(0.4) = 6400 mg/L]
2. Required VSS production (P_x)

$$P_x = \frac{[(Q)(8.34)(S_o - S_e)Y]}{[1 + k_d(SRT)]}$$

$$= \frac{(0.008)(8.34)(16,000 - 6400)(0.04)}{[1 + (0.03)(10)]}$$

= 19.7 Ib VSS/d

3. Required COD stabilization

= (1516 – 606 – 19.7) lb S-COD/d

= 890 lb S-COD/d

4. Required digester volume (V)

MCRT = HRT = SRT = 10 days

V = (SRT)(Sludge flow)

= (10)(8000 gal/d) = 80,000 gal.

5. Required gas production

Methane generation at 32°F = (5.62 $ft^3/lbCOD_r$)(890 lb S-COD/d)

and 1 atm = 5002 ft^3/d at 32°F and 1 atm

$$\text{Methane production} = \left(5002\ \text{ft}^3/\text{d}\right)\frac{460+95}{460+32} \text{ at } 95^\circ \text{F and 1 atm}$$

= 5642 ft^3/d at 95° F and 1 atm

$$\text{Total gas production } = \frac{5642\ \text{ft}^3/\text{d}}{0.6} = 9404\ \text{ft}^3/\text{d}$$

[Methane = 60%]

6. Required mixing levels

- Mechanical mixing = (0.3 hp/1000 ft^3)(80,000 × 0.1337 ft^3)

  = 3.2 hp

- Gas recirculation = (5 ft^3/1000 ft^3.min)(80,000 × 0.1337 ft^3)

  = 53.5 ft^3/min.

7. Required digester dimensions

Volume	=	80,000 gal
No. of digester	=	1
Working depth	=	20 ft
Surface area	=	535 ft^2
Digester diameter	=	26 ft
Free board	=	2 ft
Total depth	=	(20 + 2) = 22 ft.

8. Required heat balance

Table 74 : Heat losses

Parameters	*Area (ft^2)*	*H ($BTU/ft^2.hr\ ^oF$)*	*Heat ($BTU/hr.^oF$)*
Roof	535	0.50	270
Floor	535	0.12	65
Walls	1796	0.35	629
		Total	964

$$H_{loss} = UA\ (T_2 - T_1)$$

$$H_{loss} = [(UA)_{flow} + (UA)_{roof} + (UA)_{walls}][T_2 - T_1]$$

$$= [964]\ [95^oF\ (Digester) - 40^oF\ (Ambient)]$$

$$= 53{,}020\ BTU/hr$$

Heat required for the feed (H_r)

$$H_r = W\ Cp\ (T_2 - T_1)$$

$$= (8000\ gal/d)(1)\left(\frac{8.34}{24}\right)(95 - 40)$$

$$= 152{,}900\ BTU/hr$$

Total heat required = ($H_{loss} + H_r$)

$$= (53020 + 152{,}900) = 205{,}920\ BTU/hr$$

Heat available :

Methane heating value = 960 BTU/ft^3 at STP

Methane generated at 32°F = 5002 ft^3/d

Total heat generated $= \left(\frac{5002}{24} ft^3/hr\right)\left(960\ BTU/ft^3\right)$

= 200,080 BTU/hr

Heat to be added = (205,920 – 200,080) BTU/hr

(Not sufficient) = 5840 BTU/hr.

Note :

1. Reactor concentration (Well mixed reactor)
   - Equilibrium reactor concentration:

$$X = \frac{Y(S_o - S_e)}{1 + k_d\,(SRT)} \quad [\text{Without recycle}]$$

   - Batch Reactor : Initial solids level is a direct function of the net residual solids maintained. This is a function of the quantity of solids generated within the batch treatment cycle and the quantity wasted for sludge management.

- Contact stabilization

$$X = \frac{Y(S_o - S_e)}{[1 + k_d (SRT)]}\left[\frac{SRT}{HRT}\right]$$

- Separators can handle influent concentrations upto 5000 mg/L, thickening to a maximum 10,000 mg/L
- Reactor concentration (X) is typically maintained at 3000 to 5000 mg/L, with a provisions for 2 to 4:1 recycle (recycle to influent flow)
- Recycle range must be adequate to allow the operator process control to :
  - Compensate for an expected increase in substrate loading
  - Increase the loading during periods of low biological activity
  - Increase the inventory of fresh micro-organism during surges of toxic influent.
- Other systems :

  Reactor concentration (X) is not always directly controlled, but is an inherent property of the system's operating or physical properties.

2. Reactor volume (dispersed growth)

$$\text{Digester volume} = (\text{Washing rate})\left[\frac{SRT}{X_v}\right]$$

$$= \frac{(\text{Wasting rate})(\text{Cell growth rate})^{-1}}{(X_v)}$$

$$\text{Digester volume} = \frac{[SRT][Q(S_o - S_e)Y]}{[X][1 + k_d (SRT)]}$$

3. Mixing levels

Mechanical mixing	:	0.2 to 0.3 hp/1000 ft^3
Gas recirculation	:	5 to 7 ft^3/1000 ft^3.min (Confined gas system)
	:	4.5 to 5 ft^3/1000 ft^3.min (Unconfined gas system)
Velocity gradient (G)	:	50 to 80 s^{-1}
Turn over time	:	$\frac{\text{Digester working volume}}{\text{Internal flow rate(influenced by mixing)}}$
	:	20 to 30 min.

4. Sludge growth

$$\text{lb VSS/d} = Y(F_r) - k_d\, M_v$$

where Y is the gram of VSS produced/kg COD_r, k_d is the decay coefficient (d^{-1}), F_r is the COD destroyed (lb/d), and M_v is the MLVSS (lb) in the reactor

$$A = \frac{YF}{1 + k_d(SRT)}$$

where A is the lb of VSS produced per day, F is the lb of ultimate BOD (COD) added per day, and k_d is the decay constant (d^{-1}).

5. Methane production

$$C = C_f\,(e.F - 1.42A)$$

where C is the ft^3 of CH_4 produced per day at STP (0°C and 1 atm), e is the fraction of the ultimate BOD removed (0.80 to 0.95), F is the lb of ultimate BOD added per day, A is the VSS (lb) produced per day, and C_f is equal to 0.35 (theoretical methane produced per kg ultimate BOD or if British units are used 5.62, and 1.42 is the conversion factor to convert VSS to ultimate BOD

6. Typical reactions

$$C_nH_aO_bN_c + \left[n - \frac{a}{4} - \frac{b}{2} + \frac{3c}{4}\right]H_2O = C - NH_3 + \left[\frac{n}{2} - \frac{a}{8} + \frac{b}{4} + \frac{3c}{8}\right]CO_2$$

$$+ \left[\frac{n}{2} + \frac{a}{8} - \frac{b}{4} - \frac{3c}{8}\right]CH_4$$

Example 3.90

Anaerobic Digester

Determine the mean cell residence time, the effluent quality, the reactor volume, the biogas production, the CO_2 partial pressure, the alkalinity, the mixing requirement, and the heating requirements for an anaerobic treatment process. Assume the following data :

Wastewater flow	:	1 MGD
Influent COD	:	4000 mg/L
Effluent COD	:	500 mg/L
Nitrogen required	:	0.12 lb-N/lb VSS
Phosphorus required	:	0.02 lb-P/lb VSS
Fraction biodegradable	:	0.12
Ultimate BOD solids	:	1.42

Design temperature	: 35°C
Digester depth	: 20 ft
Decay coefficient (k_d)	: 0.015 d^{-1} at 35°C
Maximum specific growth (k_m)	: 6.67 kg COD/kg VSS.d at 35°C
Half substrate saturation constant (K_s)	: 2500 mg COD/L at 35°C
Yield coefficient (Y)	: 0.04 kg VSS/kg COD at 35°C

Solution

1. Required MCRT

$$(SRT)^{-1} = \left[\frac{Yk_{max}s_c}{K_s + S_c} - k_d\right]$$

$$= \left[\frac{(0.04)(6.67)(500)}{2500 + 500} - 0.015\right]$$

SRT = 34 days

- Minimum SRT (washout) $S_e = S_o$

$$(SRT)^{-1} = \left[\frac{Yk_mS_o}{k_s + S_o} - k_d\right]$$

SRT = 6.7 days

- Limiting SRT ($S_o >> S_s$)

$$[SRT\ (limiting)]^{-1} = Y\ k_m - k_d$$

$$= (0.04)(6.67) - 0.015$$

SRT (limiting) = 4 days

- Design SRT

SRT (design) = (SF)[SRT (min)]

= 5(6.7) = 33.5 days

2. Required effluent quality (S_e)

$$S_e = \frac{K_s(1 + k_d\ SRT)}{SRT(Yk_m - k_d) - 1}$$

$$= \frac{2500\left[1 + 0.015(33.5)\right]}{(33.5)\left[(0.04)(6.67) - 0.015\right] - 1}$$

= 505 mg/L COD [OK]

3. Required solids production (VSS)

   - Applied COD load :

     (1 MGD)(8.34)(4000) = 33360 lb COD/*d*

   - COD utilized :

     (1 MGD)(8.34)(4000 – 500) = 29190 lb COD/*d*

   - Solids produced :

$$P_x = \frac{[Q(8.34)(S_o - S_e)Y]}{[1 + k_d SRT]}$$

$$= \frac{[1(8.34)(4000 - 5000)(0.04)]}{[1 + 0.015(33.5)]} = 777 \text{ lb VSS/d}.$$

   - P_x(in terms of COD) = (1.42)(777) = 1103 lb COD/d

   - Amount of COD stabilized :

     = (29190 – 1103) = 28,087 lb COD/d

   - Mass balance for COD

     Input S – COD Output COD

     = 33,360 lb COD/d = 28,087 lb COD/d (stabilized)

     = 1103 lb COD/d (VSS)

     = 4170 lb COD/d (effluent)

     Total = 33,360 lb S-COD/d.

4. Calculations for contact stabilization volume

   - MLVSS inventory

     = (SRT)(Solid wasted)

     = (33.5)(777 lb VSS/d) = 26,029.5

   - Reactor volume (V)

$$V = \frac{\text{Total MLVSS}}{(3500 \text{ mg/L})(8.34)} = \frac{26029.5 \text{ lb}}{(3500 \text{ mg/L})(8.34)}$$

     = 0.892 MG

5. Conventional treatment reactor volume (V)

     MCRT = HRT = 33.5 days

     V = (HRT)(Flow) = 33.5 MG

6. Required biogas production, alkalinity, and CO_2 partial pressure

   - Methane gas production

= [5.62 SCF at 32°F/lb COD_r at STP (32°F and 1 atm)]

$$= (5.62)\frac{(460+95)}{(460+32)}(28087 \text{ lb/d})$$

= 178,061 ft^3/d at 35° C and 1 atm

- Total gas production (assume CH_4 = 60%)

$$= \frac{178061 \text{ ft}^3/\text{d}}{0.60} = 296{,}769 \text{ ft}^3/\text{d at } 35^\circ\text{C and 1 atm}$$

- CO_2 partial pressure

  - 25% CO_2 concentration in gas

  pCO_2 = (0.25)(1 atm) = 0.25 atm

  - 40% CO_2 concentration in gas

  pCO_2 = (0.40)(1 atm) = 0.40 atm [p = partial pressure]

- Required alkalinity

$$\text{Alkalinity } (CaCO_3, \text{ mg/L}) = \frac{0.00063 \, pCO_2}{10^{-pH}}$$

Assume pH = 7.0

– Alkalinity at pCO_2 = 0.25 atm :

$$\text{Alkalinity } (CaCO_3) = \frac{(0.00063)(0.25)}{10^{-7}} = 1575 \text{ mg/L}$$

– Alkalinity at pCO_2 = 0.40 atm

$$\text{– Alkalinity } (CaCO_3) = \frac{(0.00063)(0.40)}{10^{-7}} = 2520 \text{ mg/L}$$

[It is necessary to increase the pH to 7.2 in order to bring alkalinity within the range of 2500 to 3500 mg/L].

7. Required mixing levels

- Contact stabilization :

  – Mechanical mixing : (0.3 hp/1000 ft^3)(0.892 × 10^6 gal)(0.1337 ft^3/gal) = 36 hp

  – Gas recirculation : (5 ft^3/1000 ft^3.min)(0.892 × 10^6 gal)(0.1337 ft^3/gal) = 596 ft^3/min

8. Required nutrient

Nitrogen requirement = (0.12 lb N/lb VSS)(777 lb VSS/d)

= 932 lb N/d

Phosphorus requirement = (0.02 lb-P/lb VSS)(777 lb VSS/d)

= 15.54 lb P/d

9. Required digester size

Volume = 0.892 MG

No. of digesters = 2

Liquid depth = 20 ft

Surface area = $\frac{0.892 \times 10^6 \times 0.1337}{20} = 5963\ ft^2$

[2982 ft^2 for each digester]

Free board = 2 ft

Total depth =z (20 + 2) = 22 ft

Diameter of digester = 62 ft (each)

10. Required heat loss

Ambient temperature = 40°F (Winter)

Waste temperature = 30°F (Winter)

Design ambient temperature = 73°F (Summer)

Design waste temperature = 68°F (Summer)

Table 75

Parameters	*Approx. area (ft^2)*	*H (BTU/ft^2.h.°F)*	*Heat (BTU/h.°F)*
Floor	5963	0.12	716
Roof	5963	0.50	2982
Side walls [2 × pD × 22]	8566	0.35	2998
		Total	6696

$H_{loss} = UA\ (T_2 - T_1)$

$H_{loss} = [(UA)_{roof} + (UA)\ )_{wall} + (UA)\ )_{flow}]\ (T_2 - T_1)$

where H_{loss} is the heat loss (BTU/hr), A is the total heat transfer area (ft^2), T_2 is the reactor temperature (°F), T_1 is surface temperature (°F), and U is the overall heat transfer coefficient (BTU/ft^2.h.°F)

H_{loss} = [6696 BTU/hr.°F][95°F (digester temperature) – 40°F (ambient)]

= 368,280 BTU/hr [Winter]

H_{loss} = [6696 BTU/hr.°F][95°F (digester) – 73°F (ambient)]

= 147,312 BTU/hr [Summer].

11. Required feed heat

Feed heat requirements (H_r)

$$H_r = W\ C_p(T_2 - T_1)$$

where H_r is the feed heat input (BTU/hr), W is the wastewater flow (lb/hr), C_p is the heat capacity [= 1], T_2 is the reactor temperature (°F), and T_1 is the feed temperature.

Winter conditions :

$$H_r = (1\text{ MGD})\left(\frac{8.34}{24}\right)\ (1)\ (95 - 30)$$

$$= 22.6 \times 10^6\ \text{BTU/hr}$$

Summer conditions :

$$H_r = (1\text{ MGD})\left(\frac{8.34}{24}\right)\ (1)(95 - 68) = 9.4 \times 10^6\ \text{BTU/hr}$$

Total heat requirement :

Winter conditions :

= Heat loss + Heat required for feed

= (0.37 + 22.6) = 22.96 × 10^6 BTU/hr

Summer conditions

= (0.15 + 9.40) = 9.55 × 10^6 BTU/hr

12. Required heat availability

Methane heating value = 960 BTU/ft^3 at STP

Methane produced at STP (32°F, 1 atm)

= (5.62 ft^3/lb COD stabilized)(28,087 lb COD/d)

= 6577 ft^3/hr at 32°F and 1 atm

Heat available = (960 BTU/ft^3)(6577 ft^3/hr)

= 6.314 × 10^6 BTU/hr

Required heat = (22.96 – 6.314) × 10^6 BTU/hr [Winter]

= (16.65 × 10^6 BTU/hr [Winter]

= (9.55 – 6.314) × 10^6 BTU/hr [Summer]

= 3.24 BTU/hr [Summer]

[Waste does not generate enough methane to sustain heating requirements. A COD of 12,000 mg/L would produce enough methane to sustain the digester temperature of 95°F].

Note :

1. Reactor kinetics

$$\frac{\frac{dX}{dt}}{X} = \mu = SRT$$

$$\frac{dS}{dt} = \frac{\mu_{max}}{Y}\left(\frac{SX}{K_s + S}\right) = \frac{k_{max}SX}{K_s + S}$$

$$\frac{dX}{dt} = Y\frac{dS}{dt} - k_d X$$

$$\frac{\frac{dX}{dt}}{X} = \frac{1}{SRT} = \frac{Y\,k_m\,S_e}{K_s + S_e} - k_d$$

- At Washout ($S_e = S_o$)

$$[SRT]_{min}^{-1} = \frac{Y\,k_{max}S_o}{K_s + S_o} - k_d$$

- If $S_o >>> k_s$

$[SRT]^{-1}$(limiting) $= Yk_m - k_d$

Table 76 : Anaerobic treatment kinetic constants

Waste Type	*Magnitude*
Domestic sludge	k_m = (6.67 kg COD/kg VSS.d)$(1.035)^{T-35}$
	K_s = 1800 mg COD/L $(1.112)^{T-35}$
	k_d = (0.03 d^{-1})$(1.035)^{T-35}$
	Y = 0.04 to 0.054 kg VSS/kg COD
General wastes	General values for methane fermentation
	k_m = 6.67 kg COD/kg VSS.d $(1.035)^{T-35}$
	K_s = 2224 mg COD/L $(1.112)^{T-35}$
	k_d = (0.010 to 0.040 d^{-1})$(1.035)^{T-35}$
	Y = 0.04 to 0.054 kg VSS/kg COD
Organic classes	
• Carbohydrates	Y = 0.01 to 0.18 mg VSS/mg COD_r;
	k_g = 0.02 to 0.03 kg VSS_d/kg VSS.d
• Acetic acid	Y = 0.04 to 0.06 mg VSS/mg COD_r
	k_d = 0.011 to 0.015 mg VSS_d/mg VSS.d
• Mixed organic acids	Y = 0.04 to 0.06 mg VSS/mg COD_r
	k_d = 0.015 mg VSS_d/mg VSS.d

- Application of aerobic system equations for anaerobic systems

$$\frac{F_r}{M_v} = K\frac{S_e}{S_o}$$

where F_r is substrate removal rate (lb/d), M_v is the biomass quality (lb), S_e is the substrate (effluent), concentration, S_o is the influent substrate concentration, and K is the constant.

Aerobic system :

$$\frac{F_r}{M_v} = \frac{Q(S_o - S_e)}{XV} = \frac{(S_o - S_e)}{X\theta}$$

$$[SRT]^{-1} = Y\left(\frac{F_r}{M_v}\right) - k_d$$

$$= \frac{k_m S_e}{K_s + S_e}$$

2. Effect of temperature and mixing on safety factor (SF)

Table 77 : Effect of temperature

Temp.(°C)	*SRT*			*Effective Safety Factor (SF)*		
	Theoretical (SRT_e)	*Min. (SRT_m)*	*Limiting (SRT_l)*	$\frac{SRT_d}{SRT_l}$	$\frac{SRT_d}{SRT_m}$	$\frac{SRT_d}{SRT_e}$
25	47.5	9.7	6.0	3.4	2.1	0.4
30	20.6	6.8	5.0	4.0	2.9	1.0
35	11.6	5.1	4.2	4.7	3.9	1.8

SRT_d : SRT – design

Table 78 : Effect of mixing on safety factor (SF)

Temp. (°C)	*% Vol. used*	*Real SRT_r*	*Effective SF*		
			$\frac{SRT_r}{SRT_l}$	$\frac{SRT_r}{SRT_m}$	$\frac{SRT_r}{SRT_c}$
25	100	20	3.4	2.1	0.4
25	50	10	1.7	1.0	0.2
25	25	5	0.8	0.5	0.1
30	100	20	4.0	2.9	1.0
30	50	10	2.0	1.5	0.5
30	25	5	1.0	0.7	0.2
35	100	20	4.7	3.9	1.8
35	50	10	2.4	2.0	0.9
35	25	5	1.2	1.0	0.5

Critical temperature and mixing conditions producing washout are indicated by SFs of less than 1. Too high a solids residence time increases reactor volumes and costs, but does not significantly improve the process efficiency :

Table 79

SRT (days)	*Theoretical removals (%) at 35°C*	
4.2	Limiting SRT	
5.1	Washout SRT	
6	50	Waste properties
7	67	S_o = 10 g/L COD
8	75	Y_o = 0.04
9	80	k_d = 0.03 d^{-1}
10	83	K = 6.67 mg COD/mg VSS.d
11	85	K_s = 1800 mg/L
15	90	
20	92	
25-30	94	
35-40	95	
45-50	96	

3. pH and alkalinity
   - Optimum pH range : 6 to 8
   - Methane production will drop below pH 6.5 to 6.8, and completely ceases below 6 and above 8.5
   - CO_2 concentration of 25 to 40% by volume (OK)
   - Bicarbonate alkalinity: 1000 to 5000 mg $CaCO_3$/L

     – $CO_2 + \text{Cation} = HCO_3^-$

     – $\text{Org} - \text{N compound} \rightarrow NH_3 + CO_2 = NH_4^+ + HCO_3^-$

     – $H_2O + CO_2 \leftrightarrow H_2CO_3 \leftrightarrow H^+ + HCO_3^-$

     $$\left[H^+\right] = K\frac{[H_2CO_3]}{\left[HCO_3^-\right]}$$

     – Biocarbonate alkalinity :

     $$\text{Bi} - \text{alk}\left(\text{mg/L as } CaCO_3\right) = 0.00063\frac{pCO_2}{10^{-pH}}$$

     The factor 0.00063 to 0.00128 accounts for ionic strength of the constituents. In a balanced system volatile acids formed are rapidly converted to methane, resulting in minimum volatile acid concentration, so that the total alkalinity is essentially the bicarbonate alkalinity in the system.

Under unstable conditions, the VA formed are neutralized, reducing the bicarbonate alkalinity according to the stoichiometric relations :

$$HCO_3^- + HA_C = CO_2 + H_2O + VA\overline{C}$$

[HA_C = unionized acetic acid; A_C = acetate ion]

Under unstable conditions, the resulting bicarbonate alkalinity can be determined by analytically measuring the total alkalinity and volatile acid content :

Bicarbonate alkalinity (mg $CaCO_3$/L) = (Total alk) – (0.85)(0.833) TVA

[TVA = Total volatile acids (mg/L)]

The factor 0.85 accounts for the analytical limitation in that the titration analysis to pH-4 accounts for only 85% of the volatile acids, and 0.833 converts the mg/L of VA to equivalent $CaCO_3$.

- Chemicals used for alkalinity

  Lime :

  - Inexpensive
  - Easy control
  - Slow solution and reaction rate
  - Increase inert solids contents
  - Carbonate formation
  - Ash content of the biomass increases

  Sodium hydroxide :

  - Expensive
  - Easier to handle than lime
  - No inherent buffering capacity
  - If not handled properly, pH can rapidly increase to 10

  Sodium bicarbonate :

  - Excellent buffering capacity
  - Relatively inexpensive
  - Ideal for pH control
  - Can increase bicarbonate alkalinity, eliminating the potential for a partial vacuum (without reacting with CO_2).

Example 3.91

Conventional Anaerobic Digester for Wastewater Sludges

A digester has been designed for 14,000 lb/d of VS. Calculate the volume of the digester at 100°F. During critical period Winter (40°F), the digester has been receiving 20,000 gal/d of VSS at 8% solids. Check for heat balance.

Solution

1. Digester volume (V)

   Assume a loading rate of 0.25 lb VS/ft^3.d

$$V = \frac{\text{Amount of VS feed/day}}{\text{VS loading rate}}$$

$$= \frac{14,000}{0.25} = 56,000 \text{ ft}^3$$

$$= 418,936 \text{ gal.}$$

2. Solids retention time (θ) at high through put of VS

$$\theta = \frac{\text{Volume}}{\text{VS feed/d}} = \frac{418,936}{20,000} = 21 \text{ days}$$

3. Reduction of volatile solids (at 60%)

$$= 14,000 \times 0.60 = 8400 \text{ lb/d}$$

4. Biomass generation at 15 ft^3/lb VS

$$= 8400 \times 15 = 1,26,000 \text{ ft}^3/\text{d}$$

   4.1 Assume methane content of 60% in the biogas

   Total methane produced = 1,26,000 × 0.60 = 75,600 ft^3/d

5. Heat Balance : Heat the raw sludge to digester temperature of 100°F. Assume heat loss of 1°F/day and heat transfer efficiency of 40%

   5.1 Required energy

$$\text{BTU} = \frac{20,000 \text{ gal/d} \times 9 \text{ lb/gal}(100 + 21 - 40)}{0.4}$$

   5.2 Methane gas required for heating purposes

$$= \frac{3,64,50,000 \text{ BTU/d}}{960 \times \text{BTU/ft}^3 \times 0.60} = 63,281 \text{ ft}^3/\text{d}$$

   5.3 Methane available (75,600 ft^3/d) is greater than the methane required (63,281 ft^3/d) for heating purposes.

Example 3.92

High Rate Anaerobic Digestion

Calculate the digester volume for a single-stage floating cover digestion system and for a two-stage high rate digestion system operating at 35°C. Sludge drying sand beds are used for drying the digested sludge. Assume the following information :

Municipal wastewater flow	:	0.50 MGD
Water content in raw sludge	:	95%
Volatile solids in raw sludge	:	75%
Water content of digested sludge	:	92%
Volatile solids reduction	:	45%
Sludge flow rate	:	900 lb/d

Solution

1. Single-stage digestion

 1.1 Volume of raw sludge

$$V = \frac{900}{\left[\frac{(100-95)}{100}\right] \times 62.4} = 288 \text{ ft}^3/\text{d}$$

 1.2 Digested sludge volume (V)

$$V = \frac{[\text{Inert solids} + \text{Solids remaining after digestion}]}{\left[\left(\frac{100-p}{100}\right)\right] \gamma S}$$

where p is the percent of water in the digested sludge

$$V = \frac{[0.25(900) + (0.75)(0.45)(900)]}{\left[\left(\frac{100-92}{100}\right)\right] \times 62.4 \times 1}$$

$$= \frac{(225+304)}{0.08 \times 62.4} = 106 \text{ ft}^3/\text{d}$$

 1.3 Volume required for single-stage floating cover digester (V)

$$V = \left(\frac{V_1 + V_2}{2}\right) T_1 + V_2 \times T_2$$

where V is the total digester capacity (ft^3), V_1 is the volume of raw sludge feed (ft^3/d), V_2 is the digested sludge volume accumulated in tank (ft^3/d), T_1 is the time required for digestion (25 days at a temperature of 85-95°F), and T_2 is the period of digested sludge storage (30-120 days)

$$V = \left[\frac{(288+106)}{2}\right] \times 25 + (106 \times 60)\ ft^3$$

$$= 197 \times 25 + 6360$$

$$= 4925 + 6360 = 11{,}285\ ft^3.$$

1.4 Volatile solids loading

$$= \frac{0.65 \times 5000 \times 0.18}{11{,}285} = 0.052\ lb\ VS/ft^3.d.$$

1.5 Digester capacity per capita

$$= \frac{11{,}285\ ft^3}{5000} = 2.26\ ft^3/capita$$

This value is less than 5 ft^3/capita, therefore, digester capacity has to be increased by 2.2 times the value given above or it has to be used for a population of 2500 (if population is 5000 and 0.18 lb-VS/capita)

2. High rate anaerobic digestion (Two-stage)

Assume the following :

Maximum allowable loading	:	0.15 lb VS/ft^3.d
Minimum detention period	:	15 day
Required period for thickening	:	10 day
Quantity of sludge (volume)	:	288 ft^3/d
Water content of digested sludge	:	94%
Volatile solids reduction	:	45%

2.1 First-stage digester value

$$V_I = V_1 \times T$$

where V_I is the digester volume required for first-stage high rate (ft^3), V_1 is the feed sludge volume/day (ft^3/d), and T is the digestion time (d)

$$V_I = 288\ ft^3/d \times 15 = 4320\ ft^3$$

2.2 VS loading

$$= \frac{0.75 \times 900\ lb/d}{4320\ ft^3} = 0.1563\ lb\ VS/ft^3.d$$

2.3 Digested sludge volume (V_2) per day

$$V_2 = \frac{0.25 \times 900 \text{ lb/d} + 0.75 \times 50.45 \times 900 \text{ lb/d}}{\left[\frac{(100-94)}{100}\right] \times 62.4 \times 1}$$

$$V_2 = \frac{529}{0.06 \times 62.4} = 212 \text{ ft}^3/\text{d}$$

2.4 Second-stage digester volume

$$V_{11} = \left(\frac{V_1 + V_2}{2}\right) T_1 + V_2 \times T_2$$

where V_{II} is the required volume for second stage digested sludge thickening and storage (ft^3) V_1 is the volume of digested sludge feed = volume of average daily raw sludge (ft^3/d), V_2 is the volume of daily digested accumulation in tank, ft^3/d T is the time required for thickening (day), and T_2 is the time of digested sludge storage

$$V_{11} = \left(\frac{288 + 212}{2}\right) \times 10 + (212 \times 90)$$

$$= (2500) + 19{,}080 = 21{,}580 \text{ ft}^3$$

Note :

1. Single-stage floating cover digester has three functions : (a) anaerobic digestion of VS (b) gravity thickening, and (c) storage of sludge.
2. The unmixed operation of the tank permits gravity thickening of sludge solids and withdrawal of the supernatant.
3. The two stage digester-the first-stage is complete mixing for anaerobic digestion of VS and the second-stage is used for gravity thickening and storage of the digested sludge.
4. Table 80, Loadings and detention times for heated anaerobic digesters :

Table 80

Parameters	*Conventional Single-Stage (Unmixed)*	*First-Stage High Rate (Completely Mixed)*
Loading (lbVS/ft^3.d)	0.02-0.05	0.1-0.2
Detention time (days)	30-90	10-15
VS reduction	50-70	50

5. Table 81, General conditions for sludge digestion

Table 81

1. Temperature	
Optimum	98°F (35°C)
General range of operation	85-95°F
2. pH	
Optimum	7.0-7.1
General range of operation	6.4-7.4
3. Gas production	
Per lb of VS added	6-8 ft^3
Per lb of VS destroyed	16-18 ft^3
4. Gas composition	
Methane	65-69%
CO_2	31-35%
H_2S	Traces
5. Volatile acids (as acetic acid)	
Normal operation	200-800 mg/L
Maximum	2000 mg/L
6. Alkalinity as $CaCO_3$	
Normal operation	2000-3500 mg/L

Example 3.93

High Rate Anaerobic Digester

High rate sludge digestion is accomplished by a) continuous mixing of the digester contents. This makes use of the total tank capacity for active digestion as compared to approximately one third of the tank capacity in conventional tank, b) use of a high feed concentration, and c) use of continuous feed. Sludge concentration and separation of supernatant is accomplished in secondary digestion units. The following design procedure is taken from ASCE recommended procedure.

1. Total solids remaining in the digester at the end of t days

$$= W_d.t - \frac{W_d \ t \ V_a}{2}$$

$$= W_d.t\left(1 - \frac{V_a}{2}\right)$$

where W_d is the lb dry solids added daily, V is the volatile fraction, t is the digestion period (days), and a is the fraction of volatiles digested in t-days.

2. Daily withdrawal

$$= W_d \left(1 - \frac{V_a}{2}\right)$$

3. Capacity of digester

$$= \frac{W_d\ t\left(1 - \frac{V_a}{2}\right)}{\left[62.5\left(1 - \frac{W}{100}\right)\right]}$$

where W is the average moisture content of the sludge.

4. Daily additions and periodic withdrawals

Maximum capacity = The accumulation of sludge for t_o days plus accumulation of digested sludge for t_1 days

$$= W_d\ t_o\left(1 - \frac{V_a}{2}\right) + W_d\ t_1 (1 - V_a)$$ = Maximum capacity in Ibs of dry solids.

5. Consider a high rate digester with a secondary digester for sludge storage to handle primary sludge from a sewage flow of 1 MGP. The initial suspended solids in sewage is 150 mg/L of which 50 percent are volatile and 50 percent removal can be expected from PST. Sludge will be pumped at 5% solids from PST under flow.

5.1 Sludge loading

$$= 1 \times 150 \times 0.5 \times 8.34 = 625.5 \text{ lb/d}$$

5.2 Solids remaining at the end of 10 days period in the digester

$$= W_d \times t\left(1 - \frac{V_a}{t}\right);$$ [a = 50% solids reduction in 10 days]

$$= 625.5 \times 10\left(1 - \frac{0.5 \times 0.5}{2}\right)$$

$$= 5473 \text{ lb}$$

5.3 Sludge volume (assume density of sludge is same as water)

$$= \frac{5473 \text{ lb}}{0.05 \times 62.4} = 1754 \text{ ft}^3$$

5.4 Secondary digester (sludge thickening + storage)

Assume 30 days; storage at average solids concentration of 8 percent and 1 ft^3 of sludge stored at 8 percent contains 6 lbs of total solids.

5.4.1 Total solids to the secondary digester = 547.3 lb/d

5.4.2 At 8% solids which contains 6 lb of total solids

$$= \frac{547.3 \text{ lb/d}}{6 \text{ lb/ft}^3} = 91.2 \text{ ft}^3/\text{d}$$

5.4.3 Total digester (secondary) capacity = 91.2 × 30

= 2770 ft^3.

Example 3.94

Gas Production from Anaerobically Decomposed Sludge

Determine the amount of gas that would be produced under anaerobic conditions in a landfill per unit weight of sludges. Assume the following data :

Empirical formula of the sludge	:	$C_{60}H_{90}O_{40}N$
Moisture content of the sludge	:	20%
Inert content of the sludge	:	30%
Decomposable content of the sludge	:	50%
Efficiency of conversion	:	95%

Solution

1. Required biological reactions
   - Partial conversion :

$$C_aH_bO_cN_d \rightarrow nC_wH_xO_yN_z + mCH_4 + sCO_2 + rH_2O + (d - nx)NH_3$$

where : $s = a - nw - m$

$r = c - ny - 2s$

   - Complete conversion :

$$C_aH_bO_cN_d + \left(\frac{4a - b - 2c - 3d}{4}\right)H_2O \rightarrow \left(\frac{4a + b - 2c - 3d}{8}\right)CH_4$$

$$+\left(\frac{4a - b + 2c + 3d}{8}\right)CO_2 + dNH_3$$

   - Methane content = 50 – 60% (typical)
   - Gas generation = 10 – 16 ft^3 of gas/lb of biodegradable VS
2. Required gas generation [Assume total weight of sludge = 100 lb]

a = 60; b = 90; C = 40; d = 1

   - Complete conversion

$$C_{60}H_{90}O_{40}N + 16.75H_2O = 30.87CH_4 + 29.125CO_2 + NH_3$$

- Methane weight

$$= \frac{30.87(16)(50)(0.95)}{[60(12)+90(1)+40(16)+1(14)]}$$

= 16.0 Ib

- CO_2 weight

$$= \frac{29.125(44)(50)(0.95)}{1464}$$

= 41.6 lb

- Methane volume

$$\text{Specific weight of } CH_4 = 0.0448 \text{ lb/ft}^3$$

$$\text{Methane volume} = \frac{16.0 \text{ lb}}{0.0448 \text{ lb/ft}^3} = 357 \text{ ft}^3$$

- CO_2 volume :

Specific weight of CO_2 = 0.1235 lb/ft^3

$$CO_2 \text{ volume } = \frac{41.6 \text{ lb}}{0.1235 \text{ lb/ft}^3} = 337 \text{ ft}^3$$

- Methane percent

$$= \frac{357 \text{ ft}^3}{(357+337)\text{ft}^3} \times 100 = 51.4\%$$

- CO_2 percent

= (100 – 51.4) = 48.6%

- Amount of gas produced based on dry organic decomposable sludge

$$= \frac{694 \text{ ft}^3}{50(0.95) \text{ Ib}} = 14.6 \text{ ft}^3/\text{lb}$$

- Amount of gas produced based on 100 lb of sludge

$$= \frac{694 \text{ ft}^3}{100 \text{ lb}} = 6.94 \text{ ft}^3/\text{lb}.$$

Example 3.95

Conventional Mesophilic Digester

Calculate the retention time for a conventional anaerobic digester (without recycle). Assume the following for this example

k at 35°C = 8.7 d^{-1} ; k_s = 165 mg/L as COD;

k_d at 35°C = 0.019 d^{-1} ; Y = 0.044, and S_e = 150 mg/L as COD

Solution

A biomass solids balance may be written as :

$$V\frac{dX}{dt} = QX_o + (\mu - k_d)XV - QX$$

At steady-state and no influent biomass into the system, the equation becomes

$$QX = (\mu - k_d)V \quad \text{or}$$

$$\frac{1}{t} = \frac{1}{\theta_c} = \mu - k_d$$

where t is the hydraulic retention time and q_c is the mean cell residence time

$$\mu = -\frac{ds}{dt}\frac{Y}{X} = \frac{\mu_{max}S_e}{(k_s + S_e)} = \frac{1}{X}\frac{dX}{dt}$$

$$\theta_c = (\mu - k_d)^{-1}$$

$$= \left[\frac{\mu_{max}S_e}{(k_s + S_e)} - k_d\right]^{-1}$$

$$= \left[\frac{Y\,kS_e}{(k_s + S_e)} - k_d\right]^{-1}$$

$$k = \frac{\mu_{max}}{Y}$$

$$\theta_c = \left[\frac{8.7 \times 0.044 \times 150}{165 + 150} - 0.019\right]^{-1} = 6.1\text{ d}$$

Example 3.96

Gas Production and Heat Budget in a Mesophilic Anaerobic Digester

Gas Production

The rate of gas evolution is commonly around 27 L per capita per day or approximately 1 m^3 (at STP) per kg volatile material destroyed. The ratio of methane to CO_2 varies with the substrate; however for domestic sewage a ratio of 1.9:1 is commonly accepted, giving methane concentration of approximately 66%. Some variation with temperature is found, as also are variations due to toxic substance inhibiting the methane fermentation process, this latter process producing reductions in methane down to 55% as total gas for heavy metal concentration of less than 400 mg/L. Some variation is found, depending on the type of sewage treatment process. (e.g., gas production of 16, 22 and 26 L per capita per day have been suggested for Imhoff tanks, primary settlement only, primary settlement +

activated sludge, and primary settlement + percolating filter installations, respectively).

The gas also contains approximately 2 – 3% H_2, similar amounts of N_2 and traces of O_2 and H_2S, with variable amounts of SO_2 and SO_3. The calorific value of the gas shows considerable variation about a mean of 22,400 kJ m^{-3} (5350 kcal/m^3) at STP.

Heat Budget

In essence a mesophilic digester has a heat budget as follows :

Losses

- Loss to the atmosphere and soil from the digester contents through the concrete/brickwalls.
- Losses due to removal of sludge to dewatering tank. This is equal to the heat required to raise similar (inlet) mass of sludge from ambient to operating temperature.
- Losses associated with heat exchangers, pipelines, boilers, etc.

Gains

- The major gain is from the gas itself, which may be burned.
- Reaction heat gains. These are generally neglected in calculation of overall heat requirements. The specific heat of sludge is 4.18 kJ $kg^{-1}/°C$.

  Consider a population of 200,000 and combined sludge (primary + activated produced at 0.09 kg dry SS per cap per day and 70% volatile).

$$\text{Total sludge production} = 200{,}000 \times 0.09$$
$$= 18{,}000 \text{ kg/d}$$

  Assuming 96% moisture content (MC) for the sludge (40 kg/m^3), the daily sludge volume is

$$\frac{18{,}000}{40} = 450 \text{ m}^3\text{d}^{-1}$$

  A retention time of 25 d gives a total digester capacity of $450 \times 25 = 11{,}250$ m^3 (If an average loading of 1.55 kg VSS per m^3 capacity had been chosen, a capacity of

$$\frac{18{,}000 \times 0.70}{1.55} = 8129 \text{ m}^3$$

  would have been required, which corresponds roughly to an MC of 94.5% i.e., 8200 m^3).

  A choice of three digesters each of approximately 3800 m^3 would possibly be made in order to keep down machinery costs and heat loss and enable still reasonable mixing within the tanks.

In order to reduce problems of foaming, it is advisable to allow the sludge depth to exceed 11 m. Therefore, the digester diameters are

$$d = \left(\frac{3750}{11} \times \frac{4}{\pi}\right)^{1/2} = 21 \text{ m}$$

In practice, somewhat greater sludge depths are used, giving reduced (diameter/depth) ratios. This reduces the effect of top cover heat losses.

The heat budget can be calculated from the heat losses and gas heat production, assuming a specific heat for the sludge of 4.18 kJ kg^{-1}/°C (10^3 cal kg^{-1}/°C). Each daily sludge input must be raised to the operating temperature of 33°C. Assuming an overall sludge yearly mean temperature of 10°C, the heat required is $4.18 \times 10^3 \times 450\ (33 - 10) = 43.26 \times 10^6$ kJ/d. The heat loss from the digester can be calculated; however, an approximate guide has been given in terms of an equivalent temperature drop of 0.4-0.8°C per day. With efficient design a large digester could operate with an average value of 0.6°C/d. The heat loss to air/soil is, therefore :

$$4.18 \times 0.6 \times 10^3 \times 11{,}250 = 28.22 \times 10^6 \text{ kJ/d}$$

The total heat requirement is 71.48×10^6 kJ/d. This total heat requirement must be offset by the gas production. An average value of 1.1 m^3 per kg VM destroyed for gas production with approximately 50% organise destruction would give :

$$18{,}000 \times 0.7 \times 0.5 \times 1.1 = 6930 \text{ m}^3 \text{ gas/d}$$

With an average calorific value of, say, 22,360 kJ/m^3, this would give 155 $\times$ 10^6 kJ/d. Even if the gas boiler/pipe losses were to represent 33% overall loss, a net value of 108×10^6 kJ/d would be adequate to provide for the above heat losses.

In practice, however, the gas would run a dual fuel engine producing approximately 1.9 kWh/m^3 gas (6.8 MJ/m^3 gas) is electrical power and approximately 30% of the net gas calorific value as hot water in the engine jacket and exhaust calorifiers. A split of gas, therefore, between boiler and dual fuel engine can be calculated to give sufficient heat to satisfy the heat loss requirements. If Z is the volume of gas burned in the boiler, then (boiler efficiency $\times$ gas calorific value $\times$ Z m^3 gas) + (water jacket efficiency $\times$ gas calorific value $\times$ gas volume left) = å heat requirement

$$0.7 \times 22{,}360 Z + 0.3 \times 22{,}360\ (6930 - Z) = 71.5 \times 10^6 \text{ kJ}$$

$$Z = 2790 \text{ m}^3$$

Thus, if 2790 m^3 is used as boiler feed at 70 percent efficiency, the remaining 4140 m^3 gas would provide approximately 4140×1.9 or 7860 kWh as electrical power (28.3 GJ).

The effect of varying the operating conditions can now be considered. If the original sludge input were consolidated to a moisture content of 95%, the sludge input heat requirement would be

$$4.18 \times \frac{18,000}{50} \times 10^3 (33 - 10) = 34.61 \times 10^6 \text{ kJ/d}$$

and (still assuming an equivalent heat loss of 0.6 °C/d for the total contents), the digester – ambient heat loss would be

$$4.18 \times \frac{18,000}{50} \times 25 \times 10^3 \times 0.6 = 22.57 \times 10^6 \text{ kJ/d}$$

giving a total of 57.18×10^6 kJ/d. The value of Z then becomes 1198 m^3 required for boiler feed; the electrical output from dual fuel engine/ generator would increase to (6930 – 1198)×1.9 = 10890 kWh (39.2 GJ). If the original sludge were only drawn off at 98% MC with no consolidation, the heat requirement would be

$$4.18\left(\frac{18,000}{20} \times 10^3 (33 - 10) + \frac{18,000}{20} \times 25 \times 10^3 \times 0.6\right) = 143 \text{ kJ/d}$$

In this case a critical situation could possibly arise if the gas production were reduced (owing to a slug of heavy metals being discharged to the walls). The effect of increased heat loss of digester contents is not critical provided the total digester volume is minimised by sludge consolidation- *e.g.;* at 95% MC input an increase of equivalent temperature drop to 0.8°C/d increases heat requirements by

$$4.18 \times \frac{18,000}{50} (0.8 - 0.6) \times 10^3 \times 25 = 143 \text{ kJ/d}$$

The most critical factor which is immediately controllable by the plant operator is, therefore, the consolidation of the input sludge.

Note :

Major causes of digester failure in increasing order of importance are :

- Stratification and solids loss
- Overloading
- Temperature control
- Scum formations, and
- Poor mixing.

The preponderance of causes linked to poor mixing if the contents is an indication of how badly many digesters deviate from the kinetic model of a completely mixed reactor. Modern digesters usually have some scum drawoff and mixing systems in order to alleviate the above problems but very few operate with efficient mixing.

Example 3.97

Anaerobic Contact Process

The required reactor volume for contact anaerobic process is calculated from the kinetic relationship :

$$t = \frac{S_r}{X_v k S_e}$$

The VSS is generally in the range of 3000-5000 mg/L at SRT (sludge age) should be in the excess of 20 days in the digester to insure the growth of the methane fermentation micro-organisms. The sludge yield and methane (CH_4) production (G) are calculating using the following relationships

$$\Delta X_v = aS_r - bX_v$$

$$G = 5.62(S_r - 1.42\Delta X_v)\text{ft}^3/\text{d}$$

Heat loss is calculated based on 1.0° F/d of detention time. The sludge digester should be designed based on a SRT of 10 days and for an organic loading of 0.4 lb VSS/ft^3.d the larger volume of the two should be selected for design. Anaerobic filters can be designed using a relationship developed by Young and McCarty.

$$t = \frac{S_o - S_e}{kS_e}$$

Design an anaerobic contact process to achieve 85% removal of COD. Assume the following information :

Wastewater flow	:	150,000 gal/d
Total influent COD	:	8000 mg/L
Recalcitrant COD	:	2000 mg/L
SRT	:	20 days (minimum)
Temperature	:	35°C
'a' and 'b'	:	0.140 and 0.020 d^{-1}
Reaction rate constant (k)	:	0.00045 L/mg .d
X_v	:	4500 mg/L

Solution

1. Digester value

$$t = \frac{S_r}{X_v\ kS_e} = \frac{S_o - S_e}{X_v\ kS_e} = \frac{6000}{4500 \times 0.00045 \times 600}$$

$$V = 0.15 \text{ mil} - \text{gal/d} \times 4.9 \text{ d} = 0.74 \text{ mil-gal}$$

$$\text{SRT} = \frac{X_v}{\Delta X_v} = \frac{X_v \cdot t}{aS_r - bX_v \cdot t}$$

$$= \frac{4500 \times 4.9}{0.14 \times (6000 - 600) - 0.020 \times 4500 \times 4.9}$$

$$= 70 \text{ day}$$

SRT of 70 days is greater than 20 days (minimum) to insure methane growth.

2. Sludge production rate

$$\Delta X_v = a\ S_r - b\ X_v$$

$$= 0.14(5400 \times 0.15 \times 8.34) - 0.020 \times (4500 \times 0.45 \times 8.34)$$

$$= (946 - 338) = 608 \text{ lb/d}$$

3. Methane production rate

$$G = 5.62(S_r - 1.42\ \Delta\square X_v)$$

$$= 5.62(5400 \times 0.15 \times 8.34 - 1.42 \times 608)$$

$$= 5.62(6755.4 - 863.4)$$

$$= 33{,}113 \text{ ft}^3\text{/d of } CH_4$$

4. Heat balance

Critical winter temperature for wastewater flow is 70°F

Heat transfer efficiency is 40%

4.1 Heat required (BTU/d)

$$= \frac{W(T_i - T_e)}{E}$$

$$= \frac{0.15 \text{ MGD}(8340000 \text{ lb/MG})(100^\circ + 4.9^\circ - 70^\circ)}{0.4}$$

$$= 109150000 \text{ BTU/d}$$

4.2 Heat available = 33,113 $\text{ft}^3\text{/d-}CH_4 \times 960 \text{ BTU/ ft}^3 - CH_4$

$$= 31788480 \text{ BTU/d}$$

4.3 Application of external heat to maintain the anaerobic digester contents at 95°C

$$= (10{,}91{,}50{,}000 - 3{,}17{,}88{,}480) = 7{,}73{,}61{,}520 \text{ BTU/d}$$

5. Required nutrients (N and P)

$$N = 0.11 \times 608 \text{ lb/d} = 67 \text{ Ib/d}$$

$$P = \frac{67}{7} \text{lb/d} = 9.6 \text{ lb/d}$$

6. Clarifier with a vacuum degasifier or flash aerator should be provided between the digester and the clarifier to purge gas for the sludge.

Note :

1. *Mixing requirement :* 30 – 50 hp/MGD (0.0059 – 0.0099 kW/m^3)
2. Aerobic digestion

   2.1 Batch or plug flow

$$\frac{(X_d)e}{(X_d)d} = \exp(-k_d\ t)$$

or

$$\frac{X_e - X_n}{X_o - X_n} = \exp(-k_d\ t)$$

   2.2 Completely mixed reactor

$$\frac{X_e - X_n}{X_o - X_n} = \frac{1}{1 + k_d\ t}$$

or

$$t = \frac{X_o - X_e}{k_d (X_e - X_n)}$$

   2.3 Multiple reactors in series

$$\frac{X_e - X_n}{X_o - X_n} = \frac{1}{(1 + k_d\ t_n)^n}$$

   2.4 Oxygen requirement

   = 1.4 lb O_2 considered for each lb of VSS destroyed

   2.5 Temperature dependence for k_d

$$k_d\ (T) = k_d(20)1.023^{T-20}\ ;\ k_d = 0.10\text{-}0.2\ d^{-1}$$

   2.6 *Notations :* $(X_d)_e$ and $(X_d)_o$ are the degradable VS after time, t and initial degradable VS respectively, k_d is the decay reaction rate constant (d^{-1}), t is the time of aeration (day), X_o, X_e, X_n are the concentrations of initial VS, effluent VS and non-degradable VS, respectively.

Example 3.98

Anaerobic Contact Plant

Determine daily sludge accumulation based on Speece's formulation, digester volume, gas production, and nutrient requirements for an anaerobic contact process. Use the following information :

Daily waste flow : 0.25 MGD

Influent COD	:	6000 mg/L
Solids retention time (SRT)	:	15 days
BOD reduction	:	85 percent at 35°C
a	:	0.08
b	:	0.01

Solution

1. Solids accumulation (Speece Formula)

$$A = \frac{a\,F}{[1 + b(SRT)]}$$

where A is the lb volatile biological solids produced per day, F is the, lb COD removed per day, SRT is the solids retention time (days), a is the growth constant, and b is the endogenous respiration rate (d^{-1}).

This expression is similar to the expression

ΔS (lb VSS_r/d) = a lbBOD_r/d – b lbMLVSS

Applied COD (lb/d) = 6000 × 8.34 × 0.25 = 12,510 lb/d

$$A = \frac{0.08 \times 12{,}510 \times 0.85}{[1 + 0.01 \times 15]} = 740 \text{ lb/d}$$

2. Digester volume

Assume MLSS of 6000 mg/L

$$SRT = 15 \text{ days} = \frac{\text{Total sludge in the digester}}{\text{Sludge produced/d}}$$

$$= \frac{TS}{740 \text{ lb/d}}$$

TS = 15 × 740 lb/d = 11,100 lb

$$\text{Required volume} = \frac{11{,}100 \text{ lb}}{6000 \times 8.34} = 0.22 \text{ mil-gal}$$

$$= 29{,}403 \text{ ft}^3$$

3. Process loading

$$= \frac{12{,}510 \text{ lb/d}}{29{,}403} = 0.425 \text{ lb/ft}^3\text{.d}$$

4. Hydraulic detention period

$$\frac{0.22 \times 10^6 \text{ gal}}{0.25 \times 10^6 \text{ gal/day}} = 0.88 \text{ day}$$

5. Methane production

$$G = 5.62(e\ F - 1.42\ A)$$
$$= 5.62(0.85 \times 12{,}510 - 1.42 \times 740)$$
$$= 53{,}856\ ft^3/d - CH_4$$

6. Available heat energy

$$= 53{,}856\ ft^3/d \times 960\ BTU/ft^3 = 5{,}17{,}0220\ BTU/d$$
$$= 5.2 \times 10^6\ BTU/d$$

7. Nutrient requirements

Biomass formula	:	$C_5H_7NO_2$
Active biomass contains:		
• Nitrogen percent	:	12.3
• Phosphorus percent	:	2.6
Nitrogen requirements	:	740 lb/d × 0.123 = 91 lb/d
Phosphorus requirements	:	740 lb/d × 0.026 =19.3 lb/d

Note :

1. The cellular residue after endogenous respiration has been reported to contain 7% nitrogen and 1% phosphorus.

Example 3.99

Electric Power Generation from Anaerobic Digester

Calculate the electric power generation from a 10 MGD wastewater facility. Assume the following :

Total sludge fed to anaerobic digester on dry basis	:	40,000 lb/d
Percent volatile solids	:	80%
Percent destruction of volatile solids	:	60%
Engine efficiency	:	30%
Engine generator efficiency	:	30%

Solution

Total volatile solids destroyed	:	40,000 × 0.80 × 0.60
	:	19,200 lb/d
Assume standard biogas generation	:	10 SCF/lb of VSS destroyed
Total biogas generation	:	19,200 × 10 = 1,92,000 SCF
Total methane generation	=	1,92,000 × 0.6
	=	1,15,200 SCF/d

Heat to be generated by combustion of methane

(Heat value of methane = 960 BTU/SCF)

$$\text{Total heat to be generated} = 1{,}15{,}200 \times 960$$
$$= 1105920000 \text{ BTU}$$
$$= 111 \times 10^6 \text{ BTU}$$

Power available (assume 2500 BTU/hp/hr at 30% efficiency or 8333 BTU/hp/hr)

$$\frac{111 \times 10^6}{8333} = 540 \text{ hp}$$

Electric power generated

$$= \frac{2500 \text{ BTU/hp/hr}}{0.746 \text{ kWh/hp/hr} \times 0.30} = 11{,}171 \text{ BTU/kWh}$$

Power generated

$$\frac{111 \times 10^6}{11{,}171 \times 24} = 414 \text{ kW}$$

Example 3.100

Sludge Lagoon Disposal System

Determine the lagoon surface area, the sludge depth accumulation, and the air flow requirements for a lagoon to stabilize the sludge generated from an ASP.

Assume the following data :

Biomass formula	:	$C_5H_7NO_2$
Wastewater flow to an ASP	:	2.0 mil – gal/d
Influent BOD	:	300 mg/L
ASP-sludge age	:	35 days
Average temperature	:	20°C
Coefficients		
• Yield coefficient (a)	:	0.5
• Decay coefficient (b)	:	0.1 d^{-1}
Hydraulic detention time	:	0.7 d
Mixed liquor volatile suspended solids (MLVSS)	:	3000 mg/L
Volatile fraction of MLSS	:	80%
Effluent BOD	:	20 mg/L

Solution

1. Degradable fraction of biological VSS (X_d)

$$X_d = \frac{X_d^*}{1 + b\, X_n^* \theta_c}$$

where X_d is the degradable fraction of the biological VSS, X_d^* is the degradable fraction of the biological VSS at generation, X_n^* is the non-degradable fraction of the biological VSS at generation [average of 0.2 $\left(X_d^* + X_n^*\right) = 1\right]$, b is the endogenous rate coefficient (d^{-1}), and q_c is the sludge age (d)

$$X_d = \frac{0.8}{1 + (0.1)(35)(0.2)} = 0.47$$

2. Excess sludge production (DX_v)

$$\Delta X_v = a\,[S_r + f_d\, f_x\, X_i] - b\, X_d\, f_b\, X_v\, t + (1 - f_d)\, f_x\, X_i + (1 - f_x)\, X_i$$

where a is the fraction of organics removed that is synthesized to cell mass, S_r is the soluble substrate removed, f_d is the fraction of degradable influent VSS degraded, f_x is the fraction of influent VSS that is degradable, X_i is the influent VSS (mg/L), b is the fraction per day of degradable biomass oxidized, X_d is the fraction of the biological VSS, f_b is the fraction of MLVSS that is biomass, X_v is volatile SS concentration (mg/L), and *t* is the hydraulic detention time (d).

Therefore,

$$\Delta X_V = \left[0.5(300 - 20) - 0.1 \times 0.47 \times 0.7 \times 3000\right] \times 2.0 \times 8.34$$

$$= 688 \text{ lb VSS/d}$$

$$= \frac{688}{0.8} = 860 \text{ lb SS/d}$$

Assume 80% of VSS will be degraded in the lagoon, therefore, lagoon area is :

$$A = \frac{R_a}{AS}$$

where R_a is the loading rate of VSS (lb/d) to be degraded, and AS is the average stabilization ratio (600 lb VSS/acre.d)

The required surface area of the lagoon is :

$$A = \frac{688 \text{ lb VSS/d}}{600 \text{ lb VSS/acre.d}}$$

$$= 1.147 \text{ acre}$$

3. Required sludge depth accumulation

Residual sludges = [860 – 688 × 0.80] lbs residue SS/d

= 310 × 365 lb residual SS/yr

= 113,004 lb residual SS/yr

$$\text{Sludge volume at 4\% solids} = \frac{113{,}004 \text{ lb SS/yr}}{(0.04)\left(62.5 \text{ lb/ft}^3\right)}$$

$= 45{,}202 \text{ ft}^3\text{/yr}$

Therefore, sludge depth accumulation is :

$$= \frac{45202 \text{ ft}^3/\text{yr}}{(1.147 \text{ acre})\left(43{,}560 \text{ ft}^2/\text{acre}\right)} = 0.905 \text{ ft/yr}$$

4. Required air flow

Assume oxygen requirement of 760 lb O_2/acre.d

Total oxygen requirement = 760×1.147 = 872 lb O_2/d

$$\text{Airflow requirement}^* = \frac{872 \text{ lb } O_2/\text{yr}}{(144 \text{ min/d})(0.1)(0.232 \text{ lb } O_2/\text{lb air})\left(0.0746 \text{ lb air/ft}^3\right)}$$

- 10 ft depth = 350 Standard ft^3/min (transfer efficiency ofr 10%)

Assume each diffuser operates at 5 Standard ft^3/min, the No. of diffusers will be :

$$= \frac{350}{5} = 70$$

Note :

1. In general, lagoons should be considered where large land areas are available and the sludge will not present a nuisance to the surrounding environment.
2. Benthal stabilization occurs in a lagoon as a result of anaerobic or a combination of aerobic and anaerobic mechanisms.
3. Below the surface aerobic layer, anaerobic conditions prevail resulting in methane production as well as other products of anaerobic decomposition. Ammonia as well as some of the less-reduced products generated in the anaerobic layer (organic acids) diffuse into aerobic layer in which they are oxidized.
4. Rich [Water Res. 16 [9(1)], 399 (1982)] has reported an average benthal stabilization rate of 80 gm/m^2.d of bio-degradable solids at 20°C. Under these conditions, as much as 63% of total carbon stabilization can occur view methane fermentation. Assuming, oxidation of all ammonia and

BOD released to the water, the oxygen uptake would be 86 gm O_2/ m^2.d. The sludge should concentrate in the bottom of the lagoon to 2 to 3% solids.

5. Assuming continuous input to the sludge lagoon and withdrawal once a year, the average annual stabilization rate is estimated as 68 gm biomass/ m^2.d.
6. Only sludges can be successfully disposed of on land :
   - Oil degradation rate is directly related to the percentage of oil in soil.
   - Fertilization improved the degradation rate.
   - Aeration (tilling) frequencies vary (1 week to 3 months).
   - 380 and 400 m^3 of oil per ha should be degraded in an 8 months growing season.
   - Sludge farming is about one-fifth as expensive as incineration.

Example 3.101

Aerobic Sludge Digestion : Primary and Secondary Sludges

Determine the following for an aerobic digestion to process sludge from mixed ASP :

- Digester retention time and volume
- Oxygen requirement
- Horsepower requirements and power level (P/V)

Assume the following data :

Sludge parameters

– Primary sludge

Flow	:	0.03 MGD
Temperature	:	20°C
Total Kjeldhal nitrogen	:	6000 mg/L
Ultimate BOD (BOD_u)	:	30,000 mg/L
Total suspended solids (TSS)	:	30,000 mg/L

– Wasted activated sludge :

Concentration	:	1.0%
Flow	:	0.1 MGD
Temperature	:	30°C
Ammonia-N [NH_4^+– N]	:	15 mg/L
SRT (θ_c)	:	8 days
Total suspended solids [TSS]	:	10,000 mg/L

Solution

1. Required mathematical expressions for aerobic digesters

 1.1 Mass balance for biodegradable solids [MLVSS] in a CSTR at steady-state :

$$Q(X_d)_o - \left[\frac{dX_d}{dt}\right]V - Q(X_d)_e = 0$$

$$t_d = \frac{V}{Q} = \frac{(X_d)_o - (X_d)_e}{k_b (X_d)_e}$$

$$\left[\frac{dX_d}{dt} = -k_b X_d\right]$$

$$(X_d)_o = X_o - X_n$$

$$(X_d)_e = X_e - X_n$$

Therefore, $t_d = \dfrac{(X_o - X_e)}{k_b (X_e - X_n)}$

Simplifying it further, t_d becomes :

$$t_d = \frac{(X_o - X_e)}{k_b D(X_{oad}) X_o}$$

or
$$t_d = \frac{(X_o - X_e)}{k_d (0.77) D(X_{oa}) X_o}$$

or
$$t_d = \frac{(X_o)_m - (X_e)_m}{(k_d)_m (0.77) D(X_{oa}) (X_o)_m} \qquad (1)$$

$$X_{om} = \frac{(X_o)_p Q_p + Q_A X_o}{Q_p + Q_A}$$

 1.2 Assumptions for the equations are :

- Total suspended solids are composed of an active fraction and an inactive fraction.
- Inactive fraction of the influent total suspended solids is non-degradable [material that cannot be oxidized/solubilized through microbial activity].
- Active fraction of the influent TSS comprises of non-degradable and degradable fractions. Degradable material can be oxidized/solubilized through microbial activity.

- Degradable active fraction of the total suspended solids decreases during digestion

$$(k_d)_m = \left[\frac{\text{Active degradable biomass lost through microbial activity/time}}{\text{Active degradable biomass in the system}}\right]$$

$$- \left[\frac{\text{Biomass saved from destruction / time because of presence of external food source}}{\text{Active degradable biomass in the system}}\right]$$

$$= \frac{k_d\left[D(X_{od})_m (X_{oad})V\right]}{D(X_{oad})(X_o)_m\ V} - \frac{Y_t\ Q\ S_a}{D(X_{oad})(X_o)_m\ V}$$

or
$$(k_d)_m = k_d - \frac{Y_t\ S_a}{0.77\ D(X_{oa})(X_o)_m\ t_d} \qquad (2)$$

where
Q = Volumetric flow rate $[L^3/T]$

$(Xd)_o$ = Degradable VSS concentration in the influent $[M/L^3]$

$(Xd)_e$ = Degradable VSS concentration in the effluent $[M/L^3]$

V = Digester volume $[L^3]$

t_d = Digester retention time [T]

X_n = Non-biodegradable VSS, assumed to be constant throughout the digestion period $[M/L^3]$

k_b = Degradable solids lost rate as a result of endogenous respiration $[T^{-1}]$

X_d = Degradable solids remaining at time, t $[M/L^3]$

X_{oa} = Fraction of the total SS concentration in the influent which is active

X_{oad} = Fraction of the active mass which is degraded [active mass which can be oxidized or stabilized through biological activity]

D = Fraction of degradable active biomass in the influent that appears as degradable active biomass in effluent [0.1 to 3]

k_d = Decay rate of the degradable fraction of the active biomass approximated from the decrease in total SS

[0.77] = Degradable fraction of a cell [only for the active biomass, not for the total or VSS]

Y_t = True yield coefficient of the organic content of primary sludge

S_a = $\dfrac{Q_P}{Q_A + Q_P} S_o$ = BOD_u of the primary sludge in the feed [M/L^3]

S_o = Primary sludge BOD_u [M/L^3]

Q_p = Flow rate of primary sludge [L^3/T]

Q_A = Flow rate of activated sludge [L^3/T]

$(X_o)_p$ = TSS concentration in primary sludge [L^3/T]

$(X_{oa})_m$ = $\dfrac{X_o}{(X_o)_m} X_{oa}$ = Fraction of total solids in the digester feed that is active

$(k_d)_m$ = Overall decay rate of the degradable fraction of the active biomass [T^{-1}], and takes into account the presence of external food source in the form of primary sludge and assumes that all the external food is utilized [T^{-1}]

$(X_e)_m$ = Total SS concentration [M/L^3] in the effluent from aerobic digester treating primary and wasted activated sludge

Substituting for $(k_d)_m$ in equation (1) from equation (2):

$$t_d \left[k_d - \frac{Y_t S_a}{0.77(D)(X_{oa})_m \, t_d} \right] = \frac{(X_o)_m - (X_e)_m}{0.77\, D (X_{oa})_m \, (X_o)_m}$$

or
$$t_d = \frac{(X_o)_m + Y_t S_a - (X_e)_m}{k_d \left[0.77\, D (X_{oa})_m \, (X_o)_m \right]}$$

- Temperature effect on decay coefficient (k_d) :

  $k_d(T) = k_d(20)(1.023)^{T-20}$

1.3 Oxygen requirement

- Carbon oxidation [Primary and activated sludge]

$$\text{lb } O_2/d = 1.42\begin{bmatrix}\text{Biological solids}\\ \text{destroyed (lb/d)}\end{bmatrix} + \begin{bmatrix}BOD_u \text{ satisfied from}\\ \text{primary sludge (lb/d)}\end{bmatrix}$$

$$= 1.42(8.34)\ [Q_oR(0.77)\ X_{oa}X_o] + 8.34\ Q_p\ S_o$$

where R = Reduction in active degradable biomass [fraction]

X_{oa} = TSS fraction in digester feed which is active biomass [fraction]

X_o = TSS concentration in digester feed [mg/L]

Q_o = Digester feed [MGD]

S_o = BOD_u of primary sludge [mg/L]

Q_p = Primary sludge flow rate [MGD]

- Nitrogenous oxygen demand [NOD] :

$$NOD = \begin{bmatrix}\text{oxygen required in the}\\ \text{biological oxidation of}\\ \text{ammonia contained in}\\ \text{the influent digestion}\end{bmatrix} + \begin{bmatrix}\text{oxygen required in the}\\ \text{biological oxidation of}\\ \text{ammonia released during}\\ \text{oxidation of cellular}\\ \text{proteins}\end{bmatrix}$$

$$+\begin{bmatrix}\text{oxygen required in the biological oxidation}\\ \text{of ammonia released during oxidation of}\\ \text{primary sludge}\end{bmatrix}$$

or $NOD[\text{lb } O_2/d] = (8.34)(4.57)\{(Q_A)\left(NH_4^+ - N\right)_A + Q_oR(0.77\ X_{oa}X_o)\ 0.122 + Q_p\ (TKN)_p\}$

where $[NH_4^+ - N]$ = Ammonia concentration in the waste activated sludge [mg – N/L]

$[R(0.77X_{oa}X_oQ_o)]$ = Biological solids destroyed per day

[0.122] = Nitrogen released as a result of biological solids destroyed based on empirical cell composition: $C_{60}H_{80}O_{23}N_{12}\ P$

$[TKN]_p$ = Total Kjedahl nitrogen in primary sludge feed [mg – N/L]

Q_A = Wasted activated sludge flow rate [MGD]

- Total oxygen requirement becomes :

  lb O_2/d = 1.42(8.34)[Q_o R(0.77) X_{oa} X_o] + 8.34Qp S_o + NOD

  [For wasted activated sludge = 15 to 20 scfm/1000 ft^3 of tank capacity

  For primary and secondary sludge = 25 to 30 scfm/1000 ft^3]

1.4 Mixing requirement :

- Power level :

$$\frac{P}{V} = 0.00475(\mu)^{0.3}(X)^{0.298}$$

where $\frac{P}{V}$ = Power level [hp/1000 gal]

μ = Water viscosity [cp]

X = Digester TSS concentration at steady state [mg/L]

- Required compressed air flow :

$$\frac{G_s}{V} = 50.5\frac{\frac{P}{V}}{\log\left[\frac{h+34}{34}\right]}$$

$$\frac{P}{V} = \text{Power level}\left[\text{hp/1000 gal}\right]$$

h = submergence depth to the diffuser [ft]

where = Cfm/1000 ft^3 at operating air temperature

= Power level [hp/1000 gal]

h = submergence depth to the diffuser [ft]

2. Required digestion temperature (T) primary to secondary sludge mass ratio (M_P/M_A) effluent TSS concentration and active fraction of the sludge mixture (X_{oa})m

2.1 Required T, M_P/M_A, and (X_{oa})m

- Digester temperature (T) :

$$T = \frac{20(0.30)+30(0.1)}{(0.03+0.1)} = 28^\circ C$$

- Sludge mixture concentration $(X_o)_m$:

$$= \frac{30000(0.30)+10000(0.1)}{(0.03+0.1)}$$

$$= 14600 \text{ mg/L}$$

- Primary to secondary sludge ratio $[M_p/M_A]$

$$\left[\frac{M_P}{M_A}\right] = \frac{(0.03)(30{,}000)}{(0.10)(10{,}000)}$$

$$= 0.9$$

It has been experimentally find that TSS reduction = 34.6% at M_P/M_A of 0.9 [when activated sludge portion of the mixture is from a culture approaching 100% activity/unit of biomass]

- Active fraction (X_{oa}) at SRT (θ_c) = 8 days :

= 0.8 [Experimental value]

2.2 Required digester detention time (t_d) :

- Effluent TSS concentration:

$$(X_e)_{design} = m\ X_e$$

where $(X_e)_{design}$ = Effluent TSS concentration to be used in design

X_e = Effluent TSS concentration possible only when the activated sludge solids are nearly 100% active

m = Adjustment factor necessary because the activated sludge solids are below 100% of their activity

The relationship is :

$$\frac{X_o - X_e}{1.0} = \frac{X_o - m\ X_e}{X_{oa}}$$

or

$$m = \frac{X_o - X_{oa}(X_o - X_e)}{X_e}$$

$$= \frac{14600 - 0.8\left[14600 - 14600(1 - 0.346)\right]}{14600(1 - 0.346)}$$

$$= 1.1058$$

Therefore, $[X_e]_{design}$ = 1.1058[14600(1 – 0.346)]

= 10560 mg/L

- Active fraction of sludge mixture $(X_{oa})_m$

$$(X_{oa})_m = \frac{X_o}{(X_o)_m}(X_{oa})$$

$$= \left(\frac{10{,}000}{14600}\right)(0.8) = 0.55$$

- Ultimate BOD_u of the primary sludge in the feed to the digester

$$BOD_u(S_e) = \left[\frac{Q_p}{Q_A + Q_p}\right] S_o$$

$$= \left[\frac{0.03}{0.1+0.03}\right](30,000)$$

$$= 6920 \text{ mg/L}$$

- Decay coefficient (k_d) at 28°C

$k_d(T) = k_d(20)(1.023)^{T-20}$

$= 0.20\ (1.023)^{28-20}$

$= 0.24\ d^{-1}$

- Digester detention time (t_d) :

$$t_d = \frac{(X_o)_n + Y_t S_a - (X_e)_m}{k_d\left[0.77\,D(X_{oa})_m (X_o)_m\right]}$$

Assume the following :

Y_t = 0.5

D = 0.2 [Fraction of degradable active biomass in the influent that appears as degradable active biomass in effluent (active degradable biomas must be reduced by 80%)]

$$t_d = \frac{(X_o)_n + Y_t S_a - (X_e)_m}{k_d\left[0.77\,D(X_{oa})_m (X_o)_m\right]}$$

= 25.3 days

- Digester volume (V)

V = [(0.1 + 0.03) MGD][25.3 days]

= 3.3 MG

2.3 Required amount of oxygen for carbon and nitrogen oxidation

lb O_2/d = $1.42(8.34)[Q_o\ R(0.77)X_{oa}X_o] + 8.34Q_p\ S_o$ + NOD

Assume : R = 0.9 or 90% reduction in active degradable biomass

= 1.42(8.34)[(0.10 + 0.03)(0.8)(0.55)14600] + [8.34(0.03)(30000) + 8.34(4.57) {0.10(15) + [(0.1 + 0.3)(0.9)(0.77)(0.55)(14600)](0.122) + 0.03(6000)}

lb O_2/d = 11.84[835] + [7506] + 38.1{1.5 + [88.26 + 180]}

= 9886 + 7506 + 10288

= 27680 lb O_2/d

2.4 Required horsepower for oxygenation and mixing

- Power for oxygen transfer :
  - Assume oxygen transfer rate of 1.5 lb O_2/hp.h
  - Total horsepower required $= \dfrac{27680 \text{ lb } O_2/\text{d}}{1.5 \text{ lb } O_2/\text{hp.h}\,(24 \text{ h/d})}$

    $= 769$ hp
  - Power level (P/V) $= \dfrac{(769 \text{ hp})(1000)}{3.3 \times 10^6 \text{ gal}}$

    $= 0.23$ hp/1000 gal
- Power for mixing :

$$\frac{P}{V} = 0.00475(\mu)^{0.3}\left(X_{e,\,\text{design}}\right)^{0.298}$$

μ of water at 28°C = 0.8363 cp

$$\frac{P}{V} = 0.00475(0.8363)^{0.3}\,(10560)^{0.298}$$

$= 0.00475(0.95)(15.8)$

$= 0.07284$ hp/1000 gal

P = (0.07284 hp/1000 gal)(3.3×10^6 gal)

= 240 hp

Oxygen transfer controls the power requirements.

Note :

Table 82 : Specific oxygen utilization rate

Sludge Type	*Oxygen Utilization Rate [mg O_2/g VSS.h]*
Primary	20-40
Waste sludge for ASP	
• Conventional	10-15
• Extended aeration	5-8
• Contact stabilization	10
• Single stage digester	2-4 [aerobic]
• Two stage	0.5-2.4 [aerobic]

Example 3.102

Aerobic Digestion of Sludges

Determine the HRT, the digested sludge concentration, the maximum oxygen demand (Summer), the digester sludge product, and the digester mixing requirements for an aerobic digester. Assume the following data:

Wasted sludge	:	8000 gal/d
Concentration of thickened sludge (SS)	:	2% (20,000 mg/L)
Volatile SS (VSS)	:	16,000 mg/L
Fraction of feed VSS biodegradable	:	0.9
Decay constant	:	0.05 d^{-1} at 20°C
Destruction efficiency of VSS	:	40%

Solution

1. Required fixed solids, biodegradable VSS, non-biodegradable VSS (X_n)

Influent SS	=	20,000 mg/L
Influent VSS	=	16,000 mg/L (S_o)
Fixed solids	=	(20,000 – 16,000 = 4,000 mg/L
Biodegradable VSS	=	(0.9)(16,000) = 14,400 mg/L
Non-biodegradable VSS	=	(0.1)(16,000) = 1600 mg/L (X_n)

$$\text{Influent lb SS/d} = \frac{(8000\text{ gal/d})(8.34)(20,000)}{10^6} = 1334\text{ lb/d}$$

$$\text{Influent lb VSS} = \frac{(8000\text{ gal/d})(8.34)(16,000)}{10^6} = 1067\text{ lb/d}$$

$$\text{Influent biodegradable VSS (lb/d)} = \frac{(8000)(14,400)(8.34)}{10^6}$$

$$= 961\text{ lb//d}$$

$$\text{Influent non – biodegradable VSS (lb/d)} = \frac{(8000)(1600)(8.34)}{10^6}$$

$$= 107\text{ lb/d}$$

$$\text{Influent fixed solids (lb/d)} = \frac{(8000)(4000)(8.34)}{10^6}$$

$$= 267\text{ lb/d}$$

Effluent VSS biodegrable = (14,400 mg/L)(0.6) = 8640 mg/L

Total effluent VSS (S_w)

S_w = (8640 + 1600) = 10,240 mg/L

2. Required hydraulic retention time (HRT)

At 20°C (design) :

$$HRT = \frac{(S_o - S_W)}{k_d(S_W - X_n)}$$

$$= \frac{(16{,}000 - 10{,}240)}{(0.06)(10{,}240 - 1600)} \quad \text{at } 20°C \text{ (design)}$$

= 11 days

At 30°C (Summer) :

$$k_d(T) = k_d(20)(1.04)^{T-20}$$

$$k_d(30) = (0.06)(1.04)^{10}$$

$$= 0.09 \text{ d}^{-1}$$

$$HRT = \frac{(16{,}000 - 10{,}240)}{(0.09)(10{,}240 - 1600)}$$

3. Required oxygen demand

Effluent soluble BOD_5 = 5 mg/L

Effluent soluble ultimate $BOD_u = \frac{5 \text{ mg/L}}{0.68} = 7.35 \text{ mg/L}$

lb O_2 required/lb VSS destroyed	: 1.42
lb Nitrogen released/lb VSS destroyed	: 0.12
lb O_2 required/lb-N-nitrified	: 4.57

- Total required O_2 (lb/lb VSS destroyed) :

  = 1.42 + (0.12)(4.57) = 1.97 lb O_2/lb VSS destroyed

- S – BOD oxygen demand

$$\frac{(8000 \text{ gal/d})(8.34)(7.35)}{10^6} = 0.49 \text{ lb } O_2/\text{d}$$

- Maximum VSS daily oxygen demand at 30°C (Summer)

$$[\text{Influent VSS} - \text{Effluent VSS}][8.34]\left[\frac{Q}{10^6}\right] [1.97 \text{ lb } O_2/\text{lb VSS}]$$

$$= \left[\frac{(8000)}{10^6}\right] [14{,}400 - 5753][8.34][1.97]$$

= 1136 lb O_2/d.

- Total daily maximum oxygen demand

  $= (1136 + 0.49) = 1136.5 \text{ lb } O_2/d$

4. Required digester products

   Non-biodegradable feed VSS = 107 lb/d

   Fixed feed solids = 267 lb/d

   Maximum digester degradable VSS per day (Winter) :

   $$= \left(\frac{8000}{10^6}\right)(8.34)(8633) = 576 \text{ lb/d}$$

   Total sludge produced (dry)/d = (107 + 267 + 576)

   = 950 lb/d

   At T = 10°C (Winter) :

   $k_d(10) = (0.06)(1.04)^{-10}$

   $= 0.04 \text{ d}^{-1}$

   $$\text{HRT} = \frac{(16000 - 10240)}{(0.04)(10240 - 1600)}$$

   = 16.7 days

   Winter temperature controls, therefore digester volume (V) :

   V = (Sludge flow)(HRT)

   = (8000 gal/d)(16.7) = 133,333 gal

5. Required digested sludge concentration (S)

   At T = 20°C

   $$S = \frac{S_o + [k_d + (\text{HRT})(X_n)]}{1 + k_d(\text{HRT})}$$

   $$= \frac{16{,}000 + [0.06(16.7)(1600)]}{[1 + (0.06)(16.7)]} = 8793 \text{ mg VSS/L}$$

   Effluent degradable VSS = (8793 – 1600) = 7193 mg/L

   At T = 30°C

   $$= \frac{16{,}000 + [0.09(16.7)(1600)]}{[1 + (0.09)(16.7)]} = 7353 \text{ mg VSS/L}$$

   Effluent degradable VSS = (7353 – 1600) = 5753 mg/L

   At T = 10°C

$$= \frac{16{,}000 + [0.04(16.7)(1600)]}{[1 + (0.04)(16.7)]} = 10233 \text{ mg/L}$$

Effluent degradable VSS = (10233 – 1600) = 8633 mg/L

6. Required digester size and mixing requirement
   - V = (133,333 gal)(0.1337 ft^3/gal) = 17,827 ft^3 (Winter)

     No. of digesters = 2

     Volume of each digester $= \frac{17{,}827 \text{ ft}^3}{2} = 8914 \text{ ft}^3$

     Working depth = 10 ft

     Area of digester = 891 ft^2

     Diameter of each digester $= \left[\frac{4}{\pi}(891)\right]^{0.5} = 34$ ft

     Free board = 2 ft

     Total depth = (10 + 2) = 12 ft
   - Mixing requirements

$$= \left(\frac{1 \text{ hp}}{1000 \text{ ft}^3}\right)(8914) = 8.9 \text{ hp for each digester.}$$

Note :

1. Reduction of digester volume possibilities.
   - Operating the process at an elevated temperature.
   - Increasing the size of the aeration basin (extended aeration basin operating under endogenous conditions).
   - Using a small sludge storage, providing minimum aeration to maintain aerobic conditions, serving to store and (batch) thicken the sludge.
   - Feeding the sludge directly to dewatering device, providing slightly increased dewatering capacity.
2. Oxygen requirements

   Carbon oxidation :

$$C_5H_7NO_2 \text{ (Sludge)} + 5O_2 = 5CO_2 + 2H_2O + NH_3 + \text{Energy}$$

   Carbon oxidation + Nitrification :

$$C_5H_7NO_2 \text{ (Sludge)} + 7O_2 = 5CO_2 + 3H_2O + H^+ + NO_3^- + \text{Energy}$$

Therefore, 1.42 kg O_2 required for oxidation of 1 kg of sludge

1.98 kg O_2 required for Carbon oxidation + Nitrification of 1 kg of sludge

3. Reaction kinetics

- Mass balance

$$Q_o X_o = Q_w X_w + Q_d X_d + k_d VX$$

where Q_o is the feed sludge (gal/d), X_o is the biodegradable feed concentration VSS (mg/L), Q_w is the waste sludge flow (gal/d), X_w is the biodegradable waste sludge concentration VSS (mg/L), Q_d is the decant flow (gal/d), X_d is the decant biodegradable concentration VSS (mg/L), V is the reactor volume (gal), and k_d is the decay constant (d^{-1})

$$V = \frac{(Q_o)(X_o)+(SRT)}{X_w(1+k_d\,SRT)}$$

$$\frac{X_w}{X_o} = \frac{1}{1+k_d SRT}\left(\frac{SRT}{HRT}\right)$$

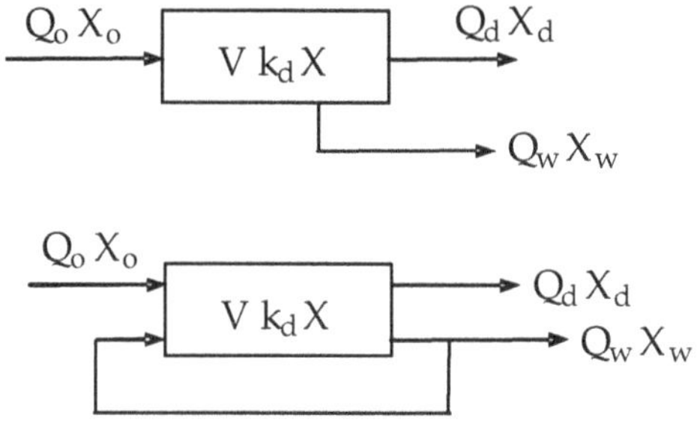

[Intermittent]

[Continuous]

Figure 30 : System configuration.

- Flow through system (none cycle)

$$\frac{X_w}{X_o} = \frac{1}{1+k_d(SRT)} \quad \text{or} \quad SRT = \frac{X_o - X_w}{k_d X_w}$$

$$X_o = S_o - X_n$$

$$X_w = S_w - X_n$$

where S_o is the total feed VSS Conc. (mg/L), X_n is the feed non-biodegradable VSS component (mg/L), S_w is the total waste VSS Conc. (mg/L),

$$SRT = \frac{S_o - S_w}{k_d(S_w - X_n)}$$ [X_n is the inactive portion of VSS]

$$S_w = \frac{S_o + (SRT)(k_d)(X_n)}{1 + k_d(SRT)}$$

Table 83 : Typical continuous aerobic digester design criteria for municipal sludge

SRT(d)	*VSS Removal*	
	40%	*55%*
[A] At 4.4°C	108	386
At 15.6°C	31	109
At 26.7°C	18	64
[B] Oxygen (kg O_2/kg VSS destroyed) 45°C or greater	::	2 at 45°C or less1.45 at
[C] Maximum sludge concentration	:	2.5 to 3.0%
[D] Mixing, mechanical diffuser	::	13-105 W/m^3 (0.5-4 hp/ 1000 ft^3)20-40 m^3/1000 m^3.min
[E] Dewatering characteristics	:	Normally poor

5. Thermophilic (auto thermal) and Cryophilic aerobic digestion
   - Thermophilic
     - Above 42°C
     - Utilising heat energy released from the reaction to sustain process conditions.
     - Applicable for sludge concentration : 2.5 to 5%
     - Destruction of VSS upto 70%
   - Cryophilic
     - Low temperature (less than 10°C)
     - Elevated SRT to compensate for low temperature [SRT $\times$ Temp. (°C) = 250 to 300 day°C] to achieve acceptable solids reduction performance.
6. Aerobic digester solids balance

$$SRT = \frac{(S_o - S_e)}{(k_d)(f)(D)(S_o)}$$

$$SRT = \frac{XV}{Q_w X_u + Q_e X_e} = \frac{\text{Biomass in reactor}}{\text{Biomass Wasted}}$$

where X_u is the thickened sludge concentration (mg/L), X is the digester suspended solids (mg/L), Q_w is the digester waste rate (gal/d), X_e is the effluent SS concentration (mg/L), Q_e is the digester decant rate (gal/d), and V is the digester working volume (gal).

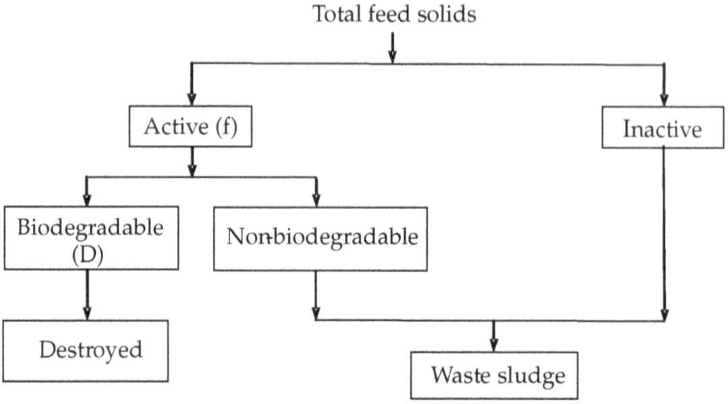

Figure 31 : Solids Balance.

7. **Electron Acceptors :** In aerobic metabolism, molecular oxygen acts as the terminal electron acceptor, *e.g.*,

$$C_7H_8 + 9O_2 = 7CO_2 + H_2O$$

3 lb of available O_2 are required to convert one pound of hydrocarbon to CO_2 and H_2O.

In anaerobic bio-remediation, alternate or substitute electron acceptors are used in place of O_2. These include in order of preference NO_3^{-2}, manganese (IV), and iron (III) oxides (MnO_2 and $Fe(OH)_3$, respectively), SO_3^{-2} , CO_2 (methanogenesis). In general, the use of a particular electron acceptor is a function of its availability, the presence of other electron acceptors, and related oxidation reduction potential of the surrounding environment. The sequence of microbially mediated redox potential processes as a function of the redox potential of the groundwater is depicted in Figure 32.

The energy yield is generally in the following order :

$$O_2 > NO_3^- > Mn(IV) > Fe^{+3} > SO_4^{-2} > CO_2$$

Nitrate reduction :

$$C_7H_8 \text{ (toulene)} + 6NO_3^- = 7CO_2 + 4H_2O + 3N_2$$

Advantages of NO_3^- electron acceptor are: (1) more soluble than O_2, (2) less expensive, and (3) it may be more economical to use nitrate than oxygen. Additionally, aerobic remediation of deeper zones of the aquifer may not be feasible. On negative side, NO_3^- is listed pollutant.

Manganese reduction :

$$C_7H_8 + 9Mn^{+2} + 7CO_2 + 4H_2O$$

Insoluble MnO_2 is reduced to dissolved Mn (II) by hydrocarbon degrading bacteria.

Iron reduction :

$$C_7H_8 + 36Fe(OH)_3 = 7CO_2 + 36Fe^{+2} + 72(OH)^{-1} + 22H_2O$$

Sulphate reduction :

$$4C_7H_8 + 18SO_4^{-2} + 12H_2O = 9(H_2S) + 9(HS)^{-1} + 28HCO_3^- + H^+$$

Methanogenesis :

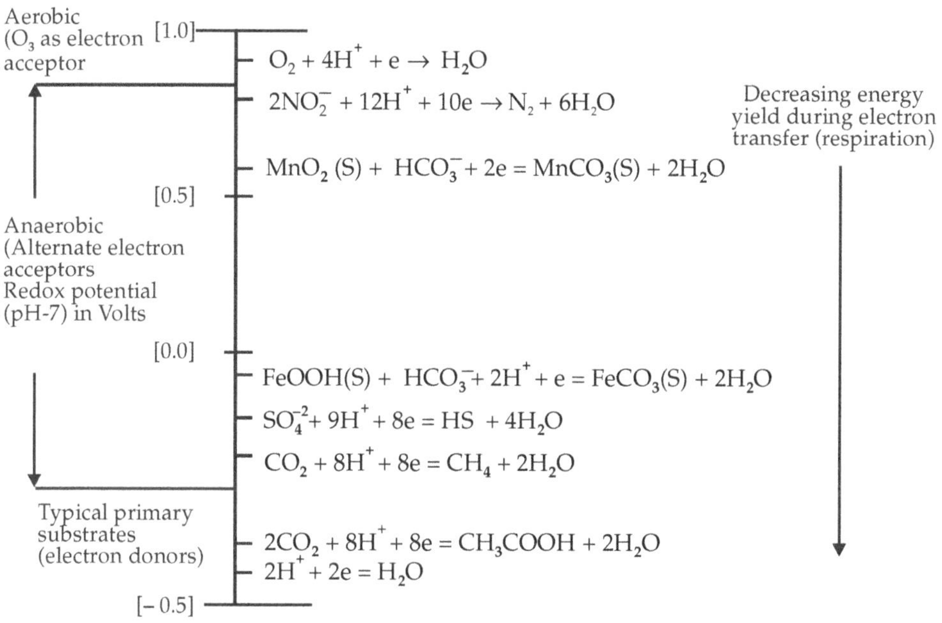

Figure 32 : Sequence of microbial-mediated redox potential as a function of redox potential.

Example 3.103

Oxygen Estimation for Aerobic Conversion of Sludges

Estimate the quantity of oxygen requirement to oxidise 5000 lb of an organic waste sludge. It is assumed that the initial composition of the sludge material to be decomposed is given by the empirical formula: $[C_6H_{10}O_5]_5$. The final composition of the aerobically oxidised sludge is given by: $[C_6H_{10}O_5]_2$; and the remaining sludge amounts to 2000 lb.

Solution

1. Required moles of initial and final sludges

- Initial moles $= \dfrac{5000 \text{ lb}}{[30(12)+50(1)+25(16)]} = 6.17$

- Final remaining moles $= \dfrac{2000}{[12(12)+20(1)+10(16)]} = 6.17$

- Moles of decomposed sludge leaving per mole of entering sludge

$$n = \frac{6.17}{6.17} = 1$$

2. Required oxygen
   - Biological reactions are :

     For partial oxidation of sludge

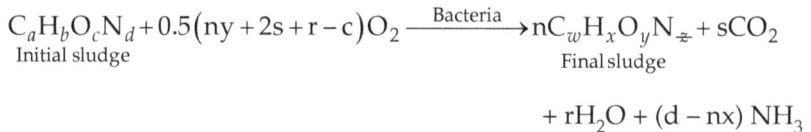

$$\underset{\text{Initial sludge}}{C_aH_bO_cN_d} + 0.5(ny+2s+r-c)O_2 \xrightarrow{\text{Bacteria}} \underset{\text{Final sludge}}{nC_wH_xO_yN_z} + sCO_2$$

$$+ rH_2O + (d - nx)\, NH_3$$

where : $r = 0.5[b - nx - 3(d - nx)]$

$s = a - nw$

For complete oxidation of sludge

$$C_aH_bO_cN_d + \left(\frac{4a+b-2c-3d}{4}\right)O_2 \rightarrow aCO_2 + \left(\frac{b-3d}{2}\right)H_2O + dNH_3$$

For nitrification of NH_3

$$NH_3 + 1.5O_2 \rightarrow HNO_2 + H_2O$$

$$HNO_2 + 0.5O_2 \rightarrow HNO_3$$

$$NH_3 + 2O_2 \rightarrow H_2O + HNO_3$$

- Oxygen requirements for decomposition of sludge $C_{30}H_{50}O_{25}$ to $C_{12}H_{20}O_{10}$

  Initial : $a = 30;\ b = 50;\ c = 25;\ d = 0$

  Final : $w = 12;\ x = 20;\ y = 10;\ z = 0$

  $r = 0.5[50 - 1(20) - 3(0 - 1\times 0)]$

  $= 15$

  $s = a - nw = 30 - 1(12) = 18$

- Oxygen requirement is

$$\text{lb } O_2 = 0.5[ny + 2s + r - c]O_2$$
$$= 0.5[1(10) + 2(18) + 15 - 25]\ (6.15)(32)$$
$$= 3540 \text{ lb}$$

- Process material balance

Input	
Organic material	= 5000 lb
Oxygen	= <u>3540 lb</u>
Total	= 8540 lb
Output	
Organic material	= 2000 lb
CO_2 = 6.15(18)(44)	= 4870 lb
H_2O = 6.15(15)(18)	= <u>1660 lb</u>
Total	= 8530 lb
Input	= Out [OK]

Example 3.104

Total Oxidation of Biological Sludge

Total oxidation is a process so designed that the biological sludge produced by synthesis is consumed by auto oxidation. The sludge balance in aeration basin can be expressed by the following expression.

1. Increase in biological solids (lb/d) DS = a lb BOD_r /d-b lb MLVSS

   Where a is the fraction of BOD synthesized to sludge. This usually varies from 50-75% of the BOD_5 removed, and b is the mean rate of auto oxidation (fraction /day). For soluble substrate, b will average between 15-25% per day at 25°C. Wastes containing SS of low oxidation rate such as domestic sewage may have endogenous rates of 6-12% per day. The above expression does not consider inert fractions of the sludge which is not oxidized.

$$\Delta S = 0, \text{ therefore, a } L_r = \text{b MLVSS}$$

   Consider, a total oxidation plant to treat a doing waste water containing 170 lb COD over a period of 20 hours. The mean auto-oxidation rate of sludge is 20% per day. The COD reduction is 85%. Determine the mean MLVSS if the waste sludge volume is 30,000 gal/d

2. COD reduction

$$\text{lb COD removed} = 0.85 \times 170 = 144.5 \text{ lb COD over 20 hours}$$

3. Required MLVSS

$$\text{MLVSS} = \frac{a\,L_r}{b} = \frac{0.45 \times 144.5}{\left(0.20 \times \frac{20}{24}\right)}$$

$$= 390 \text{ lb}$$

4. Tank volume

(*a*)	Required sludge volume	= 390 × 0.8 ft^3/lb [Assume, sludge occupies 0.8 ft^3/lb]
		= 312 ft^3
(*b*)	Waste volume $\left(\frac{30{,}000}{7.481}\text{gal/day}\right)$	= 4010 ft^3
(*c*)	Sludge free board (5% of a + b)	= 216 ft^3
	Total	= **4538 ft^3**

Note :

1. The oxygen requirements are greater than conventional system, if sludge produced by synthesis is consumed by auto-oxidation.
2. For complete sludge oxidation, the total oxygen requirements will be approximately equal the ultimate BOD.

Example 3.105

Aerobic Sludge Digestion

Determine the aerobic digester volume, and the oxygen requirements. Assume the following data :

Sludge flow rate	: 1 MGD
Influent total sludge concentration (X_o)	: 10,000 mg/L
Non-degradable sludge concentration (X_n)	: 2850 mg/L
Winter temperature of sludge	: 16°C
Summer temperature of sludge	: 30°C
Degradation rate constant for degradable VSS (k_b)	: 0.3241 d^{-1} at 20°C
Efficiency of destruction of degradable VSS	: 80%
Average oxygen utilization rate for batch reactor	: 500 mg/L.d at 20°C

Solution

1. Required design expressions

- Batch process [First order kinetics] :

$$\frac{dX_d}{dt} = -k_b X_d$$

Integrating, $(X_d)_e = (X_d)_o \exp[-k_b t]$

where $(X_d)_e$ = Degradable VSS remaining after aeration time, t

$(X_d)_o$ = Initial degradable VSS concentration

k_b = Reaction rate constant for destruction of degradable VSS (d^{-1})

$(X_d)_e = X_e - X_n$

$(X_d)_o = X_o - X_n$

where X_o and X_e = Total VSS concentration (degradable + non-degradable) in influent and effluent, respectively

Therefore, $\text{In}\left[\frac{(X_e - X_n)}{(X_o - X_n)}\right] = -k_b t$

- Continuous reactor under steady-state and well mix condition :

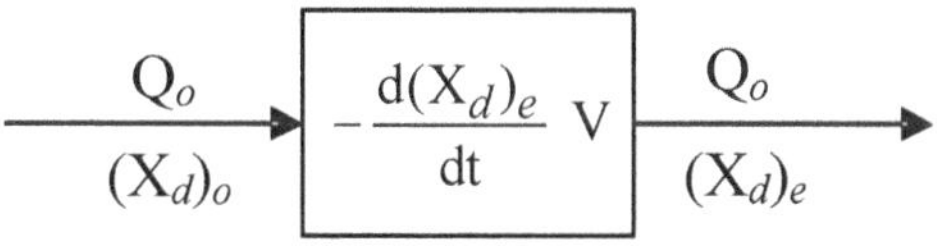

Figure 33 : Continuous reactor system

Mass balance for the degradable solids is :

$$\text{Degradable solids in} = \text{Degradable solids out} + \text{Degradable solids destroyed}$$

$$Q_o (X_d)_o = Q_o (X_d)_e - \frac{d(X_d)_e}{dt} V$$

or $$Q_o (X_o - X_n) = Q_o (X_e - X_n) - \frac{d(X_e - X_n)}{dt} .V$$

$$t = t_n = \frac{V}{Q_o} \text{ [No recycle]}$$

$$(X_o - X_e) = \left[-\frac{d(X_e - X_n)}{dt}\right] t$$

$$\frac{d(X_e - X_n)}{dt} = -k_b(X_e - X_n) \text{ [First order kinetics]}$$

$$\frac{(X_o - X_e)}{t} = k_b(X_e - X_n)$$

or $$t = \frac{(X_o - X_e)}{k_b(X_e - X_n)}$$

2. Required digester volume based on winter conditions (16°C) and percent reduction of VSS during summer conditions (30°C)

$$k_b(T) = k_b(20)(1.05)^{T-20}$$

$$k_b(16) = 0.3241(1.05)^{16-20} = 0.267 \text{ d}^{-1}$$

$$k_b(30) = 0.3241(1.05)^{30-20} = 0.528 \text{ d}^{-1}$$

Degradable influent VSS conc. $(X_d)_o$ = (10,000 – 2850) = 7150 mg/L

Degradable effluent VSS conc. $(X_d)_e$ = 0.2 (7150) = 1430 mg/L

Total influent sludge solids (X_o) = 10,000 mg/L

Total effluent sludge solids (X_e) = 1430 + 2850 = 4280 mg/L

- Digester volume during winter conditions (16°C)

$$t = \frac{X_o - X_e}{k_b(X_e - X_n)} = \frac{10{,}000 - 4280}{0.267(4280 - 2850)}$$

$$= 15 \text{ days [Continuous system]}$$

$$t = \frac{\text{In}\left[\frac{(X_e - X_n)}{(X_o - X_n)}\right]}{-k_b} = \frac{\text{In}\frac{(4280 - 2850)}{(10{,}000 - 2850)}}{-0.267}$$

$$= 6.03 \text{ days (Winter condition)}$$

$$= \frac{15 \times 10^6 \text{ gal}}{7.48 \text{ gal/ft}^3}$$

$$= 2.005 \times 10^6 \text{ ft}^3$$

- Effluent VSS during summer conditions (30°C)

$$X_e = \frac{X_o + k_b X_n t}{1 + k_b t}$$

$$= \frac{10{,}000 + 0.528(2850)(15)}{1 + 0.528(15)} = 3652 \text{ mg/L}$$

Degradable effluent VSS conc. = (3652 – 2850) = 802 mg/L

Degradable influent VSS conc. = 10,000 – 2850 = 7150 mg/L

% degradable VSS reduction $= \frac{7150 - 802}{7150}(100)$

$= 88.8\%$

3. Required oxygen consumption during summer period (30°C)
   - Batch reactor
     - Oxygen utilization $= \int_0^t (\text{OUR})\,dt$

       Plot oxygen uptake rate (mg/L.h) versus time, the area under the curve gives the oxygen utilization (mg/L)
     - Average oxygen utilization rate

$$= \frac{1}{(t-0)}\int_0^t (\text{OUR})\,dt \;\; [\text{mg/L.d}][t = 15 \text{ days}]$$

   - Continuous reactor

     The oxygen utilization may be assumed to be proportional to reduction of degradable VSS achieved in a given operation. The average oxygen utilization for batch system $\left[\frac{1}{t}\int_0^t (\text{OUR})\,dt\right]$ may be corrected for the conditions under continuous digester using:

     Oxygen utilization rate for continuous reactor

$$= \left[\int_0^t (\text{OUR})\,dt \Big/ t\right]_{\text{batch}} \frac{\left[(X_d)_o - (X_d)_e\right]_{\text{summer}}}{\left[(X_d)_o - (X_d)_e\right]_{\text{batch reactor}}}$$

$X_e = (X_o - X_n) \exp[-k_b\, t] + X_n$

Effluent sludge solids concentration at 20°C and after 15 days of digestion :

$= (10{,}000 - 2850) \exp[-0.3241\,(15)] + 2850$

$= 2905$ mg/L

$(X_d)_e = (X_e - X_n) = (2905 - 2850) = 55$ mg/L

Therefore,

oxygen utilization rate for continuous reactor during summer

$$= [500 \text{ mg/L.d}]_{\text{at } 20^\circ\text{C}} \frac{[7150 - 802]}{[7150 - 55]_{\text{at } 20^\circ\text{C}}}$$

$$= (500)\frac{(6348)}{(7095)} = 447.36 \text{ mg/L.d}$$

Total oxygen required (lb O_2/d) = (447.36 mg/L.d)(8.34)(15 MGD)

= 55964 lb O_2/d

4. Required sludge age (θ_c)

$$\theta_c = \frac{\text{lb MLVSS in the digester}}{\text{Net input of VSS to the system (lb/d)}}$$

- Winter conditions :

lb of MLVSS in the digester = (15 MG)(conc. of MLSS in the digester)

= (15 MG)(2850 + 1430)(8.34)

= 535428 lb

Net input of VSS to system is :

$Q_o(X_o - X_e)(8.34)$ = 1 MGD(10,000 – 4280)(8.34)

= 47705 lb/d

$$\theta_c = \frac{535428 \text{ lb}}{47705 \text{ lb/d}} = 11.2 \text{ days} \left[\text{Winter condition}\right]$$

- Summer conditions

lb of MLVSS in the digester = 15 MG(3652 mg/L)(8.34)

= 456865 lb of MLVSS

Net input of VSS to the system is :

$Q_o(X_o - X_e)(8.34)$ = 1 MGD(10,000 – 3652)(8.34)

= 52942 lb/d

$$\theta_c = \frac{456865 \text{ lb}}{52942 \text{ lb/d}} = 8.63 \text{ days}$$

Note :

1. Determination of non-biodegradable VSS (X_n).

   Plot the total VSS remaining versus time, the portion of the curve which is asymptotic to a VSS value defines the X_n.

2. Determination of reaction rate constant for destruction of degradable VSS (k_b).

   Plot the degradable VSS ($X_o - X_n$) remaining versus time, on the same log paper, the slope defines, k_b. Convert it to base-e [i.e., 2.303 k_b (base 10)].

Example 3.1067

Aerobic Digestion of Wastewater Sludge

Determine the temperature attained in an aerobic digester with solids feed content of 2.5% at 25°C. The daily sludge volume is 25,000 gal. VSS reduction required is 50% and endogenous rate constant (k_d) is 0.25 d^{-1}. The degradable fraction of the sludge is 0.6.

Solution

1. Required detention period

$$t = \frac{(X_o - X_e)}{k_d (X_e - X_n)}$$

$$X_o = 25{,}000 \text{ mg/L}; \; X_e = (25{,}000)(0.5)$$

$$= 12{,}500 \text{ mg/L}$$

$$X_n = (25{,}000)(0.4) = 10{,}000 \text{ mg/L}$$

$$t = \frac{30{,}000 - 12{,}500}{(0.25)(12{,}500 - 10{,}000)}$$

$$= \frac{(17{,}500)}{(0.25)(2500)} = 28 \text{ days}$$

2. Required digester volume (V)

$$V = Q \times t$$

$$= (25{,}000)\ (28) = 700{,}000 \text{ gal}$$

$$= 0.7 \text{ mil-gal}$$

3. Reduction in VSS (lb/d)

$$= (25{,}000 \text{ mg/L})(0.5)(0.025)(8.34)$$

$$= 2606 \text{ lb VSS/d reduced}$$

4. Required heat generated

$$= (9300 \text{ BTU/lb})(2606 \text{ lb VSS/d})$$

$$= 24.24 \times 10^6 \text{ BTU/d}$$

5. Required water per day

$$(25{,}000 \text{ gal/d})(8.34 \text{ lb/gal}) = 0.21 \times 10^6 \text{ lb/d}$$

6. Required rise in temperature

Heat in = Heat out

$$24.24 \times 10^6 \text{ BTU/d} = (0.26 \times 10^6)(1 \text{ BTU/lb }^{o}\text{F})(\Delta T)$$

$$\Delta T = \frac{24.24 \times 10^6}{0.26 \times 10^6} = 93^\circ F$$

[Necessary attention must be given to heat losses].

7. Required oxygen

$$\text{lb } O_2 = (1.4) \text{ lb VSS reduced}$$
$$= (1.4)(2606)$$
$$= 3648 \text{ lb } O_2/d$$

Therefore, oxygen uptake rate $= \dfrac{\text{Oxygen required}}{\text{Digester volume}}$

$$= \frac{3648 \text{ lb } O_2/d}{(0.7 \text{ MG})(8.34)(24)}$$
$$= 26 \text{ mg/L.hr}$$

Example 3.107

Aerobic Digestion of Wastewater Sludge

1. Kinetics of aerobic digestion under batch or plug flow conditions

$$(X_d)_e = (X_d)_i \exp(-k_d t)$$

or
$$\left(\frac{X_e - X_n}{X_o - X_n}\right) = \exp(-k_d t)$$

2. Kinetics of aerobic digestion under completely mixed conditions

$$\left(\frac{X_e - X_n}{X_o - X_n}\right) = \frac{1}{1 + k_d t}$$

where $(X_d)_e$ is the degradable VSS of time-t, $(X_d)_o$ is the initial degradable VSS, k_d is reaction rate constant, t is the detention time, X_o is the initial VSS, X_e is the effluent VSS, and X_n is the non-degradable VSS.

The equation for n-mixed reactor in series is :

$$\left(\frac{X_e - X_n}{X_o - X_n}\right) = \frac{1}{(1 + k_d t)^n}$$

The temperature dependence for bed is :

$$k_d(T) = k_d(20^\circ C)(1.04)^{T-20}$$

3. Determine the hydraulic detention period for an aerobic digestion system treating waste activated sludge with a SRT of 15 and 25 days. Use the following data :

Wastewater flow (Q) = 1 MGD

lb BOD removed ($S_o - S_e$) = 500 mg/L

a = 0.55

b = 0.1 d^{-1}

3.1 Sludge production (DX_V) from soluble substrate

The sludge production is expressed as :

$$\Delta X_V = a(S_o - S_e) - b\, X_d\, X_V\, t$$

In a system with sludge recycle and wastage of excess sludge, the sludge age (θ_c) is given by

$$\theta_c = \frac{X_V\, t}{\Delta X_V} = \frac{X_V\, t}{a(S_o - S_e) - bX_d X_V\, t}$$

or
$$\Delta X_V = \frac{X_V\, t}{\theta_c}$$

$$a(S_o - S_e) - b\, X_d\, X_V\, t = \frac{X_V\, t}{\theta_c}$$

$$a(S_o - S_e) = X_V\, t\left[\frac{1}{\theta_c} + b\, X_d\right]$$

$$= \frac{X_V\, t}{\theta_c}\left[1 + bX_d\theta_c\right]$$

$$= \Delta X_V\left[1 + b\, X_d\theta_c\right]$$

Therefore,
$$\Delta X_V = \frac{a(S_o - S_e)}{\left[1 + bX_d\theta_c\right]}$$

where X_d is the degradable fraction of the biological VSS, b is the endogenous rate constant (d^{-1}), θ_c is the sludge age (days), X_V is the VSS concentration (mg/L), t is the hydraulic detention period (V/Q, days), ΔX_V is the VSS wasted per day based on the influent flow (mg/L).

3.1.1 Sludge production at a SRT of 15 days

Assume $X_d = 0.6$

$$\Delta X_V = \frac{(0.55)(500)}{1 + (0.1)(0.6)(15)}$$

$$= 145 \text{ mg/L}$$

$$Q(\Delta X_V) = (1)(145)(8.34) = 1209 \text{ lb/d}$$

3.1.2 Sludge production at a SRT of 25 days

$$\Delta X_V = \frac{(0.55)(500)}{1+(0.1)(0.5)(25)}$$

$$= 116 \text{ mg/L}$$

$$Q(\Delta X_V) = (1)(122)(8.34) = 1017.5 \text{ lb/d}$$

Assume a degradable fraction of 0.4

Table 84 : Effect of SRTs on various parameters.

Item	*SRT-15 days*	*SRT-25 days*
X_o (lb/d)	1209	967.44
X_D (lb/d) [Degradable]	725.4 @ 0.6	532.1 @ 0.5
X_N (lb/d) [Non-degradable]	[1209 – 725.4 = 483.6]	[967.44 – 532.1 = 435.35]
X_e (lb/d) [Effluent portion]	[483.6/0.6 = 805]	[725.6]
X_{DN} (lb/d)	[805 – 483.6 = 321]	[725.6 – 435.35 = 290]

4. Required detention time

$$t = \frac{X_o - X_e}{k_d(X_e - X_n)}, \; kd = 0.16 \text{ d}^{-1}$$

$$t - \text{at SRT of 15 day} = \frac{(1209-805)}{0.16(805-483.6)}$$

$$= \frac{404}{0.16(321.4)} = 7.9 \text{ days}$$

$$t - \text{at SRT of 25 days} = \frac{(967-725.6)}{0.16(725.6-435.35)}$$

$$= \frac{(241.4)}{0.16(290.25)} = 5.2 \text{ days}$$

5. Required oxygen for VSS destruction

$$\text{lb } O_2/\text{d} = (X_D - X_{DN})\,(1.4)$$

$$\text{lb } O_2/\text{d @ SRT-15 day} = 1.4(1209 - 321)$$

$$= 1243.2 \text{ lb/d}$$

$$\text{lb } O_2/\text{d @ SRT-25 days} = 1.4(967 - 290)$$

$$= 947.8 \text{ lb/d}$$

6. Required nitrification

$NH_3^- - N$ released is nitrified through endogenous metabolism

Nitrogen content of VSS is dependent on SRT

Assume at SRT of 15 days N = 8% and

at SRT of 25 days N = 7%

In the effluent sludge after digestion N is 80% of the initial amount

Therefore, N-balance $(N_{ox}) = N_o X_o - N_e X_e$

at SRT-15 days, $N_{ox} = (0.08)(1209) - (0.064)(805)$

$= 96.72 - 51.51 = 45.2$ lb/d

at SRT-25 days, $N_{ox} = (0.07)(967) - (0.056)(725.6)$

$= (67.69 - 40.63) = 27.06$ lb/d

6.1 Required oxygen for nitrification

15 d – SRT, lb $O_2 = (45.2)(4.33) = 196$ lb O_2

25 d – SRT, lb $O_2 = (27.06)(4.33) = 117$ lb O_2

7. Total required oxygen

15 d – SRT : lbO_2/d = (1243.2 + 196) lb/d = 1439.2 lb/d

25 d – SRT : lbO_2/d = (947.8 + 117) lb/d = 1064.8 lb/d

8. Required alkalinity (A)

15 d – SRT : A = (45.2)(7.14)* = 323 lb/d

25 d – SRT : A = (27.06)(7.14)* = 193 lb/d

*[7.14 lb alkalinity/lb NO_3-generated]

9. Required sludge volume

(Sludge concentration)(Sludge flow rate) = Total sludge rate

$$\text{Sludge rate} = \frac{\text{Total sludge rate}}{\text{Concentration of sludge}}$$

Assume sludge concentration of 15,000 mg/L

For SRT of 15 days, sludge discharge rate

$$= \frac{1209 \times 454 \times 10^3 \times 0.2642}{15,000 \times 10^6}$$

$$= 9.7 \times 10^{-3} \text{ mil-gal/d}$$

For SRT of 25 days, sludge discharge rate

$$= \frac{967.44 \times 454 \times 10^3 \times 0.2642}{15,000 \times 10^6}$$

$= 7.74 \times 10^{-3}$ mil-gal/d

For SRT of 15 days : Volume required = 9.7 × 10^{-3} mil-gal/d)(7.9 days)

= 0.077 mil-gal

For SRT of 25 days : Volume required = 7.74 × 10^{-3} mil-gal/d)(5.2 days)

= 0.040 mil-gal

10. Required oxygen uptake (OUR) for VSS oxidation

$$\text{For SRT} -15 \text{ days, OUR} = \frac{1243.2 \text{ lb } O_2/d}{(0.077)(8.34)(24)} = 80.7 \text{ mg / L.hr}$$

$$\text{For SRT} -25 \text{ days, OUR} = \frac{947.8 \text{ lb } O_2/d}{(0.040)(8.34)(24)} = 118.4 \text{ mg / L.hr}$$

11. Required denitrification rate (mg $NO_3^- - N$ /L.hr)

It is calculated as :

For SRT – 15 days, (80.7 mg/L.hr)(0.25)* = 20.18 mg-NO_3/L.hr

For SRT – 25 days, (118.4 mg/L.hr)(0.25)* = 29.6 mg NO_3/L.hr

12. Required detention period for denitrification

For SRT – 15 days concentration of NO_3^- in the aeration basin

$$= \frac{45.2 \text{ lb } NO_3 - N/d}{0.077 \times 10^6 \text{ gal/d}}$$

$$= \frac{(45.2)(454 \times 10^3)}{0.077 \times 10^6 \times 3.785}$$

= 70.4 mg/L

For SRT-25 days concentration of NO_3^- in the aeration basin

$$= \frac{27.06 \text{ Ib } NO_3 - N/d}{0.04 \times 10^6 \text{ } 3.785}$$

$$= \frac{27.07 \times 454 \times 10^3}{0.04 \times 10^6 \times 3.785}$$

= 81.1 mg/L

Required detention period is :

$$\text{For SRT} -15 \text{ days : } t = \frac{\text{Concentration of } NO_3 \text{ in the aeration basin}}{\text{Denitrification rate}}$$

$$= \frac{70.4 \text{ mg/L}}{20.18 \text{ mg/L.hr}} = 3.5 \text{ hours}$$

$$\text{For SRT –25 days : } t = \frac{81.1 \text{ mg/L}}{29.6 \text{ mg/L.hr}} = 2.74 \text{ hr}$$

Note :

1. The oxygen requirement for aerobic digestion can be calculated as 1.4 lb O_2 consumed for each pound of VSS destroyed (1.4 kg O_2/kg VSS destroyed).
2. Nitrogen and phosphorus will be released by the oxidation process.
3. With long sludge ages and nitrifies seed, nitrification will occur under mesophilic conditions.
4. There will be build up of ammonia and non-degradable COD of nitrification does not occur, and thermophilic operation inhibits nitrification.
5. Power levels of 15 to 20 standard $ft^3/1000\ ft^3$. min [15 to 20 standard $m^3/1000\ m^3$.min using diffused air or 100 hp/mil-gal (20 W/m^3) using surface aerators are adequate for mixing purposes and oxygen requirements.
6. Pre-thickening the sludge offers a number of advantages: reducing the digester volume in particular, and increasing the temperature due to the exo-thermic heat of reaction.
7. Heat of combustion is about 9000 BTU per lb of VSS destroyed.

Example 3.108

1. Sludge Composting

Determine the fraction of organics which must degrade to produce sufficient NH_3 to neutralise acids produced from the addition of ferric chloride (neglect any neutralisation reactions which may occur during initial ferric chloride addition to the sludge). Assume the following data :

Empirical formula of sludge : $C_5 H_7 O_2 N$

Addition of 2% $FeCl_3$ (dry weight basis) as a sludge conditioner

Sludge ash content : 25% weight basis

Solution

1. Required reactions, and fraction of organic decomposition
   - Biological and chemical reactions :

     $FeCl_3$, 2% = [0.02]

     $FeCl_3 + 3H_2O = 3H^+ + Cl^- + Fe(OH)_3$

[162.5]

[y] [x]

$C_5H_7O_2N + 5O_2 = 5CO_2 + 2H_2O + NH_3$

[113] [17]

[X]

$3NH_3 + 3H^+ + 3Cl^- = 3\,NH_4^+ + 3Cl^-$

[3(17)]

- Amount of NH_3 required

$$\frac{x}{3(17)} = \frac{0.02}{162.5}$$

$x = 0.0063$ g NH_3

- Organic decomposition required for 0.0063 g NH_3 is

$$\frac{y}{113} = \frac{0.0063}{17}$$

$y = 0.042$ g

- Fraction of organic decomposition

$$= \frac{0.042\text{ g}}{0.75} = 0.056\left[\text{or } 5.6\%\right]$$

2. Air Requirements for Sludge Composting

Estimate the quantity of air required to compost 1.5 tons of mixed sludges using an in-vessel composting unit with forced aeration. Assume the following:

Empirical formula of mixed sludges	: $C_{60}H_{90}O_{40}N$
Moisture content of organic fraction of mixed sludges	: 20%
Volatile solids (VS)	: 0.90 TS (Total solids)
Biodegradable VS (BVS)	: 0.65 VS
Conversion efficiency of BVS	: 90%
Composting time	: 5 days

Solution

1. Required biodegradable volatile solids
   - Total mass of biodegradable volatile solids is

     = (1.5 tonne) (2000 lb/tonne)(0.80 day matter)(0.90)(0.65)

     = 1296 lb of BVS

- Conversion of BVS is

  = (1296 lb BVS)(0.90) = 1166 lb

2. Required air for aerobic conversion of biodegradable volatile solid (BVS)
   - Biological (aerobic) reactor for the complete conversion of BVS $[C_aH_bO_cN_d]$ is

     $$C_aH_bO_cN_d + \left(\frac{4a+b-2c-3d}{4}\right)O_2 \rightarrow aCO_2 + dNH_3 + \left(\frac{b-3d}{2}\right)H_2O$$

     a = 60; b = 90; c = 40; d = 1
   - Amount of oxygen required is

     $$= \frac{\left[\frac{4(60)+90-40-3}{4}\right]32 \text{ lb } O_2}{\left[60(12)+90(1)+40(16)+1(14)\right] \text{lb BVS}}$$

     $$= \frac{2296 \text{ lb } O_2}{1464 \text{ lb BVS}} = 1.57 \text{ lb } O_2/\text{lb BVS}$$

   - Amount of air required is

     $$= \frac{1166 \text{ lb BVS} \times 1.57 \text{ lb } O_2/\text{lb BVS}}{(0.23 \text{ lb } O_2/\text{lb air})(0.075 \text{ lb air}/\text{ft}^3 \text{ air})}$$

     = 106123 ft^3 air

     [Air contains 23% oxygen by mass, and specific weight of air is equal to 0.075 lb air/ft^3 air]
   - Consider oxygen demand as

     25, 40, 30, 20, 10% for the successive days of the 5-day composting time.

     Using the airflow rate of 40% of the total oxygen requirement (the most critical), and a safety factor of 2, the airflow rate is

     $$\text{Air} = \frac{(106123 \text{ ft}^3 \text{ air})(2-\text{SF})(0.40/\text{d})}{(1440 \text{ min/d})}$$

     = 59 ft^3 air/min.

Example 3.109

Sludge Composting

Determine the fraction of organics, which must degrade to produce sufficient CO_2 to neutralise the lime (neglect any neutralisation reactions, which may occur during initial lime addition to the sludge). Assume the following data :

Empirical formula of sludge : $C_5H_7O_2N$

Sludge ash content : 25% weight basis

Conditioning by addition of 10% lime as $Ca(OH)_2$ on dry weight.

Solution

1. Required CO_2 production by oxidizing sludge, and lime neutralisation
   - Biological and chemical reactions :

$$\begin{array}{l} \qquad\qquad\qquad\quad [x] \\ C_5H_7N_2O + 5O_2 = 5CO_2 + 2H_2O + NH_3 \\ [113] \qquad\qquad\qquad 5[44] \end{array}$$

$$\begin{array}{l} [x] \qquad [0.10] \\ CO_2 + Ca(OH)_2 = CaCO_3 + H_2O \\ [44] \qquad [74] \end{array}$$

   - Considering 0.10 g lime/g-dry solids (10%-weight basis) the CO_2 requirement is given by :

$$\frac{X}{44} = \frac{0.10}{74}$$

$$X = \left(\frac{0.10}{74}\right)(44) = 0.0595 \text{ g } CO_2$$

   - Amount of CO_2 to be produced by biological oxidation is:

$$\frac{y}{113} = \frac{0.0595}{5(44)}$$

$$y = 0.030 \text{ g}$$

Fractional organic decomposition required is estimated as :

$$y = \frac{0.030}{0.75} = 0.04 \text{ or } 4\%$$

[Therefore, only a very small percentage of organic decomposition is required to neutralise the high pH of lime conditioned sludge].

Example 3.110

Sludge Composting

1. 15 dry tonne per day of digested dewatered sludge is to be windrow-composted using recycled compost to dry the initial composting mixture. Designed mixture solids is 40% and recycled compost is 70% solids. Determine the required recycle ratio, both wet and dry basis, if the dewatered cake is 30% solids; as also the total weight of material to be processed daily.

   1.1 Required mathematical expressions

   Generalised mass balance diagram for sludge composting showing inputs of dewatered sludge cake, compost product recycle, organic amendment and bulking agent :

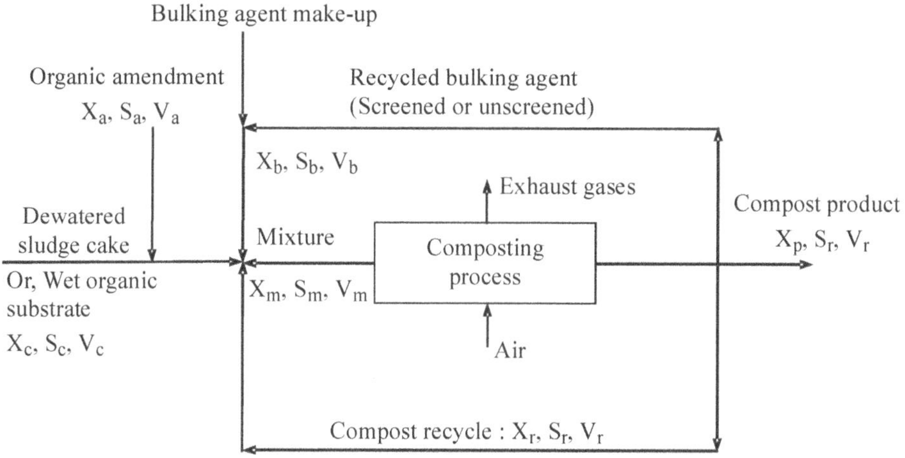

Figure 34 : Generalised Composting process.

where X_c = Total wet weight of dewatered sludge cake produced per day

X_p = Total wet weight of compost product produced per day

X_r = Total wet weight of compost product recycled per day

X_a = Total wet weight of organic amendment, other than sludge cake/compost recycle, added to mixture per day

X_b = Total wet weight of bulking agent added to the mixture per day

X_m = Total wet weight of mixed material entering the compost process per day

S_c = Fractional solids content of dewatered sludge cake

S_r = Fractional solids content of compost product and compost recycle

S_a = Fractional solids content of amendment

S_b = Fractional solids content of bulking agents

S_m = Fractional solids content of mixture before composting

V_c = Volatile solids content of sludge cake (fraction of dry solids)

V_r = Volatile solids content of compost product and recycle (fraction of dry solids)

V_a = Volatile solids content of amendment (fraction of dry solids)

V_b = Volatile solids content of bulking agent (fraction of dry solids)

V_m = Volatile solids content of mixture (fraction of dry solids)

- Mass balance on total wet solids [without amendment or bulking agent] is :

$$X_c + X_r = X_m \tag{1}$$

- Mass balance on dry solids gives :

$$S_c X_c + S_r X_r = S_m X_m \tag{2}$$

$$S_c X_c + S_r X_r = S_m(X_c + X_r) \tag{3}$$

- Recycle ratio (R_w) based on total wet weight of compost product recycled to total wet weight of dewatered sludge cake

$$R_w = \frac{X_r}{X_c} \tag{4}$$

or
$$R_w = \frac{S_m - S_e}{S_r - S_m} \tag{5}$$

- Recycle ratio (R_d) based on dry weight of compost product recycled to dry weight of the dewatered sludge cake is :

$$R_d = \frac{S_r X_r}{S_c X_c} \tag{6}$$

or
$$R_d = \frac{[S_m/S_c - 1]}{[1 - S_m/S_r]} \tag{7}$$

- Time required for evaporative drying considering a layer of wet cake of depth d_c and unit surface area. The depth of water (d_w) which must be evaporated from the unit surface is expressed as:

$$d_w = \frac{d_c\, \rho\, \gamma_c}{\gamma_w} \tag{8}$$

where d_w = Depth of water to be evaporated (cm)

d_c = Depth of sludge cake (cm)

γ_c = Bulk weight of sludge cake (g/cm^3)

γ_w = Bulk weight of water (g/cm^3)

p = Weight of water which must be evaporated to reach the desired solids per weight of original sludge cake (g-water/g-cake)

Time (t_a) required to achieve the desired cake solids is given as:

$$t_a = \frac{p d_c \gamma_c}{(E-P)\gamma_w}; (E > P) \tag{9}$$

where E = Evaporation rate (cm/d)

P = Precipitation rate (cm/d)

t_a = Time required for air drying (days)

- Area required for drying is :

$$\text{Area} = \frac{\text{Volume/d}}{\text{depth}}(t_a)$$

$$= \frac{X_c}{\gamma_c\, d_c}\frac{p d_c \gamma_c}{(E-P)\gamma_w} = \frac{\gamma_c\, p}{(E-P)\gamma_w}; (E > P) \tag{10}$$

- Volatile solids mass balance

$$V_c\, S_c\, X_c + V_r\, S_r\, X_r = V_m\, S_m\, X_m$$

Substituting equations (1) and (5), rearranging and solving for V_m:

$$V_m = \frac{V_c S_c + V_r\, S_r\, R_w}{S_m(1+R_m)};\ (\text{Wet basis}, R_w) \tag{11}$$

$$V_m = \frac{V_c + V_r\, R_d}{(1+R_d)};\ (\text{Dry basis}, R_d) \tag{12}$$

- Total mass balance (including amendment) on solids :

$$X_a = \frac{X_c\left[S_r(S_c/S_m - 1)(V_r - V_m) + S_c(S_r/S_m - 1)(V_m - V_c)\right]}{S_r(1 - S_a/S_m)(V_r - V_m) - S_a(S_r/S_m - 1)(V_m - V_a)} \tag{13}$$

$$X_r = \frac{S_c X_v (V_m - V_c) + S_a X_a (V_m - V_a)}{S_r(V_r - V_m)} \tag{14}$$

$$X_m = X_a + X_r + X_c \tag{15}$$

1.2 Required recycle ratio on dry (R_d) and wet basis (R_w)

- Dry basis :

$$R_d = \frac{(S_w/S_c - 1)}{(1 - S_m/S_r)} = \frac{(0.40/0.30 - 1)}{(1 - 0.40/0.70)} = 0.78$$

- Wet basis

$$R_w = \frac{S_m - S_c}{S_r - S_m} = \frac{(0.40 - 0.30)}{(0.70 - 0.40)} = 0.33$$

1.3 Required total weight on dry and wet basis

- Total weight = Sludge weight + Recycle weight
- Dry basis (R_d)

$$\text{Total weight} = \frac{15}{0.30} + \frac{15(0.78)}{0.70}$$

$$= 50 + 16.7 = 66.7 \text{ wet ton/d}$$

- Wet basis (R_w)

$$\text{Total weight} = \frac{15}{0.30} + \frac{15(0.33)}{0.30}$$

$$= 50 + 16.5 = 66.5 \text{ wet ton/d.}$$

2. Determine the mixture solids content if sludge, compost and amendment are blended in a 1:0.5:0.5 ratio by wet weight. Both recycled compost and an organic amendment are added to a sludge cake for moisture control. The organic amendment used is sawdust with a solids content of 70%. Dewatered cake and recycled compost are 25 and 60% solids, respectively.

2.1 Required solids and water in the mixture

For each unit weight of sludge added :

Solids = 0.25 + (0.5)(0.60) + (0.5)(0.70)

= 0.90

Water = 0.75 + (0.5)(0.40) + (0.5)(0.30)

= 1.1

Total weight = (0.9 + 1.1) = 2.0 g/g sludge cake feed

$$\text{Solids content of mixture} = \frac{0.90}{2.0} = 0.45.$$

2.2 Required quantity of amendment to achieve the same mixture solids (without recycle compost)

$$R_w = \frac{S_m - S_c}{S_r - S_m}$$

Therefore, S_a can be substituted for S_r

$$R_w = \frac{X_a}{X_c} = \frac{S_m - S_c}{S_a - S_m} = \frac{0.45 - 0.25}{0.70 - 0.45} = 0.80.$$

3. One hundred dry tonne of digested and dewatered cake is to be air dried for 25 to 40% solids per day by placing the cake in a 0.3 m layer and turning everyday to prevent formation of a hard, dry surface which would

impede evaporation. Assume rate of evaporation of about 0.6 cm/d [with no precipitation]. Determine the time and land requirements for air drying (g_c = 1.065 g/cm^3 and g_w = 1.00 g/cm^3).

3.1 Required drying time and area requirements

$$p = \frac{\text{Initial water} - \text{Final water}}{\text{Total initial weight}}$$

$$= \frac{[1/0.25 - 1] - (1/0.40 - 1)}{(1/0.25)}$$

$$t_a = \frac{p d_c \gamma_c}{(E - P)\gamma_w}; (E > P)$$

$$= \frac{0.375(30)(1.065)}{0.6(1.0)} = 20 \text{ days}$$

$$\text{Area} = \frac{X_c p}{(E - P)\gamma_w}$$

$$\text{Area} = \frac{\left(\frac{100}{0.25}\right)10^6\,(0.375)}{(0.6)(1.00)} = 2.5 \times 10^8 \text{ cm}^2$$

$$= 2.5 \text{ ha.}$$

4. Determine the amounts of recycled compost and amendment to achieve the compost comprising 60% solids with 45% volatile solids. 100 tonne/d of digested dewatered sludge is to be windrow composted. Cake solids and volatility are 25 and 50%, respectively. It is desired to maintain a mixture solids of 45% with a volatile solids content of 55%. Saw-dust is available for use as an organic amendment and is estimated to be 85% solids and 80% volatile.

4.1 Required amount of amendment (X_a)

$$X_a = \frac{X_c\left[S_r(S_c/S_m - 1)(V_r - V_m) + S_c(S_c/S_m - 1)(V_m - V_c)\right]}{S_r(1 - S_a/S_m)(V_r - V_m) - S_a(S_r/S_m - 1)(V_m - V_a)}$$

$$= \frac{\frac{100}{0.25}\left[0.6\left(\frac{0.25}{0.4} - 1\right)(0.4 - 0.55) + 0.25\left(\frac{0.6}{0.45} - 1\right)(0.55 - 0.5)\right]}{0.6\left(1 - \frac{0.85}{0.45}\right)(0.40 - 0.55) - 0.85\left(\frac{0.60}{0.45} - 1\right)(0.55 - 0.80)}$$

$$= 117 \text{ wet tonne/d.}$$

4.2 Required amount of recycle compost (X_r)

$$X_r = \frac{S_c X_c (V_m - V_c) + S_a X_a (V_m - V_a)}{S_r (V_r - V_m)}$$

$$= \frac{100(0.55-0.50)+0.85(117)(0.55-0.80)}{0.60(0.40-0.55)}$$

= 221 wet tonne/d

Total quantity of mixed material to be handled (X_m) is

$$X_m = \frac{100}{0.25} + 117 + 221$$

= 738 wet ton/d.

4.3 Required total mixed material (X_m) without amendment

Required recycle ratio (R_w) to achieve a mixture solids of 45%

$$R_w = \frac{0.45-0.25}{0.60-0.45} = 1.33$$

Using equation (11), the mixture volatility is :

$$V_m = \frac{V_c S_c + V_r\, S_r\, R_w}{S_m(1+R_w)}$$

$$= \frac{(0.50)(0.25)+(0.40)(0.60)(1.33)}{0.45(1+1.33)} @$$

= 0.424

$$@ X_m = \frac{100}{0.25} + \frac{100}{0.25}(1.33) = 932 \text{ wet ton/day.}$$

Example 3.111

Composting Processes

1. Oxygen requirement for aerobic composting

The quantity of oxygen theoretically required by aerobic composting with the following stoichiometric formulation is expressed as

$$C_aH_bO_cN_d + 0.5(ny+2s+r-c)O_2 \rightarrow nC_wH_xO_yN_z + r\,H_2O + (d-nz)NH_3 + sCO_2$$

where $r = 0.5\,[b - nx - 3(d - nz)]$

$S = a - nw$

The terms $C_aH_bO_cN_d$ and $C_wH_xO_yN_z$ represent, on an empirical mole basis, the composition of the organic matter initially and at the end of the process.

An elemental analysis is required for an evaluation of the subscripts in these terms. The oxygen requirement during aerobic compost at different temperatures within the range of 30 to 60°C is expressed as :

$$W_{O_2 = 0.07 \times 10^{0.31T}}$$

where W_{O_2} is the rate of oxygen consumption (mg O_2/gm initial volatile matter per hour, and T is the temperature (°C)).

2. Energy (heat) content (h) and degree of reduction (R)

$$h = 120\,R + 400$$

$$R = \frac{100[2.66 \times \%C + 7.94 \times \%H - \%O]}{398.9}$$

3. Determine the amount of oxygen (lb) and energy released during composting of organic matter (aerobically). Use the following data :

Initial weight of organic matter	:	2000 lb (dry basis)
Final weight of organic matter after	:	400 lb (dry basis)
Completion of composting process		
Empirical formula of the initial organic matter	:	$C_{31}H_{50}O_{26}N$
Empirical formula of the final composted matter	:	$C_{11}H_{14}O_4N$

3.1 Moles of organic matter entering and leaving the process

$$\text{Moles of matter (input)} = \frac{2000}{852} = 2.346$$

Moles of matter (final) per mole of entering matter into the process (n)

$$= \frac{400}{224 \times 2.346} = 0.76$$

3.2 Subscript number in basic composting equation

a = 31, b = 50, c = 26, d =1, w = 11, x = 14, y = 4, z = 1

$r = 0.5\,[50 - 0.76 \times 14 - 3\,(1 - 0.76 \times 1)] = 19.33$

$s = 31 - 0.76\,(11) = 22.64$

3.3 Required oxygen (W)

W = 0.5[ny + 2s + r – c][No. of moles entering the system]× 32

$= 0.5[0.76 \times 4 + 2 \times 22.64 + 19.33 - 26][2.346] \times 32$

=1566 lb

3.4 Check

Input : Organic material = 2000 lb

Oxygen	= 1566 lb
Total	= 3566 lb
Output : Organic material	= 400 lb
CO_2 $(2.346 \times 22.64 \times 44)$	= 2340 lb
H_2O $(2.346 \times 19.33 \times 18)$	= 816 lb
Ammonia $([1 - 0.76 \times 1] \times 2.346 \times 17)$	= 10 lb
Total	= 3566 lb

3.5 Percentages of elements

$$\text{Input}: \%C = \frac{31 \times 12}{852} = 43.6$$

$$\%H = \frac{50 \times 1}{852} 5.9$$

$$\%O = \frac{26 \times 16}{852} 48.9$$

$$\text{Output}: \quad \%C = \frac{11 \times 12}{224} = 59$$

$$\%H = \frac{14 \times 1}{224} = 6.3$$

$$\%O = \frac{4 \times 16}{224} = 28.6$$

3.6 Organic reduction (R)

$$\text{Input}: R = \frac{100\left[2.66(43.6) + 7.94(5.9) - 48.9\right]}{398.9} = 28.6$$

$$\text{Output}: R = \frac{100\left[2.66(59) + 7.94(63) - 28.6\right]}{398.9} = 44.8$$

3.7 Heat content (h)

Input : $h = 127 \times 28.6 + 400 = 4030$ cal/gm

Output : $h = 127 \times 44.8 + 400 = 6100$ cal/gm

3.8 Total energy released as heat

$$\text{Input}: \text{Energy} = 2000 \times 454 \times 4030 = 3860 \times 10^6 \text{ cal}$$

$$\text{Output}: \text{Energy} = 400 \times 454 \times 6100 = \frac{1106 \times 10^6 \text{ cal}}{2754 \times 10^6 \text{ cal}}$$

or (10.12 BTU)

Note :

1. When composting is performed in the open, ground refuse is stacked in long piles (windows-10 ft wide and 5 to 6 ft deep). Piles are turned periodically to ensure aeration and composting. In dry weather, the time required for open air composting is 2 to 3 weeks.
2. Mechanical composting
   - Horizontal drum (6 to 12 ft diameter).
   - Length (100 ft)
   - Rotational speed of drum (0.2 to 1 rpm)
   - Provision for air and water along the length of the drum

     As the drum rotates, the refuse is lifted and cascaded by flights fixed along the interior of the drum (promots mixing and aeration).
   - Temperature of 140 to 160°F are reached in 7 to 12 days.
3. Particle size
   - Biological activity takes place predominantly at the surface of the organic matter.
   - Finer the particle size, greater will be the rate at which material composts.
   - If particle size is too fine, exchange of O_2 and CO_2 between the atmosphere and decomposing surface is retarded as resistance to air flow increases and the voids tend to become filled with moisture.
   - Higher the moisture content, the coarser should be the particle size and, therefore, size control should be exercised through proper grinding.
4. Moisture

   Desiccation will result in cessation of microbial activity. Optimum moisture content falls within 50 to 60%.
5. C/N ratio
   - Optimum C/N = 30 to 35 (Weight basis)

     At greater C/N ratios, biological activity is hampered and composting time is increased.
   - At C/N ratio less then the optimum, NH_3 is released
   - C/N ratio of the completed compost falls within the range of 10 to 20.
6. Seeding : These sources include animal and poultry manures, rich soils, and composts.

Example 3.112

Sludge Composting : Material and Energy Balances

Determine mass and energy balance for the case of composting and drying. Assume the following data :

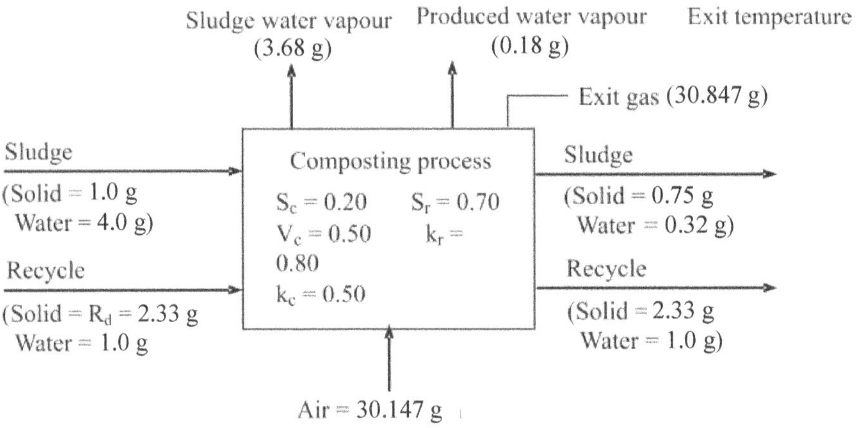

Figure 35 : Composting process.

where S_m = Fractional solids content of mixture before composting

S_c = Fractional solids content of dewatered sludge cake

S_r = Fractional solids content of compost product and compost recycle

V_c = Volatile solids content of sludge cake (fraction of dry solids)

k_c = Coefficient of decomposition for dewatered sludge (fraction, biodegradability coefficient)

k_r = Biodegradability coefficient for compost product $C_{10}H_{19}O_3N$ = Sludge formula

Solution

1. Required mass flux for input sludge and recycle solids
   - Input sludge solids = 1 g (given)
   - Input sludge water $= \left[\frac{1}{0.20} - 1\right] = 4$ g
   - Required recycle ratio (R_d) on dry basis is :

$$R_d = \frac{(S_m / S_c - 1)}{(1 - S_m / S_r)}$$

$$= \frac{(0.40/0.20-1)}{(1-0.40/0.70)} = 2.33$$

- Recycle solids input = 2.33 g

- Recycle water input $= \left(\frac{2.33}{0.70} - 2.33\right) = 1.00\text{ g}$

2. Required solids and water in composted sludge and recycle
   - Sludge solids output $= [1 - (1.0)(V_c)(k_c)]$
     $= 1.0 - (1.0)(0.5)(0.5)$
     $= 0.75$ g
   - Sludge water output $= \left[\frac{0.75}{0.70} - 0.75\right]$
     $= 0.32$ g
3. Required air
   - Aerobic decomposition

     [x = 0.25 g]

     $C_{10}H_{19}O_3N + 12.50O_2 = 10CO_2 + NH_3 + 8H_2O$

     ]201]   12.5(32) 10[44]   {17}   [8×18]

     $$\text{Oxygen required} = \frac{0.25(12.5)(32)}{201} = 0.498\ O_2/\text{g cake solids}$$

     [Theoretical]

     $$\text{Air required} = \frac{0.498}{0.232} = 2.147\text{g} - \text{air}/\text{g cake solid}$$

     [Theoretical]
   - Moisture production (H_2O)

     $$= \frac{0.28(8\times18)}{201} = 0.18\text{g } H_2O \text{ produced}$$
   - Water lost in vapour and produced water is :

     = (4 – 0.32) + 0.18 = 3.68 + 0.18 = 3.86 g
   - Air requirement for moisture removal

     Inlet air saturated at 20°C, specific humidity is

     = 0.0148 g H_2O/g-dry air

Outlet air saturated at 60°C, specific humidity is

= 0.152 g H_2O/g-dry air

Therefore, water removed is :

= (0.152 – 0.0148) = 0.1372 g H_2O/g-dry air

Air required is :

$$=\left(\frac{3.86\text{g H}_2\text{O}}{\text{g}-\text{solid}}\right)\left(\frac{1}{0.1372}\right)=28\text{g air / g cake solid}$$

- Total air required = (2.147 + 28) g air/g-cake solid
  = 30.147 g air/g-cake solid

4. Required quantity of gases other than water vapour

$$\text{Exit Gases}=\left[30.147-\left(\frac{(0.25)(12.5)(32)}{201}\right)+\left(\frac{(0.25)(10)(44)}{201}\right)+\left(\frac{(0.25)(17)}{201}\right)\right]$$

= (30.147 + 0.07) = 30.217

(Exit gases = Inlet air – O_2 consumed + CO_2 produced + NH_3 produced)

5. Required energy balance without heat losses to surroundings
   - Heat to solids (q_s)

     $q_s = mC_p\ \Delta T$

     = (3.33g)(0.25 Cal/g. °C)(60 –20°C) = 30*
   - Heat to water (q_w)

     q_w = (5.0 gm)(1.0 Cal/g.°C)(60 – 20°C) = 200*
   - Evaporative burden (q_v)

     q_v = (540 Cal/gm)(3.86 gm) = 2080*

     [Laten heat of evaporate]
   - Heat to air (q_a)

     q_a = (30.147 g)(0.25 Cal/gm.°C)(60 – 20°C) = 280

     Total = 2590

   – Cal/gm-sludge solids

     [Heat losses to surrounding excluded]

   – Heat losses are :

   – Radiation loss $(q_r)=\sigma A\left(T_1^4-T_2^4\right)F_aF_e$

– Convective (q_c) = UA(T_1 – T_2)

where s = Stefen-Boltzmana constant (4.87×10^8 k Cal/m^2.°K.h)

F_a = Configurational factor to account for relative position and geometry of the bodies

F_e = Emissivity factor to account for non-block-body radiation

T_{a1}, T_{a2} = Absolute temperature of bodies A and B

U = Overall heat transfer coefficient [conductive + convective heat transfer][Cal/cm^2.°C.h]

A = Area perpendicular to direction of heat transfer (cm^2)

T_1 and T_2 = Temperature at points 1 and 2 (°C).

Example 3.113

Aeration Requirements for Composting

1. Determine the stoichiometric oxygen requirements per 1000 kg of sludge feed, as also the heat released during composting. Assume the following data :

Empirical formula of sludge	:	$C_{31}H_{50}O_{26}N$
Input feed	:	1000 kg
After composting	:	200 kg
Empirical formula of compost	:	$C_{11}H_{14}O_4N$

1.1 Required air for composting

- Aerobic composting :

$$C_aH_bO_cN_d + 0.5(ny + 2s + r - c)O_2 \rightarrow nC_wH_xO_yN_z + sCO_2 + rH_2O + (d - nz)NH_3$$

where $r = 0.5[b - nx - 3(d - nz)]$

$s = a - nw$

$C_aH_bO_cN_d$ = Feed composition

$C_wH_xO_yN_z$ = Final product

Air at 25°C and 1 atm pressure has a specific weight of 1.20 g/L (0.075 lb/ft^3) and contains 23.2% oxygen by weight.

- Air requirement for moisture removal

  The amount of water evaporated (neglecting organic amendment or bulking agent addition) is :

  $$\Delta H_2O = (X_c - S_c X_c) - (X_p - S_r X_p) \quad (1)$$

  Considering the quantity of ash to be conservative, a mass balance on the inorganic fraction yields :

  $$(1 - V_c)S_cX_c = (1 - V_r)S_rXC_p \quad (2)$$

  Solving for X_p and substituting in equation (2) and rearranging yields :

$$\frac{\text{Amount of moisture fremoved}}{\text{g}-\text{dry solid cake}} = \frac{\Delta H_2O}{S_cX_c} = \left(\frac{1-S_c}{S_c}\right) - \left(\frac{1-V_c}{1-V_r}\right)\left(\frac{1-S_c}{S_r}\right)$$

where X_c = Total wet weight of dewatered sludge cake produced per day

X_p = Total wet weight of compost product produced per day

S_c = Fractional solids content of dewatered sludge cake

S_r = Fractional solids content of composite product and composite recycle

V_c = Volatile solids content of sludge cake (fraction of dry solids)

V_r = Volatile solids content of compost product and recycle (fraction of dry solids)

1.2 Required amount of oxygen for composting

- Moles of organics entering the process $= \dfrac{1000}{852} = 1.173$
- Moles of organics leaving the process per mole entering the process = n

  $$= \frac{200}{(1.73)(224)} = 0.76$$

- Stochiometric coefficients :

  a = 31 w = 11

  b = 50 x = 14

  c = 26 y = 4

  d = 1 z = 1

  $r = 0.5[50 - 0.76(14) - 3(1 - 0.76 \times 1)] = 19.33$

  $s = 31 - 0.76(11) = 22.64.$

- Amount of oxygen required by the composting process is :

$$W = 0.5[0.76(4) + 2(22.64) + 19.33 - 26](1.173)(32)$$
$$= 783 \text{ kg } O_2.$$

- Summary :
  - Input :

Organic material	= 1000 kg
Oxygen	= 783 kg
Total	1783 kg

  - Output :

Organics	= 200 kg
CO_2 [1.173(22.64)(44)]	= 1170 kg
H_2O [1.173(19.33)(18)	= 408 kg
NH_3 [(1– 0.76(1)(1.173)(17)	= 5 kg
Total	1783 kg

  - C/N ratio

$$\text{In} - C/N \text{ ratio} = \frac{31(12)}{14(1)} = 26.6$$

$$\text{Out} - C/N \text{ ratio} = \frac{11(12)}{14(1)} = 9.43$$

1.3 Required amount of heat released

- Percentage elemental composition of the organics entering and leaving the process are :
  - Input :

$$\%C = 31 \times \frac{12}{852} = 43.6\%$$

$$\%H = 50 \times \frac{1}{852} = 5.9\%$$

$$\%O = 26 \times \frac{16}{852} = 48.9\%$$

  - Output :

$$\%C = 11 \times \frac{12}{224} = 59.0\%$$

$$\%H = 14 \times \frac{1}{224} = 6.3\%$$

$$\%O = 4 \times \frac{16}{224} = 28.6\%$$

- Degree of reduction (R) for any type of organic matter is :

$$R = \frac{100[2.66(C) + 7.94(H) - (O)]}{398.9}$$

where C, H, and O = Percentage of carbon, hydrogen and oxygen, respectively on ash free basis.

- Heat combustion is : Q (Cal/g) = 127R + 400
- Values of R

$$\text{Input} - R = \frac{100[2.66(43.6) + 7.94(5.9) - 48.9]}{398.9} = 28.6$$

$$\text{Output} - R = \frac{100[2.66(59.0) + 7.94(6.3) - 28.6]}{398.9} = 44.8$$

- Values of heat content (Q) :

Input Q = 127(28.6) + 400 = 4030 Cal/g

Output Q = 127(44.8) + 400 = 6100 Cal/g

- Energy (heat) released :

Energy in = $1000 \times 1000 \times 4030 = 4030 \times 10^6$ cal

Energy out = $200 \times 1000 \times 6100 = 1220 \times 10^6$ cal

Net heat released 2810×10^6 Cal

2. Determine the weight and volume of air required for biological oxidation per g of original waste. Assume the following data :

Empirical formula of sludge	:	$C_5 H_2 O_2 N$
Volatility of sludge	:	80%
Biodegradibility of sludge	:	60%

2.1 Required amount and volume of air for bioxidation of sludge

$$\underset{113}{C_5H_7O_2N} + \underset{5(32)}{5O_2} \rightarrow 5CO_2 + 2H_2O + NH_3$$

For 1 g of sludge, oxygen required for complete combustion is :

$$\text{Oxygen} = \frac{5(32)(1.0)}{113} = 1.42 \text{ g } O_2 / \text{g-sludge}$$

Oxygen required based on the original feed is :

(1.42 g/g-sludge)(0.8)(0.6) = 0.68 g-O_2/g-sludge

- Weight and volume of air required at 25°C and 1 atm are :

$$\text{Air}(\text{Volume}) = \frac{0.68}{(1.20)(0.232)} = 2.44\text{L} - \text{air} / \text{g-sludge}$$

$$\text{Air}(\text{Weight}) = \frac{0.68}{0.232} = 2.93 - \text{g} - \text{air} / \text{g-sludge}$$

3. Estimate the weight and volume of required air if the exit and ambient air temperatures are 60 and 20°C, respectively for drying the composted sludge. Assume the following data :

Dewatered cake solids content	: 35%
Dewatered cake volatile solids content	: 55%
Final composted material VS content	: 40%
Specific humidity of air at 20°C and 1 atm	: 0.015 g H_2O/g dry air
Specific humidity of air at 60°C and 1 atm	: 0.152 g H_2O/g dry air
Final composted moisture content of compost	: 30%

3.1 Required moisture evaporation

$$\frac{\Delta H_2O}{S_c X_c} = \left(\frac{1-S_c}{S_c}\right) - \left(\frac{1-V_c}{1-V_r}\right)\left(\frac{1-S_r}{S_r}\right)$$

$$= \left(\frac{1-0.35}{0.35}\right) - \left(\frac{1-0.55}{1-0.40}\right)\left(\frac{1-0.70}{0.70}\right)$$

= 1.86 – 0.32 = 1.54g –H_2O/g-sludge.

3.2 Required moisture removed from the composting sludge and air requirements

- Assume exit and ambient air to be saturated
- Moisture removed from the composted sludge :

  = (0.152 – 0.015) = 0.137 g H_2O/g-dry air

- Air requirements :

$$\text{Air (Weight)} = \frac{(1.54\text{g } H_2O / \text{g} - \text{dry cake solids})}{(0.137\text{g} - H_2O / \text{g} - \text{dry air})}$$

$$\text{Air (Volume)} = \frac{(11.24\text{g dry air} / \text{g} - \text{dry cake solids})}{1.2\text{g air} / \text{L at } 25°\text{C and } 1 \text{ atm}}$$

4. Determine the air requirements for drying if ambient air at 20°C is used, as also air requirements if it is preheated to 60°C (specific humidity at 60°C and 1 atm = 0.152 – H_2O/g-dry air). Assume the following data :

Solids content of the compost = 45%

Final solids content of compost = 85%

Ambient air specific humidity at saturation, and 20°C = 0.0152 g H_2O/g dry air

Relative humidity of air = 65%

4.1 Required water to be removed per gram of solids at 20°C

- Assume that inlet and outlet air are at the same temperature, and that exit air is saturated.
- Water to be removed

$$= \left(\frac{1}{0.45} - 1\right) - \left(\frac{1}{0.85} - 1\right)$$

$$= 1.22 - 0.18 = 1.04 \text{ g-H}_2\text{O/g-solids}$$

4.2 Required air at 20°C and 60°C

- Air (volume) at 20°C

$$= (1.04 \text{ g} - \text{H}_2\text{O/g} - \text{solids})\left(\frac{1 \text{ g} - \text{dry air}}{0.015 \text{ g H}_2\text{O}}\right)\left(\frac{1}{1.0 - 0.65}\right)\left(\frac{10^6 \text{ g}}{\text{ton}}\right)$$

$$\left(\frac{22.4 \text{ L}}{29 \text{ g} - \text{air}}\right)\left(\frac{1 \text{ m}^3}{1000 \text{ L}}\right)$$

$= 221877 \text{ m}^3$ dry – air/ton – dry compost.

- Air (volume) at 60°C
  - Moisture removal capacity of air at 60°C is :

    $= 0.152 - (1 - 0.65)(0.015) = 0.147$ g – H_2O/g-dry air
  - Air required

$$= (1.04)\left(\frac{1}{0.147}\right) \times 10^6 \times \frac{22.4}{29} \times \frac{1}{1000}$$

  - $= 5465 \text{ m}^3$ dry – air/ton – dry compost.

5. Determine the oxygen demand during composting as also the number of volume exchanges required to satisfy it. Assume the following data :

Digested dewatered solids content	:	25% solids
Digested dewatered VS content	:	55%
Recycled blending compost	:	
Solids content	:	75%

VS content	:	40%
Final mixture solids content before composting	:	45%
Specific weight of mixture to be composted	:	0.64 g/cm^3

5.1 Required recycle ratio (dry basis, R_d), and specific gravities of solids in cake and compost recycle

- Recycle ratio (dry basis), $R_d = \dfrac{(S_m/S_c - 1)}{(1 - S_m/S_r)}$

where S_m = Fractional solids content of mixture before composting (dry basis)

S_c = Fractional solids content of dewatered sludge cake (dry basis)

S_r = Volatile solids content of compost product and recycle (dry basis)

$$R_d = \frac{\left(\dfrac{0.45}{0.25} - 1\right)}{\left(1 - \dfrac{0.45}{0.75}\right)} = \frac{0.8}{0.4} = 2.0$$

- Specific gravities of solids in cake and compost recycle

$$\text{Specific gravity of solids} = \frac{1}{G_s} = \frac{V_s}{G_v} + \frac{(1 - V_s)}{G_f}$$

where G_s = Specific gravity of the total solids

V_s = Volatile fraction of total solids

G_v = Specific gravity of volatile fraction

G_f = Specific gravity of fixed or ash fraction

$$\text{Cake solid}: \frac{1}{G_s} = \frac{0.55}{1.00} + \frac{(1 - 0.55)}{2.5}$$

$$= 0.73$$

$$G_s = 1.37$$

$$\text{Compost solids}: \frac{1}{G_s} = \frac{0.40}{1.00} + \frac{(1 - 0.40)}{2.5}$$

$$G_s = 1.56$$

Mixture solids : $G_s = \dfrac{1.37(1) + 1.56(2.00)}{1.0 + 2.33}$

(Average basis) $= 1.35$

- Rate of gas volume (V_g) to total volume (V_t), f is :

$$f = \frac{V_g}{V_t}$$

$$= \frac{V_t - V_s - V_w}{V_t}$$

$$= 1 - \frac{V_s}{V_t} - \frac{V_w}{V_t}$$

f for a mixture (f_m) is :

$$f_m = 1 - \frac{\gamma_m S_m}{G_s \gamma_w} - \frac{\gamma_m (1 - S_m)}{\gamma_w}$$

where γ_m = Specific weight of a mixture

S_m = Fractional solids content of mixture (dry basis)

g_w = Specific weight of water

$$f_m = 1 - \frac{0.64(0.45)}{1.35(1.00)} - \frac{0.64(1-0.45)}{1.0}$$

$$= (1 - 0.213 - 0.352) = 0.435$$

5.2 Required air

- Degradibility of digested cake = 50%
- Degradibility of recycle cake = 10%
- Air requirements

– $$\text{Cake} = \frac{(2.5\text{ g O}_2/\text{g} - \text{organic})(0.55\text{ g VS/g solids})(0.10)}{0.23\text{ g O}_2/\text{g} - \text{air}}$$

= 2.99 g – air/g – cake – solids

– $$\text{Compost} = \frac{(2.5\text{ g O}_2/\text{g} - \text{organic})(0.40\text{ g VS/g} - \text{solids})(0.10)}{0.23\text{ g O}_2/\text{g} - \text{air}}$$

= 0.435 g – air/g – compost – solids

– Total air required for 1 L of mixture

$$\text{Air for cake} = (0.64\text{ g/cm}^3)(10^3\text{ cm}^3/\text{L})(0.45)\left(\frac{1.00}{1.00 + 2.0}\right)(2.99)$$

= 96 × 2.99 = 287 g/L

$$\text{Air for compost} = (0.64\text{ g/cm}^3)(10^3\text{ cm}^3/\text{L})(0.45)\left(\frac{2.00}{3.0}\right)(0.435)$$

= 83.5 g/L

– Total air for 1 L of mixture is :

$$= (287 + 83.5) = 370.5 \text{ g/L}$$

– Number of void volume exchanges if air density is 1.20 g/L at 25°C and 1 atm

$$= \frac{370.5 \text{ g/L}}{(f_m)(\gamma_{air})} = \frac{370.5}{(0.435)(1.20)}$$

$$= 710 \text{ exchanges}$$

Example 3.114

Composting : Heat Requirements

A compost process is operated under constant pressure conditions in which water and vapour enter at a temperature, T = 15.5°C. The reactor is maintained at a temperature T_2 = 71.1°C, and different proportions of water and vapour exit the reactor at T_2. Estimate the heat required to accomplish the mass and temperature changes assuming two different paths :

- Assume that water is vapourised at T_1 and then water and vapour are heated to T_2
- Assume inlet water and vapour are first heated to temperature T_2 and then a portion is vapourised at temperature T_2

The schematics of the process :

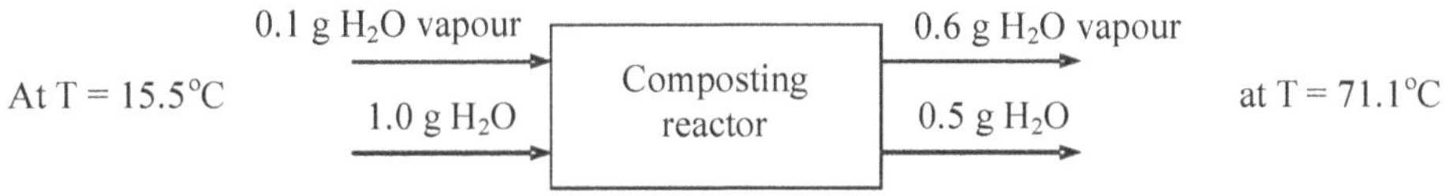

Figure 36 : Composting reactor.

Solution

1. Required enthalpy changes using path 1 :

Table 85 : Steam Table

Temperature (°C)	*Enthalpy (Cal/g)*	
	Water	*Steam (saturated)*
15.5	15.59	604.28
71.1	71.09	627.89

- Enthalpy changes (ΔH) is :

$$\Delta H = 0.5[71.09 - 15.59] + 0.5[604.28 - 15.59] + (1.0 + 0.5)[627.89 - 604.28]$$

= Liquid to T_2 + Vapourise at T_1 + All vapour to T_2

= 336.26 Cal.

2. Required enthalpy change (ΔH) using path 2

$\Delta H = 1.0[71.09 - 15.59] + 0.5[627.89 - 71.09] + 0.1[627.89 - 604.28]$

= Liquid to T_2 + vapourise to T_2 + Input vapour to T_2

= 332.26 Cal

[Enthalpy remains the same, irrespective of the path].

Example 3.115

Sludge Composting

A time temperature profile for a batch composting must be known (material to be heated, held at a constant temperature, and than cooled). Using a graphical integration approach determine log reduction of Ascaris ova. Assume the minimum lethal temperature to be 45°C. The temperature profile is approximated by construction of a number of smaller elements. The first element begins at 45°C (minimum lethal temperature). Assume the following data :

Table 86 : Temperature profile with time for various elements.

Element	*Dt (days)*	*Average T (°C)*
1	0.25	47.7
2	0.25	53.2
3	0.25	56.7
4	0.25	59.0
5	0.25	60.0
6	0.75	60.3
7	0.25	60.0
8	0.25	58.8
9	0.25	57.0
10	0.25	54.7
11	0.25	52.0

- $k_d = 0.025 \exp [0.361(T - 50)]$

Solution

1. Required mathematical expression

 - Death rate :

$$\frac{dn}{dt} = -k_d n$$

$$\frac{dn}{dt} = C \exp\left[\frac{-E_d}{R\,T_k(t)}\right] n;\; k_d = C \exp\left[\frac{-E_d}{R\,T_k(t)}\right]$$

Temperature (T_k) is a function of time and is expressed as T_k (t)

Integrating the above expressions between t = 0 and t = t_f with n = n_o at t = 0 :

$$\ln\left[\frac{n_o}{n_f}\right] = \int_0^{t_f} C \exp\left[\frac{-E_d}{R\,T_k(t)}\right] dt$$

Temperature profile must be known. Assume first a constant temperature profile T ≠ f(t),

$$\ln\left[\frac{n_o}{n_f}\right] = C\, t_f \exp\left[\frac{-E_d}{R\,T_k}\right]$$

where T_k = Absolute temperature (°K), and is constant

Similar integrations have been performed for a variety of other time-temperature profiles including hyperbolic, exponential, and linear increases. Regardless of the type of temperature-time profile, total kill is the sum of kills observed in the heating, constant (holding)-temperature, and cooling periods :

$$\left[\ln\left(\frac{n_o}{n_f}\right)\right]_{Total} = \left[\ln\left(\frac{n_o}{n_f}\right)\right]_{heating} + \left[\ln\left(\frac{n_o}{n_f}\right)\right]_{holding} + \left[\ln\left(\frac{n_o}{n_f}\right)\right]_{cooling}$$

2. Required thermal inactivation coefficient (k_d) and final microbe concentration (n_f)

   - Thermal inactivation coefficient (k_d) and average temperature (T) for each element

$$k_d\,\Delta t = \ln\left(\frac{n_o}{n_f}\right)$$

$$k_d = 0.025 \exp\,[0.361(T - 50)]$$

where T = Average temperature at each element (°C)

Table 87 : Temperature V_s kill.

Element	*1*	*2*	*3*	*4*	*5*	*6*	*7*	*8*	*9*	*10*	*11*
Δt (days)	0.25	0.25	0.25	0.25	0.25	0.75	0.25	0.25	0.25	0.25	0.25
Average T (°C)	47.7	53.2	56.7	59.6	60.0	60.3	60.0	58.8	57.0	54.7	52.0
k_d (min^{-1})	0.011	0.079	0.281	0.644	0.924	1.03	0.924	0.599	0.313	0.136	0.651
$\ln\left(\frac{n_o}{n_f}\right)$	4	29	101	232	333	1112	333	216	113	49	19

$$\ln\left(\frac{n_o}{n_f}\right) = 2541$$

$$\log\left(\frac{n_o}{n_f}\right) = \frac{\ln\left(\frac{n_o}{n_f}\right)}{2.303} = \frac{2541}{2.303} = 1100$$

Therefore, $n_f = n_o\ (10^{-1100})$

[No Ascaris oval would survive].

Example 3.116

Sludge Composting

Estimate values for thermal death coefficient (k_d) and the inactivation energy (E_d) for A. Lumbricoides during sludge composting. Assume the following data:

Table 86 : Temperature Vs Reduction and required time.

Temperature °C(°K)	Fraction Reduction (n_t/n_o)	*Time (t) for Reduction (min)*
50(323)	0.001	270
55(328)	0.001	45
60(333)	0.001	7.5

Solution

1. Required mathematical expression
   - First order decay rate model is assumed

$$\frac{dn}{dt} = -k\,n$$

$$n_t = n_o \exp(-k_d t)$$

$$\text{or} \quad t = \frac{\ln(n_t/n_o)}{k_d}$$

   - Time (t) required for one long reduction in cell concentration (90% reduction) is :

$$t_{90} = D_r = \frac{2.303}{k_d}$$

   - Temperature effect in k_d is given as :

$$k_d = C \exp\left[-\frac{E_d}{R\,T_k}\right]$$

The range of inactivation energies (E_d) for many spores and vegetable cells is between 50 and 100 kcal/mol, and it implies that heat inactivation is much more sensitive to temperature than most chemical reactions.

Or $\log k_d = \log C - \frac{E_d}{R\,T_k}$

where n = Viable cell population

k_d = Thermal inactivation coefficient

E_d = Inactivation energy

D_r = Decimal reduction factor, and is the time required to achieve a tenfold reduction in cell population

T_k = Absolute temperature (°K)

2. Required thermal inactivation coefficient (k_d) and inactivation energy (E_d)

- Inactivation thermal coefficient (k_d) is :

$$k_d = \frac{\ln(n_t/n_o)}{t}$$

  - At T = 50°C

$$k_d = \frac{\ln(1000)}{270} = 0.0256\ \text{min}^{-1}$$

  - At T = 55°C

$$k_d = \frac{\ln(1000)}{45} = 0.154\ \text{min}^{-1}$$

  - At T = 60°C

$$k_d = \frac{\ln(1000)}{7.5} = 0.921\ \text{min}^{-1}$$

- Decimal reduction factor (D_r)

$$D_r = \frac{2.303}{k_d}$$

  - At T = 50°C

$$D_r = \frac{2.303}{0.0256} = 90\ \text{min}$$

  - At T = 55°C

$$D_r = \frac{2.303}{0.154} = 15\ \text{min}$$

  - At T = 60°C

$$D_r = \frac{2.303}{0.921} = 2.5\ \text{min}$$

- Inactivation energy (E_d)

  Plot values of k_d Versus T_k and (T – 50°C)

  Therefore,

  $C = 1.81 \times 10^{49}$

  $E_d = 75200$ cal/mol = 75.2 kcal/mol

  $k_d = 1.81 \times 10^{49} \exp (3.78 \times 10^4/T_k)$; R = 1.99 cal/°K.mol

  or, $k_d = 0.025 \exp [0.361(T - 50)]$

  Final equation becomes :

$$\text{In}\left[\frac{n_o}{n_1}\right] = 0.025\,t \exp\left[0.361(T-50)\right]$$

Note :

1. Table 89, Time required for a tenfold population reduction (D_r) for various microbes during composting of sludge

Table 89 : Required time

	D_r (min)	
Microbe	*55°C*	*60°C*
Adenovirus (12 NIAID)	11	0.17
Poliovirus (type 1)	1.8	1.5
Ascaris ova	-	1.3
Histolytica cysts	44	25
Salmonella	80	7.5
Bacteriophage (f-2)	267	47

Example 3.117

Composting of Sludges

Estimate the heat of combustion for the raw sludge with the following composition :

Table 90 : Composition of raw sludge.

Element	*C*	*H*	*O*	*N*	*Cl*	*S*	*Ash*
%weight	37.51	5.54	22.56	1.96	0.33	0.37	28.06

Solution

1. Required expressions for heating value of different types of sludges
   - Empirical formulations

(a) $$Q = a\left[\frac{P_V(100)}{100 - P_c}\right]\left[\frac{100 - P_c}{100}\right]$$

where Q = Fuel (heat) value (Btu/lb dry solids)

a = Coefficient (= 131 for raw and digested primary sludge

= 107 for raw ASP sludge)

b = Coefficient (= 10 for raw and primary digested sludge

= 5 for raw ASP sludge)

P_v = Percent volatile sludge solids

P_c = Percent inert chemicals in sludge

(b) Degree of reduction for any type of organic matter

$$R = \frac{100[2.66(C) + 7.94(H) - (O)]}{398.7}$$

where C, H and O = Percentages of carbon, hydrogen and oxygen on ash free basis

Heat of combustion is :

Q (Cal/g) = 127R + 400

(c) Dulong formulation

$$Q(\text{BTU/lb}) = 145.4(C) + 620(H - O/8) + 41(S)$$

(d) Rule of thumb for heat of combustion:

- About 3.4 ± 0.2 kcal/g COD is released.
- The reason that heat released per unit of COD is relatively constant lies in the fact that COD is a measure of electron transferred. Four moles of electrons must be transferred during aerobic oxidation for each mole of oxygen transferred:

$$O_2^0 + 4_e^- \rightarrow 2.0\,O^{-2}$$

Substrate COD is proportional to the number of electrons transferred during aerobic oxidation. Heat of combustion per electron transferred to a methane type bond is taken to be about 26.05 kCal/electron equivalent or 104.2 kcal/mol O_2 (3.26 kcal/g COD).

2. Required heat of combustion

- $$Q = 131\left[\frac{71.94(100)}{100 - 0}\right]\left[\frac{100 - 0}{100}\right];\ P_V = 100 - 28.06 = 71.94\%$$

= 8114 Btu/lb TS

= 8114 Btu/lb TS

Table 91 : Considering only C, H and O component of sludge

Elements	*% Weight*	*Adjusted to 100% by Weight*
C	37.51	57.2
H	5.54	8.4
O	22.56	34.4
Total	65.61	100.0

$$R = \frac{100[2.66(C) + 7.94(H) - O]}{398.7}$$

$$R = \frac{100[2.66(57.2) + 7.94(8.4) - 34.4]}{398.9}$$

- Dulong method

$$Q(BTU/lb) = 145.4(C) + 620\left[H - \frac{O}{8}\right] + 41(S)$$

$$= 145.4(57.2) + 620\left[8.4 - \frac{34.4}{8}\right] + 0$$

= 10859 BTUu/lb

VS = (100 – 28.06) = 71.94%

Q = (10859 BTU/lb)(0.7194)

= 7812 BTU/lb TS

[Sulphur can be oxidized by autotrophic micro-organisms but it is negligible]

- Rule of thumb

VS = 0.7194

Q = 5550(VS) = 5550 (0.7194)

= 3993 cal/g TS

[Organic fraction gives 5550 cal/g of organic matter oxidized]

- COD based
  - Empirical formula of sludge is :

$$C = \frac{57.2}{12} = 4.8$$

$$H = \frac{8.4}{1} = 8.4$$

$$O = \frac{34.4}{16} = 2.2$$

Empirical formula is = $C_{4.8} H_{8.4} O_{2.2}$

Oxidation equation is :

$C_{4.8} H_{8.4} O_{2.2} + 5.8O_2 = 4.8CO_2 + 4.2H_2O$

$$\text{Amount of } O_2 \text{ required} = \frac{5.8(32)}{101.2} = 1.83 \text{ g COD/g} - \text{VS}$$

Q = (3400 Cal/g COD)(1.8 g COD/g – organics)

= 6222 Cal/g – oganics

Q = (6222 Cal/g – organics) VS

= 6222 (0.7194) = 4476 Cal/g TS.

Example 3.118

Blending Sludge with Leaves

Determine the mass proportions of leaves and wastewater sludge from a wastewater treatment plant to achieve an optimum blended C/N ratio in the range of 22-25.

Assume the following date :

C/N ratio of leaves	:	45
C/N ratio of sludge from ETP	:	7.5
Moisture content of solid sludge	:	65%
Moisture content of leaves	:	30%
Nitrogen content of solid sludge	:	5.4%
Nitrogen content of leaves	:	0.9%

Solution

1. Required mass balance for C, N and H_2O
   - Assume :

     m_s = Mass of sludge

     m_l = Mass of leaves
   - Sludge aspects :

     – $m_{H_2O} = 0.65\, m_s$

– Solid sludge = 0.35 m_s

– $m_N = 0.054\ (1 - 0.65)\ m_s$

$= 0.0189\ m_s$

– $m_c = 7.5\ [(0.054)(1 - 0.65)m_s]$

$= 0.142\ m_s\ [C/N = 7.5]$

– Inert mass in sludge (m_I)

$m_I = (0.35 - 0.0189 - 0.142)\ m_s$

$= 0.149\ m_s$

- Leaves aspects :

$m_{H_2O} = 0.65\ m_l$

$m_N = 0.009\ (1 - 0.3)\ m_l$

$= 0.0063\ m_l$

$m_C = (0.009)\ (45)(1 - 0.3)\ m_l$

$= 0.2835\ m_l$

$m_I = (0.7 - 0.0063 - 0.2835)\ m_l$

$= 0.410\ m_l$

2. Required optimum carbon to nitrogen ratio

$$\frac{C}{N} = \frac{0.142\ m_s + 0.2835\ m_l}{0.0189\ m_s + 0.0063\ m_l} = 22$$

Assume $m_l = 1$, $m_s = 0.991$, [specific gravities]

$$\frac{C}{N} = \frac{0.142\ m_s + 0.2835\ m_l}{0.0189\ m_s + 0.0063\ m_l} = 26$$

Assume $m_l = 1$, $m_s = 0.7$

Therefore, the mass ratio of sludge to leaves should be between 0.7 and 1.0.

Example 3.119

Incineration of Sludges

Determine the following :

- Inlet and outlet flow rates in lb/h on both a dry and a wet basis.
- Amount of ash remaining after combustion.
- Volatile heating value of the waste.
- Supplemental fuel that must be added to the incinerator.

Assume the following data :

Inflow of MSW	:	10,000 lb/h
Heating value of MSW	:	6500 Btu/lb, ash content = 10 wt%, and moisture content = 25 wt%
Dry gas generated	:	7.5 lb of dry gas for every 10,000 Btu of waste burned in the incinerator
Moisture produced	:	0.5 lb-moisture produced for every 10,000 Btu of waste fired
Air Cooling	:	Air cools the incinerator shell at a flow of 10,000 lb/h [and is discharged to the atmosphere at 400°F]
Influent air humidity	:	$\dfrac{0.015 \text{ lb of moisture}}{\text{lb dry air}}$

Ash produced has a heating value of 100 BTU/lb

Radiation heat loss : 1% of the total heating value of the waste

The temperature of the products of combustion (the flue gas) must not be less than 2000°F.

Solution

1. Required mass rates of moisture, ash, and volatile
   - Moisture feed rate = 0.25 (10,000 lb/h)
     = 2500 lb/h
   - Dry feed rate = (10,000 – 2500) = 7500 lb/h
   - Ash rate = 0.1 (10,000 lb/h)
     = 1000 lb/h
   - Amount of feed that is combusted (volatile)
     Volatile rate = (7500 – 1000) lb/h
     = 6500 lb/h
2. Required heating value
   - Heating value = (10,000 lb/h)(6500 BTU/lb)
     $= 65 \times 10^6$ BTU/h
   - $$\text{Volatile waste heating value} = \frac{65 \times 10^6 \text{ BTU/h}}{6500 \text{ lb/h}}$$
     = 10,000 BTU/lb

- Dry gas produced from combustion is :

  Dry gas = (7.5 lb/10,000 BTU) $(65 \times 10^6$ BTU/h)

  = 48750 lb/h

- Amount of moisture produced from combustion is :

  Combustion moisture = (0.51 lb/10,000 BTU) $(65 \times 10^6$ BTU/h)

  = 3315 lb/h

- Total combustion gas

  = 48750 lb/h (dry gas) + 3315 lb/h (combustion moisture)

  = 52065 lb/h

- Stochiometric air requirement rate
  - Air requirement = (52065 – 6500) lb/h

    = 45565 lb/h
  - Total air requirement rate

    = (45565 + 45565) lb/h

    = 91130 lb/h

- Inlet air moisture rate

  = (0.015 lb/lb dry-air)(91130 lb/h)

  = 1367 lb/h

- Total outlet moisture rate

  = (2500 + 3315 + 1367) lb/h = 7182 lb/h

- Total dry gas flow rate existing the system

  Outlet dry gas rate = (48750 + 45565) lb/h

  = 94315 lb/h

- Heat loss due to ash discharge

  = (1000 lb/h)(100 BTU/lb) = 0.1×10^6 BTU/h

- Heat loss to cool the incinerator shell

  = (10,000 lb/h) (81.8 BTU/lb)

  [Enthalpy of air at 400°F = 81.8 BTU/lb]

- Heat loss due to radiation

  = $0.01(65 \times 10^6$ BTU/h) = 0.65×10^6 BTU/h

- Amount of heat absorbed by the humidity of the inlet air supply

  = (960 BTU/lb)(1367 lb/h) = 1.31×10^6 BTU/h

- Total heat loss of the incinerator

  $= (0.1 + 0.82 + 0.65 - 1.31) \times 10^6$ BTU/h

  $= 0.26$ BTU/h

- Outlet heat content of the flue gas

  $= (65 - 0.26) \times 10^6$ BTU/h

  $= 64.74 \times 10^6$ BTU/h

- Heat content of the flue gas at 2000°F

  = (7182 lb/h)(2060 BTU/lb) + (94315 lb/h)(513 BTU/lb)

  $= 63.2 \times 10^6$ BTU/h

  [Enthalpies of dry air and moisture at 2000°F:

  - Enthalpy of dry air = 513 BTU/lb
  - Enthalpy of moisture = 2060 BTU/lb]

    Outlet heat content of 64.74×10^6 BTU/h is greater than the outlet heat content at 2000°F, supplemental fuel is not necessary.

Example 3.120

Incineration of Sludges

Estimate the heat available in the flue gas from the incineration of 1×10^5 kg/d (2.2×10^5 lb/d) of sludge. Assume the following data :

Table 92 : Composition of sludge

Component	*% of Total*	*Weight (lb/d)*
Combustible	55	121000
Non combustible	25	55000
Water	20	44000

Table 93 : Elemental composition

Element	*C*	*H*	*O*	*N*	*S*	*H_2O*	*Inerts*
Percent	27.4	3.6	23.0	0.5	0.1	21.4	24.0

- Other data :

  Heat value of sludge : 5000 BTU/lb

  Residue contains 10% unburnt carbon

  Entering air temperature : 80°F

  Grate residue temperature : 880°F

  Latent heat of water : 1040 BTU/lb

Radiation loss : 0.5% of total heat input

All oxygen in waste is boundwater

Net hydrogen available for combustion is equal to percent hydrogen minus 1/8 the percent oxygen [Accounts for the boundwater in the day combustible material]

Carbon heating value : 14000 BTU/lb

Air moisture : 1.0 percent

Solution

1. Required mass balance

Table 94 : Elemental mass balance

Element	%	*Mass (lb/d)*
C	27.4	$0.274\ (2.2 \times 10^5) = 0.6028 \times 10^5$
H	3.6	$0.036\ (2.2 \times 10^5) = 0.079 \times 10^5$
O	23.0	$0.230\ (2.2 \times 10^5) = 0.506 \times 10^5$
N	0.5	$0.005\ (2.2 \times 10^5) = 0.011 \times 10^5$
S	0.1	$0.001\ (2.2 \times 10^5) = 0.0022 \times 10^5$
Water	21.4	$0.214\ (2.2 \times 10^5) = 0.4708 \times 10^5$
Inerts	24.0	$0.240\ (2.2 \times 10^5) = 0.440 \times 10^5$

- Amount of residue is :

  Inerts = 44000 lb/d

  $$\text{Total residue with unburnt carbon} = \frac{44000}{0.90} = 49000 \text{ lb/d}$$

  Unburnt carbon = 49000 – 44000

  = 5000 lb/d

- Available hydrogen and boundwater are :

  $$\text{Net available hydrogen (\%)} = \left[3.6\% - \frac{23.0}{8}\%\right]$$

  $$= 0.725\%$$

  $$= \frac{0.725\left(2.2 \times 10^5\right)}{100}$$

  $$= 1595 \text{ lb/d}$$

  Hydrogen in boundwater = 3.6% – 0.728%

  = 2.875%

$$= \frac{2.875\left(2.2 \times 10^5\right)}{100}$$

= 6325 lb/d

Boundwater = Oxygen + Hydrogen in boundwater

= (50600 + 6325) lb/d

= 56925 lb/d

2. Required air for combustion

- For Carbon

  $C + O_2 = CO_2$: 11.52 lb air/lb – C

- For Hydrogen

  $2H_2 + O_2 = 2H_2O$: 34.56 lb air/lb – H

- For Sulphur

  $S + O_2 = SO_2$: 4.32 lb air/lb – S

- Air requirement :

  For Carbon combustion = (60280 – 5000) lb C/d [11.52 lb air/lb – C]

  = 636825 lb/d

  For Hydrogen combustion = (1595 lb H/d)(34.65 lb air/lb – H)

  = 55267 lb/d

  For Sulphur combustion = (220 lb S/d)(4.32 lb air/lb – S)

  = 950 lb/d

- Total theoretical air requirement is

  = (636825 + 55267 + 950) lb/d

  = 693042 lb air/d

  or $\text{lb air/lb sludge} = \frac{693042}{220000} = 3.15$

- Total dry air including 100% excess is

  = 2(693040 lb air/d) = 1386080 lb air/d

- Total moisture is

  = (1386080 lb air/d)(0.01) = 13860 lb H_2O/d

- Total air

  = (1386080 + 13860) lb/d

= 1399940 lb/d

- Amount of moisture produced from combustion of available hydrogen is

$$\text{lb } H_2O/d = 18 \text{ lb } H_2O/2 \text{ lb H } [1595 \text{ lb H/d}]$$
$$= 14355 \text{ lb/d}$$

3. Required heat balance

- Gross heat input = $(2.2 \times 10^5$ lb/d)(5000 BTU/lb)
  = 1100×10^6 BTU/d

- Heat lost :
  - Heat lost in unburnt carbon = (5000 lb/d)(14000 BTU/lb)
    = 70×10^6 BTU/d
  - Heat loss through radiation = (0.005 BTU/BTU)(1100 × 10^6 BTU/d)
    = 5.5×10^6 BTU/d
  - Inherent moisture = (47080 lb/d)(1040 BTU/lb)
    = 49×10^6 BTU/d
  - Boundwater moisture = (56925 lb/d)(1040 BTU/lb)
    = 59×10^6 BTU/d
  - Combustion produced moisture = (14355 lb/d)(1040 BTU/lb)
    = 15×10^6 BTU/d
  - Sensible heat in residue = (49000 lb/d)[0.25 BTU/lb °F (880 – 80)]
    = 10×10^6 BTU/d
  - Total heat lost = 208×10^6 BTU/d

- Heat available in the flue gas is
  = (1100 – 208) × 10^6 BTU/d = 892×10^6 BTU/d

- Efficiency of combustion $= \dfrac{892 \times 10^6 \text{ BTU/d}}{1100 \times 10^6 \text{ BTU/d}} \times 100$
  = 81%

- Overall efficiency (h) is :

  Assume boiler efficiency of 70%

  Therefore, $\eta = (0.81)(0.70)$
  = 0.567 or 56.7%.

Example 3.121

Sludge Incineration

Determine the fluidized furnace area, auxiliary fuel requirement (natural gas), air input rate, and the amount of sludge produced. Assume the following data :

Total solids fraction in the sludge	:	0.20
Volatile fraction of the sludge	:	0.80
Sludge feed rate to the fluidized bed incinerator	:	2000 lb/h
Operational cycle	:	8 hours/d and 5 days a week
Operating temperature	:	1500°F
Sludge composition :		
C	:	40%
H	:	5%
S	:	0.25%
Ash	:	15%
Oxygen	:	30%

Solution

1. **Furnace Diameter :** Using Liao's equations for fluidized bed incinerator design for determination of furnace diameter.

$$\log S_L = 2.7 - 0.0222\ M \tag{1}$$

$$\log B_r = 5.947 - 0.0096\ M \tag{2}$$

$$B = 93\ P \tag{3}$$

$$B = 145\ C + 620\left(H - \frac{O}{7.94}\right) + 45\ S \tag{4}$$

where S_L is the sludge loading rate (dry lb/hr-ft^2), M is the sludge moisture (% weight), B_r is the burning rate (BTU/hr-ft^2), B is the sludge heat value (BTU/dry lb), P is the sludge volatile content (%), and C, H, O, S are the percentage of carbon, hydrogen, oxygen and sulphur in the sludge, respectively.

$$\log S_L = 2.7 - 0.0222\ (M)$$

$$= 2.7 - 0.0222 \times 80$$

$$= 2.7 - 1.776 \approx 1.0\ \text{(say)}$$

$$S_L = 10\ \text{lb/hr-ft}^2$$

$$\text{Area of the incinerator} = \frac{2000 \text{ lb/hr}}{10 \text{ lb/hr ft}^2} = 200 \text{ ft}^2$$

2. Auxiliary fuel requirement (AFR)

AFR = [Total heat rate input] – [Sludge heat input]

Sludge heat input is given by equation (4)

$$B = 145\ C + 620 \left(H - \frac{O}{7.94}\right) + 45\ S$$

$$= 145 \times 40 + 620\left(5 - \frac{30}{7.94}\right) + 45 \times 0.25$$

$$= 5800 + 757 + 11.25$$

$$= 6568 \text{ BTU/lb}$$

Sludge heat input = 6568 BTU/lb × 2000 lb/hr

$= 13.14 \times 10^6$ BTU/hr

Using equation (2) to calculate burning rate

$$\log B_r = 5.947 - 0.0095\ M$$

$$= 5.947 - 0.0095 \times 80$$

$$= 5.947 - 0.760$$

$$B_r = 10^{5.14} \approx 165{,}000 \text{ BTU/ft}^2 - \text{hr}$$

$$\text{Total heat rate} = (165{,}000 - 6568) \frac{\text{BTU}}{\text{ft}^2 - \text{hr}} \times 200 \text{ ft}^2$$

$$= 158{,}432 \times 200$$

$$= 31 \times 10^6 \text{ BTU/h}$$

Auxiliary fuel required = $(31 \times 10^6 - 13.14 \times 10^6)$ BTU/hr

$= 17.86 \times 10^6$ BTU/hr

Heating volume of natural gas = 10,000 BTU/ft^3

$$\text{Therefore, amount of natural gas required} = \frac{17.86 \times 10^6}{10{,}000} = 1786 \text{ ft}^3/\text{h}$$

3. Air requirement (40% excess air supplied with 23% by weight of oxygen)

$C + O_2 = CO_2$ [11.5 lb air/lb carbon]

$2H_2 + O_2 = 2H_2O$ [34.3 lb air/lb/hydrogen]

$S + O_2 = SO_2$ [4.3 lb air/lb-sulphur]

Oxygen requirement :

Carbon : 11.5 × 0.40 × 2000 = 9200 lb/h

Hydrogen : 34.3 × 0.05 × 2000 = 3430 lb/h

Sulphur : 4.3 × 0.0025 × 2000 = <u>21.5 lb/h</u>

Theoretical air requirement = 2,651.5 lb/hr

<u>40% excess air</u> = <u>5060.0 lb/hr</u>

Total air required = 17,711.5 lb/h

Weight of air at 70°F = 0.0749 lb/ft³

$$\text{Air flow rate} = \frac{17,711}{0.0749} = 236,462 \text{ ft}^3/\text{hr}$$

4. Ash generation

[Ash input] = [Ash out]

2000 lb/hr × 0.15 = 300 lb/hr of ash generated

Example 3.122

Incineration of Sewage Sludges

The incineration of sewage sludges is difficult owing to high moisture content (MC) produced in conventional dewatering processes. The principle of auto-thermic or self supporting combustion is that the heat produced from the sludge burnt should be sufficient to allow for the heat losses and to raise the sludge moisture up to the exit temperature.

Assume a sludge to consist of 1 kg combustible material (CM) plus, say, 0.33 kg inert inorganics and an unspecific mass of moisture content. The calorific value (gross) of organic matter varies between 17,200 kJ kg^{-1} for carbohydrates, 23,600 kJ kg^{-1} for proteins and 39,500 kJ kg^{-1} for lipids. For calculation purposes a compound will be chosen of composition C (60%), H (9%), O (25%) and N (6%) with gross calorific value of 23,000 kJ kg^{-1} per kg CM. Combustion of 1 kg CM is as follows :

$$C + O_2 \rightarrow CO_2$$

$$0.6 + 1.6 \text{ kg} \rightarrow 2.2 \text{ kg}$$

$$2H_2 + O_2 \rightarrow 2H_2O$$

$$0.09 \text{ kg} + 0.72 \text{ kg} \rightarrow 0.81 \text{ kg}$$

The oxygen required is 2.32 kg less the 0.25 kg in CM, i.e., 2.07 kgO_2. This is equivalent (using air) to 2.07 × 3.31 kg N_2, i.e., – 6.85 kg N_2 after O_2 utilization. It is impossible to run an incinerator with this amount of air. Thus excess air is required. Assuming 50% excess air, the air requirement is 0.5 × 2.07 × 4.31 = 4.46 kg air.

Heat required to raise 4.46 kg air to an exit temperature of 430°C at a specific heat of 0.988 kJ per dry air is (assuming input at 15°C).

$0.998 \times 415 \times 4.46 = 1829$ kJ per kg CM

Heat required to raise 6.85 kg N_2 to the exit temperature at a specified heat of 1.186 kJ $kg^{-1}/°C$ is

$1.186 \times 415 \times 6.85$ kJ = 3372 kJ per CM

Heat required to raise 2.2 kg CO_2 by 415°C at 1.266 kJ $kg^{-1}/°C$ is

$1.266 \times 415 \times 2.2 = 1155$ kJ per kg CM

Heat required to raise 0.06 N_2 (for the sludge) is

$0.06 \times 415 \times 1.186 = 29.5$ kJ per kg CM

Σ heat in gaseous products $\approx$ 6390 kJ per kJ per CM

Since the calorific value was quoted on gross basis (including the heat gained from condensation of water formed during combustion), a loss of water formed by combustion is made at the rate of 2454 kJ per kg water. Thus water heat loss for 0.81 kg water produced per kg CM is

$2454 \times 0.81 = 1988$ kJ per kg CM

Additional losses will be due to the ash and the wall heat loss.

Assuming 1 kg CM is associated with 0.33 kg inert (75% VM in the original compound) leaving the furnace at 415°C above ambient temperature with specific heat of 0.84 kJ $kg^{-1}/°C$, the heat loss in the ash is

$0.84 \times 0.33 \times 415 = 115$ kJ per kg CM

Wall heat losses can only be determined if the size of the incinerator is known; however, assuming 100 m^2 with a heat loss rate of 1.5 kJ m^2/s at a burning rate of 0.5 kg CM/s, the heat loss is

$$\frac{100 \times 1.5}{0.5} = 300 \text{ kJ per kg CM}$$

Many other contributory factors may be evaluated; however, the foregoing, calculations give an estimate of the heat budget.

Thus, heat loss per kg CM = 6390 kJ
1988 kJ
115 kJ
300 kJ

Total = 8793 kJ per kg CM

Heat input = 23,000 kJ per kg CM

Heat available to dry cake and raise moisture to 430°C is 23,000 – 8793 = 14,207 kJ per kg CM. One kg of water requires 3270 kJ to be raised to 415°C, thus sufficient excess heat is present to heat (14,207/3270) kg (= 4.34 kg) water.

One kg CM is associated with (1/0.75) kg (i.e. 1.33 kg) dry solids at 75% volatile matter in the sludge. Thus, the solids content of a sludge in which 1 kg CM, 0.33 kg inert matter and 4.34 kg water is present is

$$\frac{1.33}{(1.33+4.34)} = 0.23 \text{ or } 23\% \text{ solids}$$

Gales has carried out calculations for different exit temperatures, excess air values, calorific values and compositions showing the minimum solids content of the cake required for self supporting combustion. The dewatering processes do place a limit to it. Since, the greater calorific values of sludge are due to lipids normally destroyed in anaerobic digestion, there is little reason for prior digestion.

Example 3.123

Incineration : Industrial Waste Sludges

Determine the following :

- Heat released
- Fuel requirement for heating upto 1800°F with 175% excess air
- Cooling water required to exhaust to 90°F
- Material balance
- Heat balance
- Flue gas discharge for incinerating

Sugar waste containing $Ca(OH)_2$.

Assume the following data :

Feed waste	:	1000 lb/h
$Ca(OH)_2$	:	10% of the feed
Excess air	:	175% [Total air = (100 + 175)]

Solution

1. Required heat of combustion for sugar and $Ca(OH)_2$
   - Heat released from sugar waste ($C_6H_{12}O_6$) :
     - Combustion reaction for sugar waste ($C_6H_{12}O_6$)

       $C_6H_{12}O_6 + 6O_2 = 6CO_2 + 6H_2O$

       MW : 180.18

       Heat of formation [ΔH_f, kcal/mol] : (– 203.8) 0 (– 94.8)(– 57.8)

       Heat of combustion of $C_6H_{12}O_6$ (ΔH_f)

       = $\Sigma\Delta H_f$ (Products) – $S\Delta H_f$ (Reactants)

$= 6 \times (-94.8) + 6(-57.8) - (-203.8)$

$= -346.8 - 564.6 + 203.8 = -707.6$ kcal/mol

ΔH_f ($C_6H_{12}O_6$) = – 707.6 kcal/mol (Exothermic reaction)

$$= \frac{(-707.6 \text{ k cal/mol})}{180.18 \text{ lb/mol}}(1802 \text{ BTU/kcal}) = -7077 \text{ BTU/lb}$$

- Heat of combustion of $Ca(OH)_2$ waste
  - Combustion reaction for $Ca(OH)_2$

$$Ca(OH)_2 \xrightarrow{(+\text{ heat})} CaO + H_2O \text{ (steam)}$$

Molecular Weight : 74.28 : 56.08 : 18.02, [$Ca(OH)_2$: CaO : H_2O]

Heat formation (ΔH_f, kcal/mol): – 239.7 – 151.8 – 57.8

Heat of combustion of

$Ca(OH)_2$ (ΔH_f, kcal/mol) $= \Sigma\Delta H_f$ (Products) – $\Sigma\Delta H_f$ (Reactants)

$= -151.8(1) + (-57.8)(1) - (-239.7)$

$= -151.8 - 57.8 + 239.7$

$= +30.1$ kcal/mol

$$= +\left(\frac{30.1 \text{ kcal/mol}}{74.28 \text{ lb/mol}}\right)(1802 \text{ BTU/kcal})$$

= + 730 BTU/lb (Endothermic reaction)

It is necessary to supply 730 BTU/lb of heat for this reaction to occur. Enthalpy of steam (H_2O) produced in included in 730 BTU/lb

  - Latent heat of steam = 970 BTU/lb

Therefore, latent heat of the moisture produced

$$= \frac{18.02 \text{ lb } H_2O}{74.28 \text{ lb } Ca(OH)_2} \times 970 \text{ BTU/lb } H_2O$$

= 235 BTU/lb $Ca(OH)_2$

  - Net heat desired to produce lime

= (730 – 235) BTU/lb $Ca(OH)_2$ = 495 BTU/lb

(Heat is supplied by the sugar)

2. Required mass balance
   - Total $Ca(OH)_2$ = 1000 × 0.1 = 100 lb/h
   - Ash (CaO) generated is from the $Ca(OH)_2$ heating

$$= \frac{\text{lb CaO}}{\text{lb Ca(OH)}_2} \times 100 \text{ lb of Ca(OH)}_2/\text{h}$$

$$= \frac{56.08(\text{CaO})}{74.28\left[\text{Ca(OH)}_2\right]} \times 100 \text{ lb Ca(OH)}_2/\text{h}$$

= 75.50 lb CaO/h

- Heat absorbed by ash (CaO) is :

$$= \left[495 \text{ BTU/lb Ca(OH)}_2\right]\left[\frac{74.28 \text{ lb Ca(OH)}_2}{56.08 \text{ lb CaO}}\right]$$

= 656 BTU/lb ash

- Amount of moisture within the caustic is :

$$= \frac{18.02 \text{ lb } H_2O}{74.28 \text{ lb Ca(OH)}_2} \times 100 \text{ lb Ca(OH)}_2/\text{h}$$

= 24 lb H_2O/h

3. Required total mass balance, heat balance, and flue gas emission

Table 95 : Required mass balance

Component	*Magnitude*
Wet feed (lb/h)	1000
Moisture (%)	2.4
Moisture (lb/h)	24
Dry feed (lb/h)	976
Ash (%)	7.8
Ash (lb/h)	76
Volatile (lb/h)	900
Heating value of volatiles (BTU/lb)	7077
Total heating value (MBTU/h)	6.37
Dry gas (lb/10 k BTU)	7.89
Dry gas (lb/h)	5026
Combustion H_2O (lb/10 k BTU)	0.57
Combustion H_2O (lb/h)	363
Dry gas + Combustion H_2O (lb/h)	5389
100% Air (Zero O_2 in flue gas, lb/h)	4489
Total air (excess air = 275% lb/h)	(2.75 × 4489) = 12345
Excess air (lb/h)	(12345 – 4489) = 7856
Humidity/dry gas (air) (lb/lb)	0.01
Humidity (lb/h)	(0.01 × 12345) = 123
Total water (lb/h)	510
Total dry air	12,882

Table 96 : Required heat balance

Component	*Magnitude*
Ash (lb/h)	76
Ash heat content (BTU/lb)	656
Ash heat content (MBTU/h)	0.05
Radiation loss (%)	3
Radiation heat (MBTU/h)	0.19
Humidity (lb/h)	123
Correction (@ 970 BTU/lb, MBTU/h)	- 0.12
Total losses (MBTU/h)	0.12
Input (MBTU/h)	6.37
Outlet (MBTU/h)	6.25
Dry gas (lb/h)	12,882
H_2O (lb/h)	510
Temperature (°F)	1651
Desired temperature (°F)	1800
Desired heat (MBTU/h)	6.83
Net heat (MBTU/h)	0.58
Fuel oil (air fraction)	1.1
Net heat content of fuel oil (BTU/gal)	70746
Net fuel oil (gal/h)	8.20
Air [lb/gal] for fuel oil (lb/gal)	114.64
Air necessary (lb/h)	940
Dry gas (lb/gal)	115.92
Dry gas (lb/h)	944
H_2O (lb/gal)	8.62
H_2O (lb/h)	71
Dry gas with respect to fuel oil (lb/h)	13826
H_2O with respect to oil (lb/h)	581
Air with respect to oil (lb/h)	13285
Outlet heat (MBTU/h)	7.40
Reference temperature (°F)	60
Cooling air wasted (lb/h)	
* °F	–
* BTU/lb	–
* MBTU/h	–

Table 97 : Required flue gas emission

Component	*Magnitude*
Inlet temperature(°F)	1800
Dry gas (lb/h)	13826
Heat (MBTU/h) [BTU/lb dry gas]	7.40 [535]
Adiabatic temperature (t, °F)	171
H_2O saturation (lb/lb dry gas)	0.4493
H_2O saturation (lb/h)	6212
H_2O inlet (lb/h)	581
Quench H_2O (lb/h) [gal/min]	5631 [11.3]
Outlet temperature (°F)	90
Raw water temperature (°F)	70
Sump water temperature (°F)	143
Temperature difference (°F)	73
Outlet heat (BTU/lb dry gas) Total heat (MBTU/h)	48.2120.67
Required cooling (MBTU/h)	6.73
H_2O (lb/h) [ft^3/min]	92192 [184]
Outlet (ft^3/lb dry gas) [ft^3/min]	14.547 [3352]
Fan pressure (in. water column)	20
Outlet (actual ft^3/min)	3525
Outlet (H_2O/lb dry gas)	0.03115
H_2O (lb/h)	431
Recirculation (ideal, gal/min)	11.3
Recirculation (actual, gal/min)	91
Cooling H_2O (gal/min)	184

Example 3.124

Incineration of Municipal Solid Waste (MSW)

Calculate the waste heat available for incineration of a municipal solid waste.

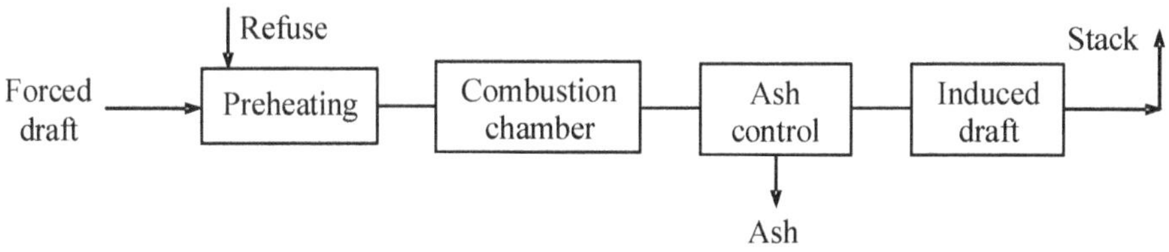

Figure 37 : Incineration of MSS.

Solution

1. Required mass balance on refuse (one pound moist refuse)

Mean reaction: $C_6H_{10}O_5 + 6O_2 = 6CO_2 + 5H_2O$ [Cellose]

Table 98 : Mass balance based on 1 lb of refuse

Input mass	*lb mass*
• Refuse	
(1) Organics	0.7
(2) Inerts (inorganics)	0.2
(3) Moisture	0.1
• Air	
1. **Grass :** Oxygen requirement for = $(0.7/162)(6 \times 32)$ = for combustion of $C_6H_{10}O_5$ (MW = 162) Assuming 200% excess oxygen, dry air requirement =	0.83 lb
[2(0.83/0.23)] =	0.722 lb
2. Moisture : 50% relative humidity at 60°F = 0.005 lb moisture/lb dry air Moisture = (7.22 lb dry air) (0.005 lb/lb dry air) =	0.036 lb
Outputness	lb mass

- Stack discharge
  1. Gas

$$CO_2 \text{ produced (combustion)} = \left(\frac{0.70}{162}\right)(6 \times 44) = 1.14 \text{ lb}$$

Dry gas discharge = Air input + CO_2 produced – O_2 utilized

= 7.22 lb + 1.14 lb – 0.83 lb

= 7.53 lb

  2. Moisture produced by reaction (combustion)

$$= \left(\frac{0.70}{162}\right)(5 \times 18) = 0.39 \text{ lb}$$

Total moisture in stack discharge :

= Refuse moisture + Air moisture + Combustion moisture

= 0.20 lb + 0.036 lb + 0.39 lb = 0.626 lb

- Ash
  1. Inerts : 0.1 lb

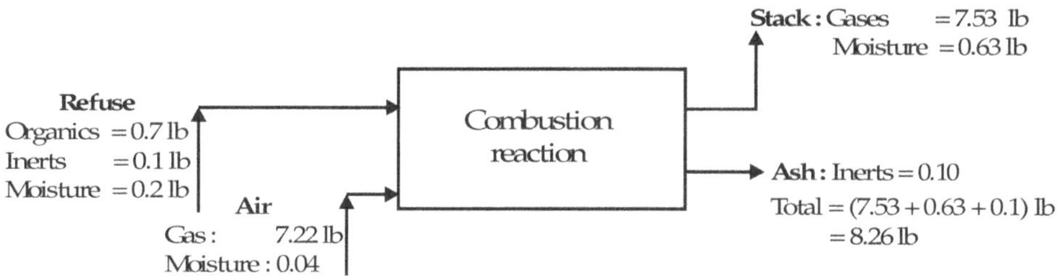

Figure 38 : Mass Balance

2. Required energy balance

A. Heat of combination

B. Heat of refuse and air from 60°F to 1300°F

1. Moisture

Sensible heat (Q_s) = 1(0.20 + 0.036)(212 – 60)

= 31 BTU

Latent heat (Q_l) = 975(0.20 + 0.036) = 230 BTU

Further sensible heat (Q_s) = 0.5(0.20 + 0.036)(1300 – 212)

= 128 BTU

Total moisture heating (Q) = Q_s(water) + Q_l(water) + Q_s(steam)

= 31 + 230 + 128 = 389 BTU

2. Gas

Q_s = 0.25(7.22)(1300 – 60) = 2238 BTU

3. Solids

Q_s = 0.25(0.1 + 0.7)(1300 – 60) = 248 BTU

C. Stack discharge

1. Moisture (60°F to stack temperature 700°F)

= Q_s (water) + Q_l (water) + Q_s (steam)

= 1(0.626)(212 – 60) + 975(0.626) + 0.5(0.626)(700 – 212)

= 95 + 610 + 153 = 858 BTU

2. Gas

Q_s = 0.25(7.53)(700 – 60) = 1205 BTU

D. Ash (ash temperature = 500°F)

Q_s = 0.2(0.1)(500 – 60) = 9 BTU

E. Refractory loss (5%)

Q_s = (5500) (0.05) = 275 BTU

Available heat = [A] – [B + C + D + E]

= 5500 – [(388 + 2238 + 248) + (858 + 1205) + (9) + (275)]

= (5500 – 5221) = 279 BTU

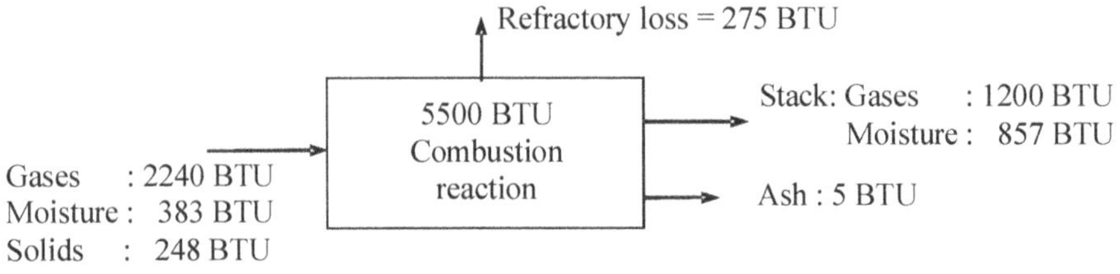

Figure 39 : Heat Balance

Note :

1. Heat of combustion

   - First method :

     h = 127R + 400 [cal/g]

     R = 0.25 (2.66% C + 7.94% H – %O)

     For any organic sources containing C, H and O; h can be expressed in BTU in BTU/lb by multiplying it with $\left(\frac{450 \times 3.97}{1000}\right)$

   - Second method

$$h = \left[\frac{\Delta O_2}{MW}\right] \times 435 \text{ kJ/g}$$

     where ΔO_2 is the number of moles of oxygen required to burn any mole of organic compound :

$$C_6H_{10}O_5 + 6O_2 = 6CO_2 + 5H_2O$$

$$\Delta O_2 = 6, \; MW = 162$$

     A solid waste (20% moisture + 10% inerts) gives heat of combustion as 5500 BTU/lb. Approximately from empirical method-1 (assuming 70% $C_6 H_{10} O_5$), the heat of combustion is 5470 BTU/lb. Approximately from empirical method-2, the heat of combustion is 4840 BTU/lb. If we assume MW to be $C_6H_{10}O_5$, method 2 will yield 6900 BTU/lb.

2. Sensible (Q_s) and latent heat (Q_l)

$$Q_s = C_p W(T_2 - T_1)$$

$$Q_l = \lambda W$$

where C_p = Specific heat

λ = Latent heat

W = Weight of material

Table 99 : Specific heat in Various Materials.

Material	*Specific Heat (BTU/lb.°F)*
Solids in refuse	0.25
Water	1.00
Steam	0.50
Air	0.25
Ash	0.20

3. Lean or rich fuel

Fuel + Oxidant + Diluent = Products

A generalised formula for the combustion hydrocarbon in air is :

$$CHy + nO_2 + n\left(\frac{79}{21}\right)N_2 \rightarrow aCO_2 + (1-a)CO + bH_2O + (y/2 - b)H_2$$
$$+ \left[n - a - \frac{(1-a)}{2} - \frac{b}{2}\right]O_2 + n\left(\frac{79}{21}\right)N_2$$

A combustible mixtures that contains excess oxidant is referred as lean. In the reverse case, where there are insufficient oxidants for complete oxidation of the available fuel, the mixture is referred as rich. Fuel to air or air to fuel ratio are the most common ways of quantitatively expressing the richness or leaneness. If n_f is the mole of fuel, M_f is the molecular weight of fuel, and n is the number of moles of oxygen per mole of fuel in the above equation, then fuel to air ratio of the mixture is either :

$$\left(\frac{F}{A}\right)_{mol} = \frac{n_f}{n\left[1 + \left(\frac{79}{21}\right)\right]}$$

or

$$\left(\frac{F}{A}\right)_{wt} = \frac{n_f M_f}{n\left[1 + \left(\frac{79}{21}\right)\right](29)}$$

Often, it is desirable to compare the richness or cleaness of combustion for different fuels. The equivalence ratio (f) is convenient for this type of comparison, and it may be defined as :

$$\phi = \frac{\left(\frac{A}{F}\right)_{actual}}{\left(\frac{A}{F}\right)_{theoretical}} = \frac{\text{Actual air to fuel ratio}}{\text{Stoichiometric air to fuel ratio}}$$

Example 3.125

Incineration : Air Requirements

Determine the amount of air required to burn completely 2 ton of wastewater treatment sludges having the following empirical chemical formula:

$$C_{50}H_{100}O_{40}N$$

Solution

1. Required amount of oxygen for the complete oxidation of organic waste
   - $CaH_bO_cN_d + \frac{[4a+b-2c-3d]}{4}O_2 = aCO_2 + \frac{[b-3d]}{2}H_2O + dNH_3$
   - $NH_3 + 2O_2 = H_2O + HNO_3$
2. Required quantity of air for the combustion (oxidation) of organic waste

   a = 50; b = 100; c = 40; d = 1 [Give by the empirical formula]

   - Chemical equations become :

$$C_{50}H_{100}O_{40}N + 54.25O_2 = 50CO_2 + 48.5H_2O + NH_3 \quad (1)$$

Molecular weights :	1354	1736	2200	873	17
Normalized weights :	1	1.28	1.62	0.65	0.0126

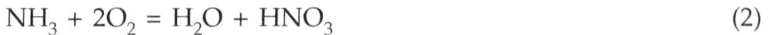

$$NH_3 + 2O_2 = H_2O + HNO_3 \quad (2)$$

Molecular weights :	17	64	18	63
Normalized weights :	1	3.76	1.06	3.70

- Oxygen requirements are :
  - Oxidation of $C_{50}H_{100}O_{40}N$ for one ton of sludge is :

    Amount of oxygen (O_2) = (1.28)(1000 kg/ton) = 1.28×10^3 kg/ton
  - Oxidation of NH_3 for one ton of sludge is :

    Amount of oxygen (O_2) = (0.0126)(3.76)(1000 kg/ton)

    = 47.38 kg/ton
  - Total oxygen required is :

    = (1.28×10^3 + 47.38) kg/ton

    = 1328 kg/ton
- Air requirements are :
  - Mass of air $= \frac{1328 \text{ kg/ton}}{0.2315}$; [Air contains 23.15% O_2 by weight]

    = 5734 kg/ton

– Volume of air $= \dfrac{5734 \text{ kg air/ton}}{1.2928 \text{ kg/m}^3}$; [Air density = 1.2928 kg/m^3]

$= 4442 \text{ m}^3/\text{ton}$

- Oxygen and air requirements for 2 tons of waste sludges are :
  – Total O_2 required = 2(1328 kg/ton) = 2656 kg
  – Total mass of air required = 2(5734 kg/ton)
  = 10468 kg
  – Total volume of air required = 2(4442 m^3/ton)
  = 8884 m^3

3. Required alternative procedure for determination of air requirements

- MW of $C_{50}H_{100}O_{40}N$ = [50 × 12] + [1 × 100] + [40 × 16] + [1 × 14]
  = 1354
- **Table 70** : Percentage distribution of elements in the waste

Element	*% by weight*
Carbon (C)	$\frac{50 \times 12}{1354} \times 100 = 44.3$
Hydrogen (H)	$\frac{100 \times 1}{1354} \times 100 = 7.4$
Oxygen (O)	$\frac{40 \times 16}{1354} \times 100 = 47.3$
Nitrogen (N)	$\frac{1 \times 14}{1354} \times 100 = 1.0$

- Net available hydrogen not bound as water

$$\text{Net available hydrogen} = \left[7.4 - \frac{47.3}{8}\right]\% = 1.49\%$$

[H_2O : 2 Hydrogen atoms are associated with one oxygen atom (therefore, number 8 is used)]

- Air required per tonne of waste sludge
  – Carbon combustion

$C + O_2 = CO_2$

Molecular weights :	12	32	44
Normalized weights :	1	2.67	3.67

Air required for 1 kg of carbon

$$= \frac{(2.67 \text{ kg/kg})}{0.2315} = [\% \text{ oxygen in air} = 23.15]$$

= 515 kg – air/ton

Total air required = (0.443)(11.53)(1000 kg/ton)

= 5108 kg/ton

– Hydrogen combustion

$2H_2 + O_2 = 2H_2O$

Molecular weights :	4	32	36
Normalized weights :	1	8	9

$$\text{Air required for 1 kg of hydrogen} = \frac{(0.0149)(1000 \text{ kg/ton})}{(0.2315)}$$

= 515 kg – air/ton

– Total air required = (5108 + 515) kg air/ton

= 5623 kg air/ton

[There is slight difference of air requirements as calculated in section 2 and 3. It is because a portion of the hydrogen combines with nitrogen to form NH_3].

Example 2.126

Incineration : Combustion Heat Balance

Estimate the heat available in the exhaust gases from the combustion of 50 tonnes/d of wastewater sludges with an energy content of 10,000 kJ/kg. The incinerator residue contain 5% carbon and the temperatures of incoming air and residue from the grate are 25 and 400°C, respectively. Assume the following composition of the sludge :

Element	C	H	O	N	S	H_2O	Inerts
% by weight	28	5	22	4	1	20	20

Solution

1. Required mass balance for various components
   - Inerts = 50(0.20) =10 tons/d

$$\text{Total residue} = \frac{10 \text{ tons/d}}{0.95} = 10.53 \text{ tons/d}$$

Residual carbon = (10.53 tons/d) (0.05)

= 0.53 tons/d

- Net available hydrogen $= \left(5 - \frac{22}{8}\right)\%$

  $= 2.25\%$

- Bound water:
  - Hydrogen in bound water = (5 – 2.25)%

    = 2.75%
  - Bound water = (22 + 2.75)% = 24.75%
- Water produced through combustion of hydrogen :

$$H_2O \text{ produced} = \left[\frac{9 \text{ kg } H_2O}{1 \text{ kg H}}\right]\left[0.025 \times 50 \text{ tons/d}\right]$$

$$= 10.13 \text{ tons/d}$$

2. Required heat balance for various components
   - Gross heat input

     = (10,000 kJ/kg)(50 × 1000 kg/d)

     = (+) 500 × 10^6 kJ/d
   - Heat lost in unburnt carbon

     = (0.53 × 1000 kg/d)(32789 kJ/kg)

     = (–) 18 × 10^6 kJ/d
   - Heat lost through inherent moisture

     = (0.2)(50 × 1000 kg/d)(2420 kJ/kg)

     = (–) 24 × 10^6 kJ/d
   - Heat lost through bound water

     = (0.2475)(50 × 1000 kg/d)(2420 kJ/kg)

     = (–) 30 × 10^6 kJ/d
   - Heat lost through oxidation of hydrogen to water

     = (10.13 × 1000 kg/d)(2420 kJ/kg)

     = (–) 24.5 × 10^6 kJ/d
   - Radiation loss

     = (–) (50 × 1000 kg/d)(0.005 kJ/kg)

     = (–) 250 kJ/d; [Negligible]
   - Sensible heat loss through inerts

     = – (10.5 × 1000 kg/d)(1.047 kJ/kg) (673 – 298)°C

     = – 4.13 × 10^6 kJ/d

Total sensible heat in hot gases is :

$$= [500 \times 10^6 \text{ kJ/d}] - [(18 + 24 + 30 + 24.5 + 4.13) \times 10^6 \text{kJ/d}]$$

$$= [500 - 100.63] \times 10^6 \text{ kJ/d} = 399.37 \times 10^6 \text{ kJ/d}.$$

Note :

1. Typical heat values of some components

   Carbon 32789 kJ/kg

   Incinerator

   - ***Unburnt carbon :*** Residue is about 4-8% carbon. Heat value is 32789 kJ/kg.
   - ***Radiation :*** Heat lost through the incinerator walls and other appurtenances to the surroundings is about 0.003 – 0.005 kJ per kg of incinerator input.

- Latent heat
  - ***Inherent moisture :*** Water content of waste. Latent heat of evaporation for is about 2420 kJ/kg.
  - Moisture in bound water ≈ 2420 kJ/kg.
  - Moisture from oxidation of net hydrogen ≈ 2420 kJ/kg.

  Sensible heat : Specific heat of residue is about 1047 kJ/kg.°K

- Stack gases.

Example 3.127

Partial Combustion of Methane Gas

Determine the amounts (weight and volume) of carbon monoxide (CO) produced during partial combustion of methane (CH_4) with oxygen and 20% excess air.

Solution

1. Required amount (weight) of carbon monoxide (CO) using oxygen
   - Partial combustion reaction of methane (CH_4) using oxygen is :

$$CH_4 + 2O_2 \rightarrow CO_2 + 2H_2O + CO \text{ (trace)}$$

Table 71 : Details of various compounds in the combustion reaction.

Molecular weight :	16.05	32.0	44.01	18.02	28.01
Weights :16.05	64.0	44.01	36.04	-	
Normalized weights :	1	3.99	2.74	2.25	-

- Stoichiometric requirements for oxygen are = 3.99 mol O_2/mol CH_4
- Assume that this reaction occurs at 2500°F, with 20% excess air supplied. The amount of CO formed per lb of stoichiometric air is 1.972×10^{-5} [experimental data]
- Amount of air required is :

  = (3.99 lb O_2)(4.3197 lb air/lb O_2)

  = 17.24 lb air

  [1 lb O_2 = 4.3197 lb dry air]
- Amount of CO produced as :

  = (17.24 lb air)(1.972×10^{-5} lb CO/lb air)

  = 0.00034 lb CO
- 1 lb of methane when combusted at 2500°F produces 0.00034 lb CO.

2. Required amount (volume) of carbon monoxide (CO)

   Perfect gas load :

   144 P = mRT

   where P = Pressure (psi)

   m = Specific weight (lb/ft³)

   T = Absolute temperature (460 + t, t in °F)

   R = Universal gas constant divided by molecular weight of gas (ft/°F)

   Universal gas constant (1545) is constant.

   The value of R is for various gases are :

Table 72 : Gas Constant for various gases

Gas	*Molecular Weight*	*Magnitude of R (ft/°F)*
Air	28.9	53.3
O_2	32.0	48.3
H_2	2.02	764.9
N_2	28.02	55.1
CO	28.01	55.2
Water vapour	18.02	85.83
SO_2	64.06	24.01

- Specific weight is :

$$W = \frac{144\ P}{RT}$$

where $W = \frac{m}{V}$

m = Weight of gas (lb)

v = Volume of gas (ft^3)

$$V = \frac{mRT}{144\ P}$$

$$V = \frac{1545\ mT}{144\ PM}; \quad \left[R = \frac{\text{Universal gas constant}}{\text{Molecular weight}} = \frac{1545}{M}\right]$$

Volume ratio of CO and CH_4 at the same temperature and pressure

$$\frac{V_{CO}}{V_{CH_4}} = \left(\frac{m_{CO}}{m_{CH_4}}\right)\left(\frac{M_{CH_4}}{M_{CO}}\right)$$

$$= \left(\frac{0.00034\ \text{Ib}}{1\ \text{Ib}}\right)\left(\frac{16.05}{28.01}\right)\left(10^6\right)$$

= 195 ppm by volume, CO to CH_4

3. Required amount of CO with 20% excess air
   - Combustion reaction of methane (CH_4) using 20% excess air is :

$$CH_4 + 2.4O_2 + 9.03N_2 \rightarrow CO_2 + 2H_2O + 0.4O_2 + 9.03N_2$$

Table 73 : Molecular and normalized weights of various gases.

Molecular weight :	16.05	32.0	28.02	44.01	18.02	32.0	28.02
Weights :	16.05	76.80	253.02	44.01	36.04	12.80	253.02
Normalized weights :	1	4.79	15.76	2.74	2.25	0.80	15.76

   - 20% excess air corresponds to (1.2 × 2) = 2.4 moles of oxygen
   - 2.4 moles of oxygen in air corresponds to (2.4 × 3.7619) = 9.03 moles of nitrogen

- Volume ratio of CO and flue gas (FG) at temperature (T) and pressure (P) is determined as :
   - Volume of flue gas (V_{FG}) at 20% excess air is :

$$V_{FG} = \frac{1545T}{144P}\left[\frac{m_{CO_2}}{M_{CO_2}} + \frac{m_{H_2O}}{M_{H_2O}} + \frac{m_{O_2}}{MO_2} + \frac{m_{N_2}}{M_{N_2}}\right]$$

– Volume ratio of V_{CO} at partial combustion and V_{FG} (at complete combustion) is :

$$\frac{V_{CO}}{V_{FG}} = \left(\frac{m_{CO}/M_{CO}}{m_{FG}/M_{FG}}\right) \quad ; \quad \left[\text{at}\frac{\text{Partial combustion}}{\text{Complete combustion}}\right]$$

$$= \frac{\dfrac{0.00034 \text{ lb}}{28.01 \text{ lb}}}{\left[\dfrac{2.74 \text{ lb}}{44.01 \text{ lb}} + \dfrac{2.25 \text{ lb}}{18.02 \text{ lb}} + \dfrac{0.80 \text{ lb}}{32.0 \text{ lb}} + \dfrac{15.76 \text{ lb}}{28.02}\right]} \times 10^6$$

= 15.7 ppm by volume CO in the exiting flue gas.

Example 3.128

Basic Combustion Calculation (with Air)

1. Determine the amounts of the following :

    Combustion products.

    Air requirements for burning of carbon and hydrogen.

    1.1 Required combustion calculation for carbon

    In most situation waste or fuel is burnt with air (N_2 and O_2 as components)

    For each mole of oxygen, 3.76 moles of N_2 $\left(\frac{79\%}{21\%}\right)$ are present in dry air. N_2 is inert gas

    Calculations for carbon oxidation are :

$$C + \underbrace{O_2 + 3.76N_2}_{\text{Dry air}} = CO_2 + 3.76N_2$$

Table 74 : Molecular and Normalized weigths of Various gases.

Molecular weight :	12.01	32.0	28.02	44.01	28.02
Weight :	12.01	32.0	105.36	44.01	105.36
Normalized weight :	1	2.66	8.77	3.66	8.77

Requirements for combustion are :

1 lb of carbon requires 2.66 lb of oxygen for combustion

1 lb of carbon requires (2.66 + 8.77) = 11.43 lb of air for combustion

Products of combustion are :

1 lb of carbon generates 3.66 lb of carbon dioxide

1 lb of carbon generates (3.66 + 8.77) = 12.43 lb of combustion products.

1.3 Required combustion calculation for hydrogen

Calculations for hydrogen combustion are :

$$2H_2 + O_2 + 3.76N_2 = 2H_2O = 3.76N_2$$

Table 75 : Molecular and Normalized weights of Various gases.

Molecular weight :	2.02	32.0	28.02	18.02	28.02
Weight :	4.04	32.0	105.36	36.04	105.36
Normalized weight :	1	7.92	26.08	8.92	26.08

Requirements for combustion are :

- 1 lb of hydrogen requires 7.9 lb of oxygen.
- 1 lb of hydrogen requires (7.9 + 26.08), 34 lb of air

Products of combustion are :

- 1 lb of hydrogen produces 8.92 lb of water.
- 1 lb of hydrogen produces (8.92 + 26.08), 35 lb of combustion products.

2. Excess air calculations for benzene (C_6H_6)

Calculate the combustion requirements if 20% excess air is used for oxidation of benzene (C_6H_6).

2.1 Required combustion calculations for benzene

$$C_6H_6 + 7.5O_2 + (7.5 \times 3.76)N_2 \rightarrow 6CO_2 + 3H_2O + (7.5 \times 3.76)N_2$$

Table 76 : Molecular and Normalized weights of Various gases.

Molecular weight :	78.12	32.0	28.02	44.01	18	28.02
Weight :	78.12	240	790.16	264.06	54.06	790.16
Normalized weight :	1	3.07	10.11	3.38	0.69	10.11

Stoichiometric requirement of oxygen for C_6H_6 is 3.07 lb O_2/lb C_6H_6 (7.5 mol O_2/mol C_6H_6).

For 20% excess air, an additional amount of oxygen (7.5 × 0.2), 1.5 mol of oxygen is necessary

Each mole of oxygen in dry air has 3.76 mol of N_2. Therefore 7.5 moles of oxygen must have 3.76 × 7.5 moles of N_2

2.2 Required revised calculations for C_6H_6 reaction (with 20% excess oxygen)

$$C_6H_6 + (7.5 + 1.5)O_2 + (9 \times 3.76)N_2 \rightarrow 6CO_2 + 3H_2O + 1.5O_2 + (9 \times 3.76)\ N_2$$

Table 77 : Molecular and Normalized weights of Various gases.

Molecular weight	78.12	32.0	28.02	44.06	18.02	32.0	28.02
Weight	78.12	288.0	948.20	264.06	54.06	48.00	948.20
Normalized weight	1	3.69	12.14	3.38	0.69	0.62	12.14

Requirements for combustion are :

1 lb of C_6H_6 requires 3.69 lb of O_2

1 lb of C_6H_6 requires (3.69 + 12.14), 15.83 lb of air.

Products of combustion are :

1 lb of C_6H_6 generates 3.38 lb of CO_2

1 lb of C_6H_6 generates (3.38 + 12.14 + 0.69), 16.21 lb of dry air (gas)

1 lb of C_6H_6 generates 0.69 lb of moisture (water).

3. Excess air calculation for cellulose ($C_6H_{10}O_5$)

Calculate the combustion requirements if 150% excess air is used for the oxidation of cellulose.

3.1 Required combustion calculations for cellulose

$$C_6H_{10}O_5 + 6O_2 + (6 \times 3.76)N_2 \rightarrow 6CO_2 + 5H_2O + 22.56N_2$$

Table 78 : Molecular and Normalized weights of Various gases.

Molecular weight	162.16	32.0	28.02	44.01	18.02	28.02
Weight	162.16	192.0	632.13	264.06	90.10	632.13
Normalized weight	1	1.19	3.90	1.63	0.56	3.90

Stoichiometric requirement of oxygen for $C_6H_{10}O_5$ is 1.19 lb O_2/lb $C_6H_{10}O_5$ (6 mol O_2/mol $C_6H_{10}O_5$).

For 150% excess air, an additional amount of oxygen (1.5 × 6), 15 mol of oxygen is necessary.

Each mole of oxygen in dry air has 3.76 mol nitrogen. Therefore, 9 mole of oxygen must have 15 × 3.76 = 56.4 mol of nitrogen.

Requirements for combustion are :

- 1 lb of $C_6H_{10}O_5$ requires 1.19 lb of oxygen.
- 1 lb of $C_6H_{10}O_5$ requires (1.19 + 3.90) = 5.09 lb of air.

Products of combustion are :

- 1 lb of $C_6H_{10}O_5$ generates 1.63 lb carbon dioxide.
- 1 lb of $C_6H_{10}O_5$ generates 0.56 lb water.
- 1 lb of $C_6H_{10}O_5$ generates (1.63 + 3.90) = 5.53 lb of dry gas.

3.2 Required revised calculation for $C_6H_{10}O_5$ reaction (with 150% excess air)

$$C_6H_{10}O_5 + (6 + 9)O_2 + (15 \times 3.76)N_2 \rightarrow 6CO_2 + 5H_2O + 9O_2 + 56.40N_2$$

Table 79 : Molecular and Normalized weights of Various gases.

Molecular weight :	162.16	32	28.02	44.01	18.02	32.0	28.02
Weight :	162.16	480.0	1580.33	264.06	90.10	288.00	1580.33
Normalized weight :	1	2.69	9.75	1.63	0.56	1.77	9.75

Finally, similar conclusion can be drawn as shown above.

Example 3.129

Incineration : Assessment of Particulate Matter in Air

Determine the amount of particulate matter generated from incinerators by burning the sewage sludges and municipal solid waste, and compare it with one day volcano erruption.

Assume the following data :

Percapita contribution for sewage sludges and emission	: 0.2 lb/Cap.d and 1.5 lb/ton
Percapita contribution for municipal solid waste (MSW)	: 3 lb/Cap.d
One day volcano particulate matter emission	: 300×10^6 tonne/d
Population	: 150×10^6
Emission discharge for MSW with 95% efficiency	: 6 lb/ton × 0.05 (= 0.3 lb/ton)

Solution

1. Required particulate matter emission rate

Sewage sludges :

$$= \frac{(0.2 \text{ lb/Cap.d})(150 \times 10^6)(365 \text{ d/yr})}{(2000 \text{ lb/ton})}$$

= 5,475,000 tons of sewage studys/yr

Assume an allowable particulate emission = 1.5 lb/dry ton

$$\text{Therefore, total emission} = \frac{(1.5 \text{ lb/dry ton})(5,475,000 \text{ d/yr})}{2000 \text{ lb/ton}}$$

= 4106 ton/yr

- Municipal solid waste :

$$= \frac{(3 \text{ lb/Cap.d})(150 \times 10^6)(365 \text{ d/yr})}{(2000 \text{ lb/ton})}$$

$$= 82125000 \text{ ton/yr}$$

$$\text{Particulate emission} = \frac{(0.3 \text{ lb/ton})(82125000 \text{ d/yr})}{(2000 \text{ lb/ton})}$$

$$= 12319 \text{ ton/yr}$$

Total particulate emission :

$$= (4106 + 12319) = 16425 \text{ ton/yr.}$$

2. Required comparison of particulate matter from sewage sludges and municipal solid waste with single day volcano erruption

$$\text{No. of years} = \frac{\text{One day volcano erruption}}{\text{Total particulate emission}}$$

$$= \frac{300 \times 10^6 \text{ tons}}{16425 \text{ ton/yr}}$$

$$= 18265 \text{ years.}$$

Note :

1. Mettalic particulate emission into the atmosphere.

   It is less than 1% of the total particulate emission.

   Important metallic components are

   - Nickel
   - Lead
   - Cadmium
   - Mercury.

Example 3.130

Incineration : Heating value for Hydrocarbons

Determine the heat value (approximate) for the combustion of hydrocarbons (C_6H_6 and C_4H_{10}).

Solution

1. Required approximate heating value of hydrocarbon (C_6H_6)

   Heating value of a hydrocarbon when the heat of formation data are not available, a figure of 184000 BTU released for every mole of stoichiometric oxygen is used.

Reaction of oxygen with benzene (combustion) is :

$$C_6H_6 + 7.5O_2 = 6CO_2 + 3H_2O$$

Molecular weight : 78.12

Heating value for benzene is :

$$= \left(\frac{7.5 \text{ mol } O_2}{\text{mol } C_6H_6}\right)\left(\frac{\text{mol } C_6H_6}{78.12 \text{ lb}}\right)\left(\frac{184000 \text{ BTU}}{\text{lb mol } O_2}\right)$$

2. Required approximate heating value of hydrocarbon (C_4H_{10})

Reaction of oxygen with isobutane (C_4H_{10}) is :

$$C_4H_{10} + 6.5O_2 = 4CO_2 + 5H_2O$$

Molecular weight: 78.14

- Heating value of isobutane (C_4H_{10}) is :

$$= \left(\frac{6.5 \text{ mol } O_2}{\text{mol } C_4H_{10}}\right)\left(\frac{\text{mol } C_4H_{10}}{78.14 \text{ lb}}\right)\left(\frac{184000 \text{ Btu}}{\text{lb mol } O_2}\right)$$

$$= 15306 \text{ BTU}$$

Note :

1. Precaution while using approximate heating value.

The method of heating value approximation is valid for hydrocarbons.

The presence of oxygen, nitrogen, halogens, or other compounds might result in erroneous values by use of this method.

Example 3.131

Incinerator : Particulate and HCl Emissions

Determine the total amount of HCl produced, as also the HCl removal, lead removal, and concentration of particulate in the incinerator flue gas. A hazardous waste incinerator is burring a waste containing solids with 50% excess air at 2100°F with a residence time of 2.5 seconds. Assume the following data :

Stack gas flow rate : 14280 dscfm

Composition of contaminant in the stack gas :

Table 80 : Composition of stake emissions.

Constituent	*Inlet (lb/h)*	*Outlet (lb/h)*
Toluene	860	0.20
Chlorobenzene	450	0.02
Dichlorobenzene	300	0.03
HCl	–	4.2
Particulate	–	9.65
Lead (Pb)	-	0.65 lb/100 lb TOC

Solution

1. Required destruction and reduction efficiency (DRE) for various components [Principal Organic Hazardous constituents (POHC)]

   - $$\text{DRE (Toluene)} = \left(\frac{860 - 0.20}{860}\right) 100 = 99.98\%$$
   - $$\text{DRE (Chlorobenzene)} = \left(\frac{450 - 0.02}{450}\right) \times 100 = 99.996\%$$
   - $$\text{DRE (Dichlorobenzene)} = \left(\frac{300 - 0.03}{300}\right) \times 100 = 99.99\%$$

2. Required amount of HCl produced in the incinerator

   Chlorobenzene

   $C_6H_5Cl + 7O_2 = 6CO_2 + HCl + 2H_2O$

   $$\text{HCI produced} = \frac{450 \text{ lb/h}}{(6 \times 12 + 5 \times 1 + 1 \times 35.45) \text{lb/mole}}$$

   $$= 4.00 \text{ mol/h}$$

   - Dichlorobenzene

     $C_6H_4Cl_2 + 6.5O_2 = 6CO_2 + 2HCl + H_2O$

     $$\text{HCI produced} = \frac{300 \text{ lb/h}}{(6 \times 12 + 4 \times 1 + 2 \times 35.45) \text{lb/mole}}$$

   Total HCl produced = (4.00 + 2.04) mol/h

   = 6.04 (36.45 lb/mol)

   = 220.2 lb HCl/h

3. Required amount of lead and particulate matter in the flue gas

   Toluene (C_7H_8)

   $$\% \text{ TOC in toluence} = \frac{7 \times 12}{7 \times 12 + 1 \times 8} = 0.913, \text{ [in terms of Carbon]}$$

   Toluene–TOC emision rate = 0.913(860 lb/h toluene)

   = 785 lb TOC/h

   - Chlorobenzene (C_6H_5Cl)

   $$\% \text{ TOC in Chlorobenzene} = \frac{6 \times 12}{6 \times 12 + 1 \times 5 + 1 \times 35.45} = 0.64,$$

   [in terms of Carbon]

   Chlorobenzene TOC emission rate = 0.64 (450 lb/h)

   = 288 lb TOC/h

Dichlorobenzene ($C_6H_4Cl_2$)

$$\% \text{ TOC in dichlorobenzene} = \frac{6 \times 12}{6 \times 12 + 1 \times 4 + 2 \times 35.45} = 0.49$$

[in terms of carbon]

Dichlorobenzene TOC emission rate = 0.49(300 lb/h)

= 147 lb TOC/h

- Total TOC emission rate in the flue gas

  = (785 + 288 + 147) lb TOC/H

  = 1220 lb TOC/h

  Total lead (Pb) emission rate :

  = (1220 lb TOC/h) (0.005 lb-Pb/lb – TOC)

  = 6.1 lb – Pb/h

- Lead concentration in the flue gas :

  $$= \frac{6.1 \text{ lb} - \text{Pb/h}(1 \text{ h}/60 \text{ min})}{14280 \text{ desfm}}$$

  = 7.12 × 10^{-6} lb/dscf

- Total Particulate concentration

  Total particulate emission = 9.65 lb/h (7000 gr/lb)

  = 67500 gr/h

  $$\text{Particulate concentration} = \frac{67500 \text{ gr/h}}{14280 \text{ dscfm}(60 \text{ min/h})}$$

  = 0.079 gr/dscfm

Example 3.132

Incineration Weight and Volume Percentages of the Products in the Flue Gas

Determine the weight and volume percentages of the products in the flue during the combustion of cellulose ($C_6H_{10}O_5$) at 100% excess air and 150% excess air.

Solution

1. Required weight and volume percentages of the products of combustion at 100% excess air (oxygen content in the flue gas is zero).

   Combustion reaction with 100% air (zero oxygen content in the products) :

$$C_6H_{10}O_5 + 6O_2 + (6 \times 3.76)N_2 = 6CO_2 + 5H_2O + 22.56\ N_2$$

Figure 81 : Molecular and normalized weights of various components.

Molecular weight :	162.15	32.0	28.02	44.01	18.02	28.02
Weights :	162.15	192.0	632.13	264	90.10	632.13
Normalized weights :	1	1.19	3.90	1.63	0.56	3.90

[lb moes of nitrogne in air in comparison to lb moles of oxygen in air:

$$= \frac{79\%\ \text{volume}}{21\%\ \text{volume}} = 3.7619\ \text{lb mol}\ N_2/\text{lb mol}\ O_2]$$

- Determination of weight percentages of products in the flue gas :
  - Total weight of the flue gas = (1.63 + 0.56 + 3.90) lb/lb cellulose
    = 6.09

$$\%\ CO_2\left(\text{total weight}\right) = \frac{1.63}{6.09} \times 100 = 26.77\%$$

$$\%\ H_2O\left(\text{total weight}\right) = \frac{0.56}{6.09} \times 100 = 9.19\%$$

$$\frac{\%\ N_2\ \left(\text{total weight}\right)}{\text{Total}} = \frac{3.90}{6.09} \times 100 = \frac{64.04\%}{100.00\%}$$

  - Total dry weight of the flue gas = (1.63 +3.90) lb/lb cellulose
    = 5.53 lb/lb cellulose

$$\%C_2O\left(\text{dry weight}\right) = \frac{1.63}{5.53} \times 100 = 29.48\%$$

$$\frac{\%\ N_2\ \left(\text{dry weight}\right)}{\text{Total}} = \frac{3.90}{5.53} \times 100 = \frac{70.52\%}{100.00\%}$$

- Determination of volume percentages of products in the flue gas
  - Total moles of flue gas = (6 + 5 + 22.56) mole/mole of cellulose
    = 33.56 moles [combustion products]

$$\%CO_2\left(\text{total volume}\right) = \frac{6}{33.56} \times 100 = 17.88\%$$

$$\%H_2O\left(\text{total volume}\right) = \frac{5}{33.56} \times 100 = 14.90\%$$

$$\frac{\%\ N_2O\left(\text{total volume}\right)}{\text{Total}} = \frac{22.56}{33.56} \times 100 = \frac{67.22\%}{100.00\%}$$

  - Total dry moles of flue gas = (6 + 22.56) mol/moles of cellulose
    = 28.56 moles

$$\% \, CO_2 \, (\text{dry volume}) = \frac{6}{28.56} \times 100 = 21.01\%$$

$$\frac{\% \, N_2 \, (\text{dry volume})}{\text{Total}} = \frac{22.56}{28.56} \times 100 = \frac{78.99\%}{100.00\%}$$

2. Required weight and volume percentages of products of combustion at 150% excess air (total air = 250%)

Combustion reaction with 150% excess air :

$$C_6H_{10}O_5 + (6 + 9)O_2 + (15 \times 3.76)N_2 = 6CO_2 + 5H_2O + 9O_2 + 56.4N_2$$

Figure 82 : Molecular and normalized weights of various components.

Molecular weight :	162.16	32.00	28.02	44.01	18.02	32.00	28.02
Weight :	162.16	480.00	150.33	264.06	90.10	288.00	1580.33
Normalizedweights :	1	2.69	9.75	1.63	0.56	1.77	9.75

- 6 moles of oxygen defines the stoichiometric requirements.
- 150% excess air, an additional 6 × 1.5 = 9 moles of oxygen must be added, and the corresponding moles of nitrogen will be = (6 + 9) 3.76 = 56.4 moles of nitrogen (inert gas).
- Determination of weight percentages of products in the flue gas :
- Total weight of flue gas = (1.63 + 0.56 + 1.77 + 9.75)

  = 13.71 lb/lb of cellulose

$$\% \, CO_2 \, (\text{total weight}) = \frac{1.63}{13.71} \times 100 = 11.90\%$$

$$\% \, H_2O \, (\text{total weight}) = \frac{0.56}{13.71} \times 100 = 4.08\%$$

$$\% \, O_2 \, (\text{total weight}) = \frac{1.77}{13.71} \times 100 = 12.91\%$$

$$\frac{\% \, N_2 \, (\text{total weight})}{\text{Total}} = \frac{9.75}{13.71} \times 100 = \frac{71.11\%}{100.00\%}$$

Total weight of dry flue gas = (1.63 + 1.77 + 9.75) lb/lb of cellulose

= 13.15 lb

$$\% \, CO_2 \, (\text{dry weight}) = \frac{1.63}{13.15} \times 100 = 12.40\%$$

$$\% \, O_2 \, (\text{dry weight}) = \frac{1.77}{13.15} \times 100 = 13.46\%$$

$$\frac{\% \, N_2 \, (\text{dry weight})}{\text{Total}} = \frac{9.75}{13.15} \times 100 = \frac{74.14\%}{100.00\%}$$

Determination of volume percentages of products in the flue gas

– Total volume of flue gas = (6 + 5 + 9 + 56.40) moles/moles of cellulose

= 76.40 mole

$$\% \; CO_2 = \frac{6}{76.40} \times 100 = 7.86\%$$

$$\% \; H_2O = \frac{5}{76.40} \times 100 = 6.54\%$$

$$\% \; O_2 = \frac{9}{76.40} \times 100 = 11.78\%$$

$$\frac{\% \; N_2}{\text{Total}} = \frac{56.40}{76.40} \times 100 = \frac{73.82\%}{100.00\%}$$

– Total dry volume of flue gas = (6 + 9 + 56.40) moles/mole of cellulose

= 71.40 mole

$$\% \; CO_2 = \frac{6}{71.40} \times 100 = 8.40\%$$

$$\% \; O_2 = \frac{9}{71.40} \times 100 = 12.60\%$$

$$\frac{\% \; N_2}{\text{Total}} = \frac{56.40}{71.40} \times 100 = \frac{80.00\%}{100.00\%}$$

Note :

1. Excess air for combustion

   Oxygen fraction will increase.

   Decrease in carbon dioxide level with an increase in excess air is that a fixed amount of carbon produces a fixed amount of carbon dioxide regardless of the amount of excess air introduced.

   As the total gas flow increases, and the excess air is increase, the proportion of carbon dioxide will decrease [amount of carbon dioxide is fixed in an increased volume of flue gas].

2. Products of combustion

   Hydrogen converts to water vapour.

   Chloride and fluoride convert to HCl and HF, respectively.

   Carbon converts to carbon dioxide.

   Sulphur converts to sulphur dioxide.

Alkali metals convert to hydroxides.

$[2Na + O_2 + H_2 = 2NaOH]$

Now alkali metals convert to oxides $[2Ca + O_2 = 2CaO]$

Nitrogen from the waste, fuel, or air will take the form of a diatomic molecule (N_2).

3. Smoke

White smoke

- Non-black (white, opaque) smoke is produced due to insufficient furnace temperature when carbonaceous materials are burnt.
- Hydrocarbons will be heated to a level where evaporation and (or) cracking will occur within the furnace. The temperatures will not be high enough to produce complete combustion of these hydro.
- At 300 to 500°F stack temperature, many of the hydrocarbons condense to liquid particulate, and with the solid hydrocarbons these will appear as non-black smoke.
- Increase of furnace or stack temperature, and increased turbulence will ensure uniformity of this higher temperature within the off-gas flow. Excessive airflow would provide excessive cooling.
- Inorganics in the exit gas may also produce non-black smoke [S and S –compounds will appear yellow in a discharge, Ca and Si oxides will appear light to dark brown].

Black smoke

- Combustion of hydrocarbons in oxygen deficient atmosphere may produce carbon particle.
- Poor atomization, inadequate turbulence, and poor air distribution within the incinerator result in production of carbon (black) particles.
- Steam injection will eliminate black smoke formation :

$$3C + 2H_2O = CH_4 + 2CO$$

$$CH_4 + 2O_2 = CO_2 + H_2O$$

$$2CO + O_2 = 2CO_2$$

Example 3.133

Incineration : Effects of Excess Air on Temperature and Composition of Flue Gas

Estimate the composition and temperature of the flue gas from burning of mixed sludges (industrial and domestic), as also the effects of excess air combustion of flue gas temperature. Assume the following data :

Table 83 : Composition of various compounds.

Component(s)	*C*	*H*	*O*	*N*	*S*	*Ash*	*Total*
Composition (lb)	27.39	3.62	22.97	0.54	0.1	3.48	58.1

The moisture content of the sludge is 21.4 lb. Energy lost through the incinerator system is 20% of the total available energy per lb of sludge incinerated.

Solution

1. Required energy content of sludges

   Dulong formula is :

$$\text{Btu/Ib} = 145\text{C} + 610\left[\text{H} - \frac{1}{8}\text{O}\right] + 40\text{S} + 10\text{N}$$

   where C, H, O, S, and N are percent by weight of C, H, O, S and N, respectively.

2. Required chemical composition of the sludge

   Percent distribution of elements without and with water is :

Table 84 : Composition of various compounds.

Component	*C*	*H*	*O*	*N*	*S*	*Ash*	*Total*
Without water (lb)	27.39	3.62	22.97	0.54	0.10	3.48	58.10
With water (lb)	27.39	6.00^{+}	41.99^{++}	0.54	0.10	3.48	79.50

$(79.50 - 58.10) = 21.4$ lb H_2O

$$^{+}\text{H} = \frac{2}{18}(21.4) = 2.38 \text{ lb}$$

Total O = 3.62 + 2.38 lb

$$^{++}\text{O} = \frac{16}{18}(21.4) = 19.02 \text{ lb}$$

Total O = 22.97 + 19.02 = 41.99 Ib

- Molar composition of the elements without ash is :

Table 85 : Composition of various compounds.

Component	*C*	*H*	*O*	*N*	*S*
Without water	2.280	3.584	1.436	0.038	0.003
With water	2.280	5.940	2.624	0.038	0.003
Atomic Weight (lb/mole)	12.01	1.01	16.00	14.01	32.07

Table 86 : Sludge composition

Component	*Mole ratio*			
	Nitrogen = 1		*Sulphur = 1*	
	Without water	*With water*	*Without water*	*With water*
C	60.0	60.0	760.0	760.0
H	94.3	156.3	1194.7	1980.0
O	37.8	69.1	478.7	874.7
N	1.0	1.0	12.7	12.7
S	0.1	0.1	1.0	1.0

The empirical formula without inerts for the sludge is :

- Without sulphur

  $C_{60}H_{94}O_{38}N$ (without water)

  $C_{60}H_{156}O_{69}N$ (with water)

- With sulphur

  $C_{760}H_{195}O_{478}N_{13}S$ (without water)

  $C_{760}H_{1980}O_{875}N_{13}S$ (with water)

3. Required energy content of the sludge

   Empirical formula for the sludge is :

   $C_{760}H_{1980}O_{875}N_{13}S$

   Percent distribution of the elements is :

Table 87 : Composition of various compounds.

Component	*No. of Atoms/Mol*	*Atomic Weight*	*Weight Distribution*	*Percent*
C	760	12	9120	36.03
H	1980	1	1980	7.82
O	875	16	14000	55.36
N	13	14	182	0.72
S	1	32	32	0.13
		Total	25314	100.00

Energy content of the sludge is :

Using Dulong's formula,

$$\text{BTU/lb} = 145C + 610\left[H - \frac{1}{8}O\right] + 40S + 10N$$

$$\text{BTU/lb} = 145(36) + 610\left[7.8 - \frac{55.3}{8}\right] + 40(0.1) + 10(0.7)$$

$$= 5772.$$

Table 88 : Required stoichiometric molar amounts of oxygen per 100 lb of sludge

Components	*Weight (%)*	*Atomic Weight*	*Atomic Units*	*Moles of O_2 required**
C	27.4	12.0	2.283	2.283
H	3.6	1.0	3.600	0.900
O	23.0	16.0	1.438	0.719+
N	0.5	14.0	0.036	-
S	0.1	32.0	0.003	0.003
H_2O	21.4	18.0	1.189	-
Inerts	24	-	-	-
Total	100			2.469

*$C + O_2 = CO_2$

$2H_2 + O_2 = H_2O$

$S + O_2 = SO_2$

+Oxygen in the sludge $= \frac{1}{2}[1.438] = 0.719$ moles

$$\text{Moles of air} = \frac{2.467}{0.2069} = 11.92 \text{ mole/100 lb sludge}$$

$$\text{lb of air} = \frac{11.92(29.7)}{100} = 3.42 \text{ lb air/lb sludge}$$

[Air (ideal gas) composition: CO_2 = 0.0003; N_2 = 0.7802, O_2 = 0.2069; H_2O = 0.0126; Volume fraction = Mole fraction; MW of air = 29.7 lb/lb-mol].

Table 89 : Required flue gas composition from 100 lb sludge based on stoichiometric combustion

Combustion Product	*Moles of Flue Gas*			
	Combustion	*Air*	*Total*	*Percent*
CO_2	2.283	0.004*	2.287	15.3
H_2O	2.00 + 1.189**	0.150	3.339	22.3
O_2	–	-	-	-
N_2	0.018***	9.30	9.318	62.3
SO_2	0.003	-	0.003	< 0.1
		Total	14.947	99.9

- Moles of air/mol of flue gas $= \frac{11.92}{14.95} = 0.80$

* : 11.92(0.0003) = 0.004

** : Moles of moisture in the original sample

*** : Moles of $N_2 = \frac{1}{2}(0.036) = 0.018$.

Assume 100 moles of flue gas from stoichiometric combustion

Table 90 : Required flue gas composition with excess air [0, 50 and 100%]

Percent Excess Air	*Mole of Excess Air*+	*Total Moles of Gas*	*Flue Gas Components*				
			CO_2	O_2	N_2	H_2O	SO_2
0	0.0	100.0	15.3	0.0	62.3	22.3	< 0.1
50	40.0++	140.0	10.9+++	5.9[a]	66.8[b]	16.3[c]	< 0.1
100	80.00	180.0	8.5	9.2	69.3	12.9	< 0.1

+ : Moles of excess air = % excess air [Moles of air/mol of flue gas]

++ : (50% excess air)(0.80) = 40 moles of excess air

+++ : $\frac{[15.3 + 40(0.0003)]}{140} \times 100 = 10.9\% \; CO_2$

a : $\frac{[0 + 40(0.2069)]}{140} \times 100 = 5.9\% \; O_2$

b : $\frac{[62.3 + 40(0.7802)]}{140} \times 100 = 66.8\% \; N_2$

c : $\frac{[22.3 + 40(0.0126)]}{140} \times 100 = 16.3\% \; H_2O$

[70% relative humidity at 60°F; H_2O content = 0.0126 mole fraction of total air].

7. Required enthalpy of flue gas at 1000°F, 1500°F, 2000°F and 2500°F

$$\frac{\text{BTU in product gas}}{\text{Ib of sludge}} = \left[\left(\frac{\text{mole of flue gas}}{\text{Ib of sludge}}\right)\right]\left(\frac{\text{total moles of gas}}{\text{moles of flue gas}}\right) \times \left[\Sigma(\text{mole fraction of gas component}) \times \left(\frac{\text{BTU}}{\text{mole of gas component}}\right)\right]$$

where Moles of flue gas at stoichiometric conditions as given in Section 4 and total moles of gas includes excess air as given in Section 5.

Enthalpy of various gas components at various temperature :

Table 91

Temperature (°F)	*BTU/lb mol at Standard Conditions*+			
	CO_2	O_2	N_2	H_2O
1000	10048	6974	6720	26925
1500	16214	11008	10556	81743
2000	22719	15191	14520	36903
2500	29539	19517	18609	42405

+ Gases except H_2O at 77 °F and 1 atm.

Enthalpy calculation at 1000°F and 50% excess is :

$$\frac{\text{Btu}-\text{flue gas}}{\text{lb}-\text{sludge}}=\left[0.1495\frac{\text{mol}-\text{flue gas}}{\text{lb}-\text{sludge}}\right]\left[\frac{140}{100}\right]$$

$$\times[(0.109)(10048\ \text{BTU/mol})+(0.059)(6974\ \text{BTU/mol})]$$

$$+[(0.668)(6720\ \text{BTU/mol})]+[(0.163)(26{,}925\ \text{BTU/mol})]$$

$$=2173\ \text{BTU/lb sludge}$$

Table 92

	BTU in flue gas/lb sludge	
Temperature (°F)	*50% Excess air*	*100% Excess air*
1000	2173	2590
1500	3064	3714
2000	3995	4884
2500	4964	6101

Temperature of flue gas at 50 and 100% excess air is :

Assume that 20% of the energy is lost through the incinerator

Energy in the flue gas = (5772 BTU/lb)(0.8) = 4617.6 BTU/lb sludge

– With 50% excess air, the flue gas temperature is

Applying unitary method,

Table 93 : Energy at various temperature.

Temperature (°F)	*BTU/lb*
2000	3995
x	4618
2500	4964

$$\frac{2000-x}{2000-2500}=\frac{3995-4618}{3995-4964}=\frac{623}{969}$$

$$2000-x=-500\left(\frac{623}{969}\right)$$

$$x=2000+500\left(\frac{623}{969}\right)$$

Temperature = 2321°F

– With 100% excess air, the flue gas temperature is :

Applying unitary method,

Table 94 : Energy at various temperatures

Temperature (°F)	*BTU/lb*
1500	3714
x	4618
2000	4884

$$\frac{x-1500}{2000-1500}=\frac{4618-3714}{4884-3714}=\frac{904}{1170}=0.77$$

$$x = 1500 + 500(0.77)$$

$$\text{Temperature} = 1886°\text{F}.$$

Example 3.134

Incinerator for Industrial Waste Sludges

Determine the following for the rotating kiln :

Required volume of the incinerator

Size of the incinerator

Thickness of the castable insulation material

Heat lost through the external surface of the furnace.

Assume the following data:

Waste sludge (dry)	:	2500 lb/h
Heat of combustion (dry)	:	5200 BTU/lb
Internal furnace temperature	:	1600°F
Ambient temperature	:	100°F
Surface temperature of the kiln	:	200°F
Kiln heat release	:	25000 BTU/ft^3.h

Solution

1. Required capacity of the furnace

Total heat released is :

$$(5200\ \text{BTU/lb})(2500\ \text{lb/h}) = 13 \times 10^6\ \text{BTU/h}$$

Volume (V) of the furnace is

$$V=\frac{\text{Total heat released}}{\text{Kiln heat release}}$$

$$=\frac{13\times10^6\ \text{BTU/h}}{25000\ \text{BTU}/\text{ft}^3.\text{h}}=520\ \text{ft}^3$$

Check kiln heat release

$$= \frac{13 \times 10^6 \text{ BTU/h}}{520 \text{ ft}^3}$$

$$= 25000 \text{ BTU/ft}^3\text{.h}$$

2. Required dimensions of the furnace

Assume length (L) to diameter (D) ratio of 4:1

$$L = 4D$$

$$V = \frac{\pi D^2 L}{4} = \frac{\pi D^2 (4D)}{4} = \pi D^3$$

$$D = \left(\frac{V}{\pi}\right)^{0.333} = \left(\frac{520}{3.14}\right)^{0.333} = 5.491 \text{ ft}$$

$$L = 4D = 4(5.491) = 22 \text{ ft}$$

Inside dimensions of the furnace are

$$D = 5.5 \text{ ft}$$

$$L = 22 \text{ ft}$$

3. Required thickness of insulating materials

Inside refractory

- Kiln is rotating continuously
- Convection heat transfer can be approximated as a vertical wall
- Thickness (X) of the insulating material = RK

  General heat conduction expression is :

$$\frac{Q}{A} = \frac{K(T_i - T_o)}{X}; \qquad R = \frac{X}{K}$$

$$\frac{Q}{A} = \frac{(T_i - T_o)}{R}$$

- If the insulating wall is composed of two or more materials, then the expression becomes :

$$q = \frac{Q}{A} = \frac{K_1 (T_i - T_1)}{X_1} = \frac{K_2}{X_2}(T_1 - T_o)$$

or

$$q = \frac{Q}{A} = \frac{1}{R_1}(T_i - T_1) = \frac{1}{R_2}(T_1 - T_o)$$

[No heat is accumulating at the interface, steady state condition prevails]

$$\text{Solving for } T_1 = T_i - \frac{QX_1}{K_1} = T_o + q\frac{X_2}{K_2}$$

or $$q = \frac{(T_i - T_o)}{\left[\frac{X_1}{K_I} + \frac{X_2}{K_2}\right]}$$

Total resistance (R) = $R_1 + R_2 = \frac{X_1}{K_1} + \frac{X_2}{K_2}$

Drop in temperature across the wall with a resistance (R_x) is :

$$T = \frac{R_X}{R}(T_i - T_o)$$

where Q = Total heat

q = Heat flux

T_i = Inside furnace temperature

T_o = Outside furnace temperature

T_1 = Interface temperature at the junction of two insulating materials

K = Conductivity of insulating material

X = Thickness of insulating material

Assume the resistance (R) of 5.3 $ft^2.^oF.h/BTU$ and K = 0.377 BTU in./$ft^2.^oF.h$ the thickness of the insulating material is :

X = RK

= (5.3 $ft^2.^oF.h/BTU$)(4.53 BTU in./$ft^2.^oF/h$)

= 24 in. (= 2 ft)

[It is not possible (practical) to have an insulation thickness of 2 ft. Therefore, composite material must be used]

- Resistance of composite insulating material is expressed as :

Assume : X_1 = 6 in. (five brick)

K_1 = 10 BTU. in./$ft^2.^oF.h$

K_2 = 1.3 BTU.in./ft^2 .$^oF.h$ (castable insulating material)

Thickness of castable insulating material is :

$$X_2 = \left[R - \frac{X_1}{R_1}\right]K_2; \quad [\text{If } R = 5.3]$$

$$= \left[5.3 - \frac{6}{10}\right]1.3$$

$$= 6 \text{ in.}$$

The kiln will be lined with 6 in. of insulating castable protected from the gas stream by 6 in of refractory brick.

Heat transfer coefficient (h) using having equation for vertical plates is

$(h) = 0.25(T_o - T_a)^{0.25}$

T_a = Ambient temperature (°F)

$(h) = 0.25(T_o - T_a)^{0.25}$

$h = 0.25(200 - 100)^{0.25} = 0.79 \text{ BTU/ft}^2 \cdot {}^\circ\text{F.h}$

where T_o = Interface temperature (°F)

T_a = Ambient temperature (°F)

Kiln's external surface area calculated as :

Diameter of kiln = 5.5 ft + 2 (6 in.+ 6 in.)

= 5.5 ft + 2 ft = 7.5 ft

Length of kiln = 22 ft + [6 + 6] in. (wall thickness)

= 23 ft

External surface area of the kiln = π DL

= 3.14(7.5)(23)

= 541 ft^2

Heat lost through the external surface (convective heat) is calculated as :

Assume : Interface temperature (T_o) = 200°F

Ambient temperature (T_a) = 100°F

$q = h(T_o - T_a)$

$= [0.79 \text{ BTU/ft}^2\text{.h}][200 - 100]$

$= 79 \text{ BTU/ft}^2\text{.h}$

where q = Heat flux

T_o = Interface temperature (°F)

T_a = Ambient temperature (°F)

Total heat lost = q(A)

$= (79 \text{ BTU/ft}^2\text{.h})(541 \text{ ft}^2)$

$$= 42739 \text{ BTU/h}$$

4. Required residence time (t) in a rotating kiln

$$t = \frac{2.28(L/D)}{S(N)}$$

where t = Residence time (min)

L/D = Internal length to diameter ratio

Slope (S) = Kiln rake slope (in./ft of length)

N = Rotational speed (rpm)

Assume : S = 0.12 in./ft of length

N = 0.75 rpm

$$t = \frac{2.28(22/5.5)}{0.12(0.75)}$$

$$= 101 \text{ min}$$

Doubling the rpm (N) would have the retention time; and halving the rake (S) would double the retention time.

The above calculations are for the retention times of solids or other materials within the kiln, and not the kiln exhaust gases.

The off gas residence time can be determined by the application of the heat balance and flue gas analyses.

Note :

1. Nature of heat transfer

Conduction (Q)

– Single wall

$$Q = \frac{KA(T_i - T_o)}{X}$$

$$q = \frac{Q}{A} = \frac{K}{X}(T_i - T_o)$$

– Composite wall

$$q = \frac{K_1}{X_1}(T_i - T_1) = \frac{K_2}{X_2}(T_1 - T_o)$$

$$q = \frac{T_i - T_o}{X_1/K_1 + X_2/K_2}$$

$$q = \frac{1}{R}(T_i - T_o)$$

$$R = \frac{X_1}{K_1} + \frac{X_2}{K_2}$$

- Convection (Q_c)

  $Q_c = h\,A(T_o - T_a)$

  $q_c = h\,(T_o - T_a)$

  $h = 0.25(T_o - T_a)^{0.25}$ [Vertical plates]

  $h = 0.38(T_o - T_a)^{0.25}$ [Plates facing up]

  $h = 0.20(T_o - T_a)^{0.25}$ [Plates facing down]

  $h = 1.0 + 0.225V$ [N = wind velocity, (ft/s)]

- Radiation (Q_R)

  $Q_R = q_R\,A$

$$q_R = 0.17\varepsilon\left[\left(\frac{T_o - 460}{100}\right)^4 - \left(\frac{T_a - 460}{100}\right)^4\right]$$

- Combined heat transfer
  - Heat balance

    Total heat transfer calculated

    = [Heat transfer by radiation] + [Heat transferred by connection]

    $q = q_R + q_c$

$$\frac{1}{R}(T_i - T_o) = 0.174\,\varepsilon\left[\left(\frac{T_o - 460}{100}\right)^4 - \left(\frac{T_a - 460}{100}\right)^4\right] + h(T_o - T_a)$$

where T_i = Temperature of hot surface (°F)

T_o = Temperature of cooler surface (°F)

= Conductivity (BTU.in./ft^2.°F.h)

R = Resistance

Q = Heat flow (BTU/h)

q = Heat flux (BTU/ft^2.h)

$$R = \frac{X}{K} = ft^2.°F.h/Btu$$

h = Heat transfer coefficient (BTU/ft^2.°F.h)

T_a = Ambient temperature (°F)

$\in$ = Emissivity factor

– Typical values of emissivity factor

Table 95 : Emissivity factor ($\in$) for various materials.

Material	*Emissivity factor* ($\in$)
Polished aluminium	0.10
Aluminium paint	0.50
Lohite pafer	0.70
Brick	0.85
Steel surface oxidized	0.90

Example 3.135

Single-stage Multiple Hearth Incinerator for Excess Sludge Destruction

Determine the material balance, the energy balance, the treatment of exhaust, the specifications for the excess sludge incinerator, and the exhaust characteristics for a single stage multiple hearth incinerator, for combustion of excess sludge generated from a biological treatment system. Assume the following data :

Table 96 : Data for Sludge and Aqueous Waste.

Parameters	*Excess Waste Sludge*	*Aqueous Waste*
Excess sludge (lb/hr)	800	20
Elemental analysis		
- Carbon (%)	5	Traces
- Oxygen (%)	3	Traces
- Hydrogen (%)	1	Traces
- Nitrogen (%)	1	Traces
- Water (%)	89	100
- Ash (%)	1	Traces
Heat of combustion (BTUs/lb)	(1800)	(-900)
Specific heat (BTU/lb.°F)	1	1
Specific gravity (mg/L)	0.8	0.9

Supplementary fuel is used as a standing (sludge is self sustaining)

Combustion temperature : 1700°F

Combustion pressure : 14.7 psi

Excess oxygen : 25%

Solution

1. Material balance

1.1 Influent mass flows through sludge

Carbon flow (lb/hr) = (800 lb/hr)(0.05) = 40 lb – C/hr

Hydrogen flow (lb/hr) = (800 lb/hr)(0.01) = 8 lb – H/hr

Oxygen flow (lb/hr) = (800 lb/hr)(0.03) = 24 lb – O/hr

Nitrogen flow (lb/hr) = (800 lb/hr)(0.01) = 8 lb – N/hr

Water flow (lb/hr) = (800 lb/hr)(0.89) = 712 lb – H_2O/hr

Ash flow (lb/hr) = (800 lb/hr)(0.01) = 8 lb – ash/hr

1.2 Material balance for combustion products

Carbon balance

$$C + O_2 = CO_2$$

1 mole of C products 1 mole of CO_2

$$\text{Moles of C} = \left(\frac{40}{12}\right) = 3.33 \text{ lb mole C/hr}$$

Weight of CO_2 produced = (3.33 lb mole – C/hr)(44)

= 146.7 lb CO_2/hr

Weight of O_2 required = (3.33 lb – mole – C/hr)(32)

=106 lb O_2/hr

Hydrogen balance

$$H_2 + \frac{1}{2}O_2 = H_2O$$

$$\text{Weight of water produced} = \left[\frac{8 \text{ lb} - \text{H/hr}}{2}\right][18]$$

= 72 lb H_2O/hr

Total weight of water = [Sludge water + Aqueous wastewater + water produced through combustion]

= (712 + 20 + 72) lb/hr = 804 lb H_2O/hr

Weight of oxygen required for H_2O production

$$= \left(\frac{8 \text{ Ib/hr}}{2}\right)\left(\frac{32}{2}\right) = 64 \text{ lb } O_2/\text{hr}$$

Required oxygen

- O_2 for CO_2 = 106 lb O_2/hr
- O_2 for H_2O = 64 lb O_2/hr
- Theoretical oxygen required = 170 lb O_2/hr

- Total oxygen required = (170 lb O_2/hr + 0.25 × 170 lb O_2/hr)
  = 212.5 lb O_2/hr
- Oxygen in the feed = 24 lb O_2/hr
- Deficit oxygen = (212.5 – 24) lb O_2/hr
  = 188.5 lb O_2/hr

Oxygen and nitrogen in the exhaust

Excess oxygen in exhaust = 188.5 lb O_2/hr

Nitrogen in the feed = 8 lb N_2/hr

Nitrogen along with excess oxygen :

$$= \left[\frac{188.5 \text{ lb } O_2/\text{hr}}{32}\right]\left[\frac{79}{21}\right][28] = 620.5 \text{ lb/hr}$$

Total nitrogen = (620.5 + 8) = 628.5 Ib/hr

Table 97 : Characteristics of the exhaust

Parameters	*Exhaust*	
	lb/hr	*lb-mole/hr*
CO_2	146.7	3.34
H_2O	804	44.67
O_2 (Excess)	188.5	5.89
N_2	628.5	22.45
Ash	8	Solid

2. Energy balance

Preheat energy input (Reference temperature = 60°F (feed temperature)

Input energy required = 0

Combustion energy released

Table 98 : Energy Balance

Parameters	lb/hr	BTU/lb	BTU/hr
Excess sludge water	800	1800	1,440,000
Aqueous waste	20	(- 900)	- 18,000
	Total	1,262,000	

Total energy released = 1,262,000 BTU/hr

Total input energy :

= [Preheat + Energy released through combustion]

= [0 + 1,582, 000] = 1,582,000 BTU/hr

Energy absorbed

Reference enthalpy temperature (RT) = 60°F

Combustion enthalpy temperature (CT) = 1700°F [based on gas phase]

Table 99 : Energy Balance

Parameters	*lb/hr*	*BTU/lb*		*BTU/hr*
		CT	*RT*	
CO_2	146.7	438	0	64,255
H_2O	804	830	0	667,320
O_2 (Excess)	188.5	404	0	76,154
N_2	628.5	436	0	274,026
Ash	8	600	0	4800
	Heat losses (16.6% at total heat absorbed)180,245			
			Total Heat absorbed	1,262,0000

3. Required exhaust treatment

Use venturi scrubbar :

- 99% removal of particulate
- Reducing incinerator exhaust temperature from 1700° to 125°F
- Utilizing 7 gal of re-circulated water/1000 ft^3 of exhaust gases

3.1 Composition of exhaust gases

Exhaust dry gas = [146.7 + 188.5 + 628.5] lb/hr dry air

= 963.7 lb dry air/hr

Humidity at 125°F = 0.0954 lb water/lb dry air

Water in exhaust = (0.0954 lb water)(963.7 lb dry air/hr)

= 92 lb water/hr

Water removed = (804 – 92) = 712 lb water/hr

Composition of various components in incinerator and venture scrubber exhausts :

Table 100 : Mass and Mole balance for incinerator and venture scrubber.

Components	*Incinerator*		*Venturi scrubber*		
	lb/hr	*lb moles/hr*	*lb/hr*	*lb moles/hr*	*Blow-down lb/hr*
Temperature (°F)	1700			125	
CO_2	146.7	3.34	146.7	3.34	
H_2O	804	44.67	92	5.11	712
O_2 (Excess)	188.5	5.89	188.5	5.89	
N_2	628.5	22.5	628.5	22.5	
Ash	8	Solid	0.08	Solid	≈ 8
Total	1775.7	76.4	1055.78	36.84	720

Gas rate in the incinerator :

$$= (76.4 \text{ mole/hr})(359 \text{ std ft}^3/\text{mole})\left(\frac{1700+460}{460+32}\right)\left(\frac{\text{hr}}{3600\text{s}}\right)$$

= 33.45 ft^3/s of gas

Venturi scrubber water :

$$= \left[7 \text{ gal}/1000 \text{ ft}^3 \text{of exhaust gas}\right]\left[33.45 \text{ ft}^3/\text{s of gas}\right]\left[\frac{60\text{s}}{\text{min}}\right]$$

= 14.649 gal/min of circulating water

- Make up water : 2 gal/min (= 1000 lb/hr)

Blown-down characteristics

Water removed = 712 lb H_2O/hr

Bleed water = 1000 lb H_2O/hr

Ash = 8 lb ash/hr

Total = 1720 lb/hr

4. Required combustion reactor specifications :
   - Loading rate : 8 lb cake/ft^2.hr
   - Hearth cake loading : 800 lb/hr

   Therefore, required surface area : $\frac{800 \text{ lb/hr}}{8 \text{ lb}/\text{ft}^2.\text{hr}} 100 \text{ ft}^2$

Hearth specifications :

- 7.5 ft hearth with 6 hearths has a total area of 100 ft^2
- 7.5 ft hearth with 7 hearths has a total area of 114 ft^2.

5. Exhaust combustion from venturi scrubber

$$\text{Gas rate} = (36.84 \text{ lb moles/hr})(359 \text{ std ft}^3/\text{lb mole})\left(\frac{460+125}{460+32}\right)\left(\frac{\text{min}}{60 \text{ s}}\right)$$

$$= 262 \text{ ft}^3/\text{min}$$

$$\text{Particulate matter} = \frac{(0.08 \text{ lb/hr})(7000 \text{ grains/lb})\left(\frac{\text{min}}{60 \text{ s}}\right)}{262 \text{ ft}^3/\text{min}}$$

$$= 0.03 \text{ grain/ft}^3 \text{ of gas.}$$

Example 3.136

Two-Stage Rotary Kiln Incinerator

Determine, Carbon, Hydrogen, Sulphur, Oxygen Nitrogen and Ash material balances, energy balance, heat recovery, exhaust treatment, exhaust characteristics, and equipment size. Assume the following data :

Table 101 : Data regarding Plant Wastes, Residue wastes, Rinses and fuel oil required.

Parameters	*Waste*			
	Plant Wastes	*Residue Wastes*	*Rinses*	*Fuel Oil (Supplementary) Oil*
Waste (lb/hr)	80	160	1500	-
Carbon (%)	60	60	15	87.4
Hydrogen (%)	15	5	3	12.4
Sulphur (%)	0.1	0.5	0.02	0.2
Oxygen (%)	15	10	4	-
Chlorine (%)	2	4.5	0.5	-
Water (%)	1	10	75.98	-
Ash (%)	6.9	5	1.5	-
Combustion heat (BTU/lb)	18,000	10,000	1500	20,000
Specific heat (BTU/lb.°F)	1	1	1	0.5
Specific gravity (mg/L)	1	1	1	0.9

Incinerator operating conditions :

Primary combustion temperature	: 1800°F
Combustion pressure	: 14.7 psia
Excess air	: 25%
Primary combustion chamber heat released	: 12,000 BTU/ft^3.hr
Secondary combustion chamber (SCC) temperature	: 2200°F

SCC residence time	: 2.5 seconds
Ash specific heat	: 0.4 BTU/lb.°F

Solution

1. Required material balances [Influent]

Carbon balance

- Plant wastes (lb/hr)	= (80 lb/hr) (0.60)	= 48 lb/hr
- Residue wastes (lb/hr)	= (160 lb/hr) (0.60)	= 96 lb/hr
- Rinses (lb/hr)	= (1500 lb/hr) (0.15)	= 240 lb/hr
Total – C		384 lb/hr

Hydrogen balance

- Plant wastes (lb/hr)	= (80 lb/hr) (0.15)	= 12 lb/hr
- Residue wastes (lb/hr)	= (160 lb/hr) (0.05)	= 8 lb/hr
- Rinses (lb/hr)	= (1500 lb/hr) (0.03)	= 45 lb/hr
Total – H		65 lb/hr

Sulphur balance

- Plant wastes (lb/hr)	= (80) (0.001)	= 0.08 lb/hr
- Residue wastes (lb/hr)	= (160) (0.005)	= 0.80 lb/hr
- Rinses (lb/hr)	= (1500) (0.0002)	= 0.30 lb/hr
Total – S		1.18 lb/hr

Oxygen balance

- Plant wastes (lb/hr)	= (80) (0.15)	= 12 lb/hr
- Residue wastes (lb/hr)	= (160) (0.10)	= 16 lb/hr
- Rinses (lb/hr)	= (1500) (0.04)	= 60 lb/hr
Total – Oxygen		88 lb/hr

Water balance

- Plant wastes (lb/hr)	= (80) (0.01)	= 0.08 lb/hr
- Residue wastes (lb/hr)	= (160) (0.1)	= 16.0 lb/hr
- Rinses (lb/hr)	= (1500) (0.7598)	= 1139.7 lb/hr
Total – Water		1156.5 lb/hr

Chlorine balance

- Plant wastes (lb/hr)	= (80) (0.02)	= 1.6 lb/hr
- Residue wastes (lb/hr)	= (160) (0.045)	= 7.2 lb/hr
- Rinses (lb/hr)	= (1500) (0.005)	= 7.5 lb/hr
Total – Chlorine		16.3 lb/hr

Ash balance

- Plant wastes (lb/hr) = (80) (0.069) = 5.52 lb/hr
- Residue wastes (lb/hr) = (160) (0.05) = 8.00 lb/hr
- Rinses (lb/hr) = (1500) (0.015) = 22.5 lb/hr

Total – Ash 36.02 lb/hr

2. Required combustion products [$C + O_2 = CO_2$]

Weight of CO_2 produced [1 mole of C produces 1 mole of CO_2]

= [lb – carbon/hr.MW] [MW of CO_2]

= (384 lb/hr) (1/12) (44) = 1408 lb/hr of CO_2 produced

Weight of oxygen required [1 mole of C requires 1 mole of O_2]

$$= \left(\frac{384}{12}\right)(32) = 1024 \text{ Ib of } O_2/\text{hr}$$

Mass balance for hydrogen [$2H + 1/2\ O_2 = H_2O$

$H + Cl = HCl$]

(lb – H/2 – MW of H.hr) = lb moles of hydrogen/hr

1 mole – Cl reacts with 1 mole atom of H

[16.3 Ib/hr/[AW – CI] = mole – atom of chlorine/hr

$$= \frac{16.3 \text{ lb/hr}}{35.5} = 0.46 \text{ mole} - \text{atom/hr}$$

- Weight of HCl produced

  [0.46 lb – mole/hr] [36.5] = 16.8 lb – HCl produced/hr

- Weight of hydrogen used

  (mole – atom of H) (M.W of H) = lb H used/hr

  (0.46 lb-mole/hr) (1) = 0.23 lb moles of H_2/hr used

- Weight of water produced

  = [Input – used for HCl production]

  $$= \left[\left(\frac{65}{2}\right)\text{lb} - \text{mole of } H_2/\text{hr} - 0.24 \text{ lb} - \text{mole of } H_2/\text{hr}\right]$$

  = (32.5 – 0.24) lb mole of H_2/hr which will react with O_2 to produce 1 mole of H_2O

  = (32.26 lb mole of H_2/hr)(18) = 581 lb H_2O/hr

  Total water produced = [Feed water] + [Water produced through reaction]

  $$= [1156.5 + 581] \text{ lb } H_2O/\text{hr} = 1737.5 \text{ lb } H_2O/\text{hr}$$

Moles of O_2 required for reaction = [65/2]

Amount of oxygen required for H_2O=[65/2] [32/2]=520lb O_2/hr

Mass balance for sulphur [$S + O_2 = SO_2$]

$$\left[\frac{1.18\text{ lb/hr}}{32}\right] = 0.037\text{ lb mole of sulphue/hr}\left[\text{input}\right]$$

SO_2 produced = (0.037) (64) = 2.36 lb SO_2 produced/hr

O_2 required for SO_2 = (0.037) (32) = 1.18 lb O_2 reacted/hr

= 1.2 lb O_2/hr

Mass balance for oxygen

- Oxygen required for CO_2 = 1024 lb O_2/hr
- Oxygen required for H_2O = 520 lb O_2/hr
- Oxygen required for SO_2 = 1.2 lb O_2/hr
- Theoretical oxygen requirements 1545.2 lb O_2/hr
- Total O_2 required [Add extra 25%] = [1545 + 386] lb O_2/hr
  = 1931 lb O_2/hr
- Influent oxygen = 88 lb O_2/hr

Therefore, oxygen requirement = (1931 – 88) = 1843 lb/hr

Oxygen and nitrogen in the exhaust = 386 lb O_2/hr

Excess oxygen in exhaust = 386 lb O_2/hr

$$\text{Therefore, Nitrogen} = (\text{Ib } O_2/\text{MW } O_2)\left(\frac{79\%\ N_2}{21\%\ O_2}\right)(\text{MW} - N_2)$$

$$= \left(\frac{1843}{32}\right)\left(\frac{0.79}{0.21}\right)(28)$$

$$= 6066 \text{ lb } N_2/\text{hr}$$

[Nitrogen from supplementary (deficit) oxygen; influent nitrogen = 0]

Required ash balance

Assume 60% of the ash is discharged with the ash

Therefore, (36 × 0.6) = 21.6 lb/hr as kiln ash; and 14.4 lb/hr is in the exhaust

Required kiln exhaust analysis

Table 102 : Mass Balance

Parameters	*lb/hr*	*moles/hr*
CO_2	1408	32
H_2O	1738	96.5
SO_2	2.36	0.074
HCl	16.8	0.46
O_2	386	12.06
N_2	6066	216.64
Ash	36	Solid
Total	9653	327

3. Required energy balance
   - Consider reference temperature = 60°F (Feed temperature)
   - Therefore, preheat requirement = 0 BTU/hr
   - Combustion energy released

Table 103 : Energy Balance.

Waste Type	*lb/hr*	*BTU/lb*	*BTU/hr*
Plant waste	80	18,000	1,440,000
Residue waste	160	10,000	1,600,000
Rinses	1500	1500	2,250,000
		Total input heat	5,290,000

Total energy released through combustion :

= [Reaction – Preheat]

(5,290,000 – 0) = 5,290,000 BTU/hr

- Heat absorbed

  [Enthalpy at reference temperature (RT)

  Enthalpy at combination temperature (CT)]

Table 104 : Energy Balance.

		Enthalpy (BTU/lb)		
Parameters	*lb/hr*	*CT*	*RT*	*BTU/hr*
CO_2	1408	469*	0	660,352
H_2O	1738	888*	0	15,43,344
SO_2	2.36	313*	0	739
HCl	16.8	365*	0	6132
O_2	386	430*	0	165,980
N_2	6066	465*	0	2,820,690
Ash	36	600	0	21,600
		Head losses (1.36%)		71,173
		Total heat absorbed =		52,90,000

HCl heat capacity is based on an average heat capacity of 0.21 BTU/hr/°F over 60 to 1800°F

SO_2 heat capacity of 0.18 for the same operating range.

3.1 **Secondary combustion chamber :** Supplementary furnace oil will have to be added to increase the kiln temperature from 1800°F to 2200°F

Assume the requirement of supplementary furnace oil = 300 lb/hr

[Pilot plant data to achieve a temperature of 2200°F]

Fuel-Carbon = (300)(0.874) = 262.2 lb/hr

Fuel-Hydrogen = (300)(0.124) = 37.2 lb/hr

Fuel-Sulphur = (300)(0.002) = 0.6 lb/hr

Total SCC fuel combustion product

$$C + O_2 = CO_2$$

1 mole of C produces 1 mole of CO_2

$$\text{Amount of } CO_2 = \left(\frac{262.2}{12}\right)(44) = 961 \text{ lb } CO_2/\text{hr}$$

$$(22 \text{ lb mole } CO_2/\text{hr})$$

$$2H + 1/2O_2 = H_2O$$

1 mole of H_2 produces 1 mole of H_2O

$$\text{Amount of } H_2O = \left(\frac{37.2}{2}\right)(18) = 333 \text{ lb } H_2O/\text{hr}$$

$$(= 18.5 \text{ lb mole } H_2O/\text{hr})$$

$$S + O_2 = SO_2$$

1 mole of S produces 1 mole of SO_2

$$\text{Amount of } H_2O = \left(\frac{37.2}{2}\right)(18) = 333 \text{ lb } H_2O/\text{hr}$$

$$(= 18.5 \text{ lb mole } H_2O/\text{hr})$$

Amount of O_2 and N_2

$$\text{Amount of } O_2 = \left[\frac{262.2}{12}\right]32 + \left[\frac{0.6}{32}\right][32] + \left[\frac{37.2}{2}\right][32]$$

$$= [699 + 0.6 + 595] = 1295 \text{ lb } O_2/\text{hr}$$

Total O_2 [Additional O_2 (25%)] = 1295 + 324 = 1619 lb O_2/hr

$$\text{Amount of nitrogen} = \left(\frac{1619}{32}\right)\left(\frac{79}{21}\right)(2-8) = 5329 \text{ lb/hr}$$

Total gas for fuel (lb/hr) = $[CO_2] + [H_2O][SO_2]$

= [961 + 333 + 1.2] = 1295.2 lb/hr

Total gas from fuel (lb – mole/hr) = [22 + 18.5 + 0.019]

= 40.52 lb – mole/hr

Energy released by burning furnance oil :

= (300)(20,000 BTU/lb) = 6,000,000 BTU/hr

Heat absorbed by SCC fuel gas exhaust:

- Fuel feed at reference to temperature (RT)
- Fuel gases at SCC temperature (combustion temperature) (CT)

Table 105 : Mass and Energy Balances.

Parameters	*lb/hr*	*CT*	*RT*	*BTU/hr*
CO_2	961	594	0	570,834
H_2O	333	1130	0	376,290
O_2 (Additional)	324	539	0	174,636
N_2	5329	582	0	3,101,478
SO_2	1.2	410	0	492
		Total heat absorbed		4,223,720 BTU/hr

Total heat absorbed in the SCC = [Heat required to heat supplementary fuel gas combustion products (60 to 2200°F) + Heat required to heat the kiln gas exhaust from 1800 to 2200°F]

Enthalpy at kiln temperature (1800°F) (Reference temperature-RT)

Enthalpy at SCC temperature (CT) at 2200°F.

Table 106 : Energy and Mass Balances at Various Temperatures.

Parameters	*lb/hr*	*CT at 2200°F*	*RT at 1800°F*	*BTU/hr*
CO_2	1408	594	469	(594 – 469) 1408 = 176,000
H_2O	1738	1130	888	421,000
O_2	386	539	430	59,000
N_2	6066	582	465	710,000
SO_2	2.36	410	313	228 (Negligible)
HCl	16.8	450	365	1400
Ash	14.4	730	600	1900
		Total heat absorbed by kiln gas		1,369,000

Total heat absorbed in SCC :

- SCC fuel gases = 4,223,720 BTU/hr

- Kiln gases = 1,369,000 BTU/hr
- Losses (7.5%) = 419,454 BTU/hr

Total 6,012,174 BTU/hr (OK)
(≈ 6,000,000 BTU/hr)

Table 107 : Total SCC mass flow rate

Parameters	*SCC Exhaust Gases*	*Moles/hr*
CO_2	1408 + 961 = 2369	54
H_2O	1738 + 333 = 2071	115
O_2	386 + 324 = 710	22
N_2	6066 + 5329 = 11,395	407
SO_2	2.36 + 1.2 = 3.46	Negligible
HCl	16.8 + 0 = 16.80	0.5
Ash	14.4 + 0 = 14.4	Solid
Total	16,579.66	598.5(= 600 moles/hr)

- Exhaust gas rate = (600 moles/hr) (359 std ft^3/mole) $\left(\frac{2200+460}{460+32}\right)$

= 1,164,561 ft^3/hr

= 324 ft^3/s

3.2 **Recovery of heat :** Waste boiler for flue gas heat recovery is used to drop the flue gas temperature from 2200°F to 500°F, and the heat recovery is given is follows :

Table 108 : Heat Recovery.

Parameters	*lb/hr*	*Enthalpy at 2200°F*	*Enthalpy at 500°F*	*Heat recovered (BTU/hr)/(2–3) × 2*
1	2	3	4	5
CO_2	2369	594	99	11,72,655
H_2O	2071	1130	200	1,926,030
O_2	710	539	100	311,690
N_2	11,395	582	110	5,378,440
SO_2	3.46	-	-	Negligible
HCl	16.80	-	-	Negligible
Ash	14.4	-	-	Negligible
Total	16,579.66	-	-	8,788,815 (» 9,000,000 BTU/hr)

Percent efficiency of heat recovery boiler = 50%

Total heat recovered = 4,500,000 BTU/hr

4. Air pollution control system

   4.1 Electrostatic precipitator

   Efficiency of particulate matter removed = 99%

   Inlet particulate matter = 14.4 lb/hr

   Outlet particulate matter = 0.144 lb/hr

   Alternatively, high energy venturi scrubber can be combined with an alkaline.

   4.2 Use of low energy acid scrubber

   Removal of remaining particulate and acid component.

   Reducing the incinerator flue gas temperature from 500°F to 125°F

   CO_2 may react with excess NaOH scrubbing solution to form $NaCO_3$ [Neglected for calculations]

   Dry incinerator exhaust gas = [CO_2 + O_2 + N_2] gas components

   [Excluding H_2O, SO_2, HCl and Ash] = [2369 + 710 + 11,395] lb/hr

   = 14,474 lb/hr

   Air humidity at 125°F = 0.0954 lb/water/lb dry air

   - Water in the incinerator exhaust = (0.0954)(14,474)

     = 1381 lb H_2O/hr

   - Water removed = (2071 – 1381) lb H_2O/hr

     = 690 lb H_2O/hr

Table 109 : Analysis of incinerator exhaust gas

Parameters	*% Removal*	*Inlet (lb/hr)*	*Scrubber Exhaust lb/hr*	*Scrubber Exhaust mole/hr*	*Blow-down (lb/hr)*	*NaOH⁺ (lb/hr)*
Temperature (°F)		500	125			
CO_2	–	2369	2369	54	–	–
H_2O	–	2071	1381	77	690	–
O_2	–	710	710	22	–	–
N_2	–	11,395	11,395	407	–	–
SO_2	90	3.46	0.346	Negligible	6.8^{++}	4.3^{+}
HCl	95	16.80	0.84	Negligible	27^{+++}	18.5^{+}
Ash	20	0.1445	0.116	Negligible	–	–
	Total	16,579.66	15,750	560	724	22.8

$^{++}Na_2SO_3$

$^{+++}$ NaCl

Salt formed : $SO_2 + 2NaOH = Na_2SO_3 + H_2O$

1 mole of SO_2 requires 2 moles of NaOH

$$\text{Weight of NaOH} = \frac{80 \text{ lb of NaOH}}{64 \text{ lb of } SO_2} = 1.25$$

Therefore, 1.25 lb NaOH/lb of SO_2 forming 1 mole of Na_2SO_3 [1.97 lb of Na_2SO_3/lb of SO_2

$$NaOH + HCl = NaCl + H_2O$$

It means : 1.1 lb of NaOH/lb of HCl [40/36.5 = 1.095]

: 1.6 lb of Na*Cl*/lb HCl [58.5/36.5 = 1.6]

Exhaust gas rate :

$$= (560 \text{ moles/hr})(359 \text{ std ft}^3/\text{mole})\left(\frac{460+125}{460+32}\right)\left(\frac{\text{hr}}{3600 \text{ s}}\right)$$

= 66.4 ft^3/s at 125°F

5. Air Pollution Control equipment specifications

Rotary kiln heat released

- Design burn rate = 15,000 BTU/ft^3.hr
- Heat released = 5,290,000 BTU/hr
- $\text{Kiln volume} = \dfrac{5,290,000 \text{ BTU/hr}}{15,000 \text{ BTU/ft}^3\text{.hr}} = 353 \text{ ft}^3$
- Dimensions : Length to diameter ratio (L/D = 3.6)

  Assume diameter of 5 ft ; L = 18 ft [OK]

$$\left[\frac{\pi}{4}D^2L = \frac{3.14}{4} \times 25 \times 18 = 353 \text{ ft}^3\right]$$

Secondary combustion chamber (SCC) volume

Gas rate = 324 ft^3/s

Gas residence time = 2 seconds

SCC volume $= (324)\ (2)$

$= 648\ ft^3$

Assume diameter of 5 ft

$$\text{Length} = \frac{(648\ ft^3)(4)}{3.14(25)} = 33\ ft$$

$$\text{Therefore, } \frac{L}{D} = \frac{33}{5} 6.6\ (OK)$$

6. Characteristics of the exhaust gas

Acid scrubber exhaust = 560 moles/hr at 125°F

Gas rate = 66.4 ft^3/s at 125°F

SO_2 at 0.346 lb/hr = 0.00540 moles SO_2/hr

HCl at 0.84 lb/hr = 0.023 moles HCl/hr

Particulates at 0.116 lb/hr

Concentration units

- $SO_2 : \dfrac{0.0054 \text{ moles } SO_2/\text{hr}}{560 \text{ moles gas rate/hr}} \times 10^6$

- $HCl : \dfrac{0.023 \text{ moles HCl/hr}}{560 \text{ moles gas rate/hr}} \times 10^6$

  : 41 ppm V of HCl

- Particulates : $(0.116\ \text{lb/hr})(7000\ \text{grams/lb})\left(\dfrac{\text{hr}}{60\ \text{min}}\right)$

- 13.5 grains/min or

  $: \dfrac{(135\ \text{grains/min})}{66.4\ ft^3/s}\left(\dfrac{\text{min}}{60s}\right)$

  : 0.0034 grains/ft^3

Note :

1. Incinerator operating function.

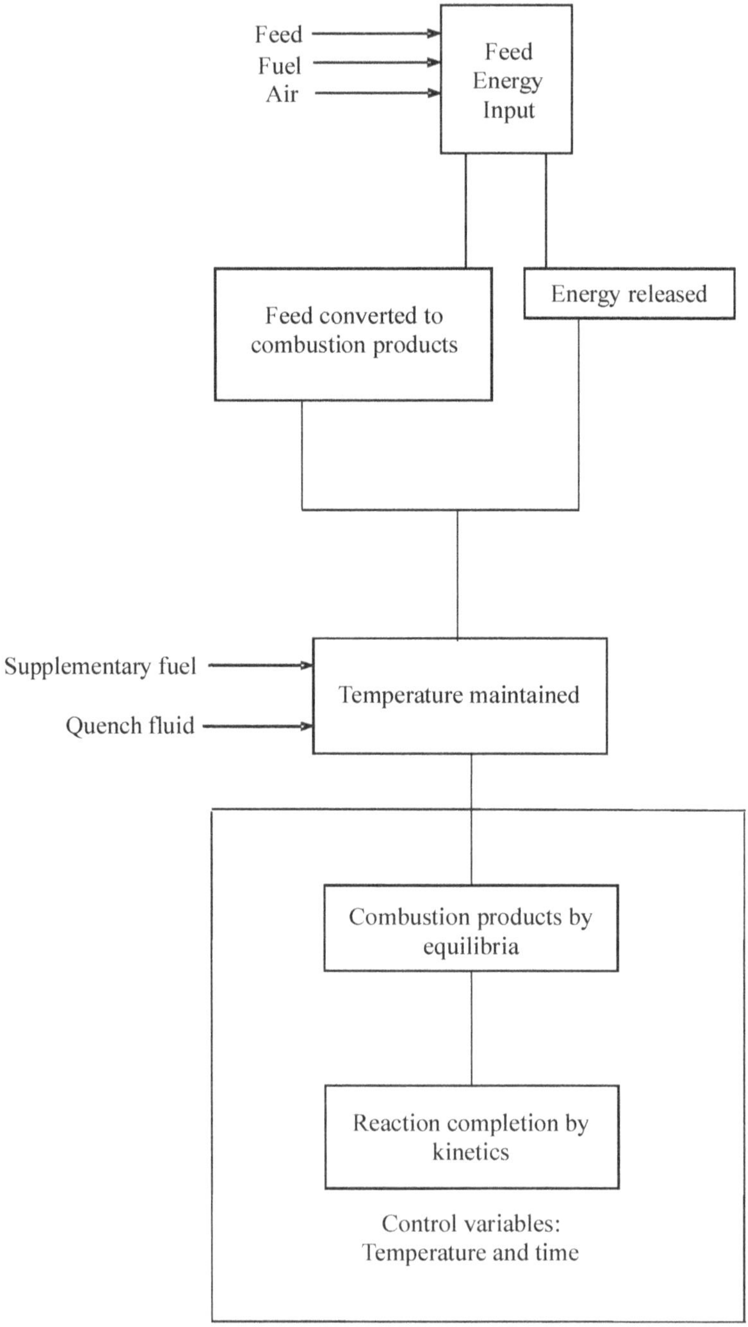

Figure 40 : Incinerator operating function.

2. Heat transfer mechanism

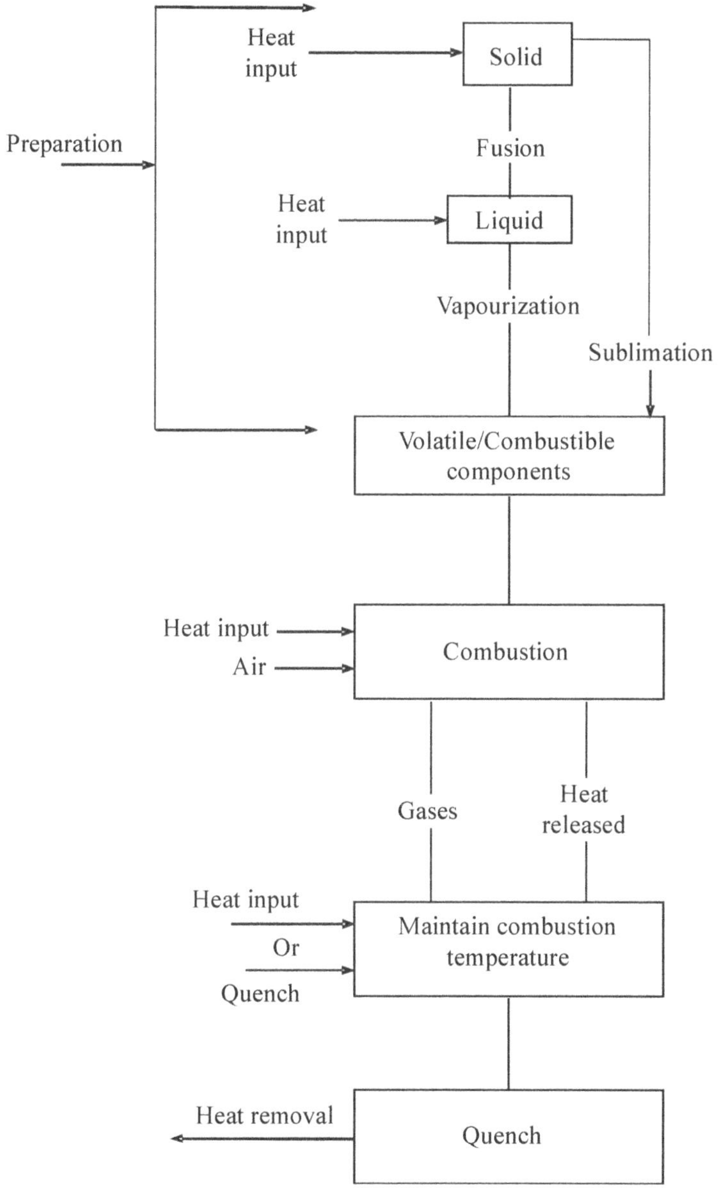

Figure 41 : Heat Transfer Mechanism.

3. Heat input (Release)

   Feed stream heat content

   Heat of combustion

Energy released → ← Energy absorbed

[Heat input] = [Heat output]

4. Heat output (absorbed)

   Heat content of the flue gases

   Latent heat for change of state

   Heat content of ash

   Heat losses from system.

5. **Heat of combustion (heat released) :** The heat released when one unit of organic is converted to combustion products is defined as the heat of combustion, e.g. $CH_4 + 2O_2 = CO_2 + 2H_2O$.

   In this case, the higher heating value at a reference temperature 20°C is 192 kcal/mole of methane. The heating value includes the heat of evaporation of the water which will be available when the steam condenses. However, since most of combustion processes operate well above the heat of condensation of water, the heat of condensation is excluded, and the lower (or net) heating value is used in combustion calculations. When hydrogen is not one of the products the higher and lower heating values are identical. Although, the heats of combustion of many organic compounds are readily available, because the composition of many wastes are usually unknown or highly variable, the heat of combustion is best determined by laboratory tests.

6. **Latent heats (heat absorbed) :** The heat required to change phase is termed as latent heat. The most common being that required to evaporate water into steam common phase changes in incineration include :

   Heat of fusion : Solid to liquid ; ash to molten form

   Heat of sublimation : Solid to vapour; release of volatiles from solids

   Heat of vapourization : Liquid to gas; evaporation of water.

   The heat absorbed in the phase change, must be accounted for in the incineration energy balance.

7. **Sensible heat (heat absorbed) :** The heat content of substances (enthalpy) at a specific temperature is based on the sensible heat content of the individual components at that temperature, relative to a base temperature.

   In case of liquids, where the specific heat is relatively constant over a temperature range encountered, the heat content is a product of the weight times the specific heat constant :

   Heat capacity = $(\text{lb/hr})(M_{Cp})(T_{out} - T_{in})$

Unlike liquids, the specific heat of gases (BTU/lb per degree change) changes significantly with temperature, so that method chosen to calculate enthalpy (heat content at a specific temperature) depends on the accuracy required. These include :

- Use of enthalpy tables for common components (air, N_2, CO_2, etc.)
- Use of sensible heat equation :

  $M_{Cp} = a + bT + CT^2 + dT^3$

- Approximate the value using an average specific heat value over the temperature range considered (OK for most incineration evaluations)
- Sensible heat can also be released from pre-heated feed steams.

8. Important variables

   Material input : Wastes, air, and fuel

   Material output : Flue gases, ash, and blow-down

9. Liquid incinerator operating data

   Combustion chamber

   - Temperature range : 815-1205°C (1500-2200°F)

     Common : 870-980°C (1600-1800°F)
   - Burn rate : 20-30,000 BTU/ft³.hr
   - Excess air (typical) : 10 to 60%

     Hazardous waste : 120 to 150%
   - Gas residence time : 0.5 – 2 seconds

10. Rotary kiln operating data

    Combustion chamber

    - Non slag : 650 – 1260°C (1200 – 2300°F)
    - Slagging : 1430 – 1540°C (2600 – 2800°F)
    - Burn rate : 15 – 40,000 BTU/ft³.hr
    - Excess air : 75 – 200%
    - Residence time (SRT) : 0.5 – 2 hours
    - Gas residence time : 1 – 5 seconds

    Secondary combustion chamber

    - Temperature : 1090 – 1370°C (2000 – 2500°F)
    - Excess oxygen : 120 – 200%
    - Gas residence time : 1 – 3 seconds

11. Fluidized bed operating date

Combustion chamber

- Bed temperature
  - * Bed below : 1095°C (2000°F);
    Typical = 705-760°C (1300-1400°F)
  - * Hazardous wastes : 760-870°C (1400-1600°F)
- Free space temperature
  - * Typical:760-870°C (1400-1600°F)
  - * Hazardous wastes:760-1095° (1400-2000°F)
- Excess air : 40 to 60% common
  100 to 150% (hazardous wastes)
- Burn rate : 100-200,000 BTU/ft^3.hr
- Gas residence time : 1-5 seconds

12. Multiple-hearth operating data

Combustion chamber

Temperature :

- After burner	:	815°C (1500°F)
- Drying zone	:	315-540°C (600-1000°F)
- Combustion range	:	760-930°C (1400-1700°F)
[Ash fusion	:	930°C (1700°F)]
- Burn rate	:	Based on the entire hearth area = 34 – 59 kg/m^2.hr or 7-12 lb/ft^2.hr
- Excess air	:	75-100%

13. Energy audit

Energy input

- Heat released from waste
- Heat released from fuel
- Sensible heat content of fuel
- Sensible heat content of air

Energy absorbed

- Sensible heat absorbed by CO_2 H_2O-vapour, SO_2 halides or acid fumes, excess oxygen, and nitrogen.
- Sensible heat absorbed by ash
- Heat losses from the system walls.

Example : 3.137

Incinerator : Lime requirements for Spray Dryer

Determine the lime requirements for HCl and SO_2 containment for the combustion of hazardous waste.

Hazardous waste flue gas	:	15,000 acfm
Flue gas temperature	:	700°F
Flue gas pressure	:	1 atm
HCl flue gas concentration	:	10,000 ppm
SO_2 flue gas concentration	:	250 ppm
HCl removal required	:	99% (or 4 lb/h)
SO_2 removal required	:	70%

Solution

1. Required HCl and SO_2 gas flow rates

$$PV = nRT$$

$$Pq = \dot{n}RT$$

$$\dot{m} = \dot{n}(MW) = Pq(MW)/RT$$

$$\dot{m}(HCl) = \frac{(1\text{ atm})(0.01)(15{,}000\text{ acfm})(36.5\text{ lb/lb}-\text{mol})}{(0.7302\text{ atm.ft}^3/\text{lb mol.°R})(460+700\text{°R})(1\text{ h}/60\text{ min})}$$

$$= 387.8\text{ lb HCI/h}$$

$$\dot{m}(SO_2) = \frac{(1\text{ atm})(0.00025)(15{,}000\text{ acfm})(64\text{ lb/lb}-\text{mol})}{(0.7302\text{ atm.ft}^3/\text{lb mol.°R})(160\text{°R})(1\text{ h}/60\text{ min})}$$

$$= 170\text{ Ib }SO_2/\text{h}$$

Control

HCl = (0.01)(387.8 lb/h) = 3.88 lb/h (OK)

SO_2 = (0.3)(170 lb/h) = 5.1 lb/h

2. Required lime doses for containing HCl and SO_2

$$1.3Ca(OH)_2 + SO_2 + 0.5O_2 = CaSO_4 + H_2O + 0.3Ca(OH)_2$$

$$1.1Ca(OH)_2 + HCI = CaCI_2 + H_2O + 0.1Ca(OH)_2$$

- Lime requirement $= \left(\frac{1.3\text{ lb mol}-\text{lime}}{\text{lb mol }SO_2}\right)\left(\frac{17\text{ lb }SO_2/\text{h}}{64\text{ lb }SO_2/\text{lb}-\text{mol}}\right) +$

$$\left[\left(\frac{(387.3-4.0)\text{ lb HCl/h}}{36.45\text{ lb HCl/lb}-\text{mol}}\right)\left(\frac{0.55\text{ lb mol lime}}{\text{lb mol HCl}}\right)\right]$$

= 6.13 lb – mol lime/h

= (6.13) (74) = 453 lb/h

Amount of $CaSO_4$ produced

$$\frac{(17)(0.7)}{64} = 0.186\text{ lb}-\text{mol/h}$$

= (0.186)(136) = 25.3 lb/h

Amount of $CaCl_2$ produced

1 mole of $CaCl_2$ = 2 moles of HCl reacted

$$= \frac{(387.3-4)(0.5)}{36.45} = 5.258\text{ lb mol/h}$$

= (5.258)(110.9) = 583.1 lb/h

Un-reacted $Ca(OH)_2$

= (6.12 – 5.258 – 0.186) = 0.676 lb mol/h

= (0.676)(74) = 50.0 lb/h

Total solids generated

= $CaCl_2$ + $CaSO_4$ + $Ca(OH)_2$ (Unreacted)

= 583.1 + 25.3 + 50.0

= 658.4 lb/h

3. Required size of the spray dryer

$$V = q \times \theta = \pi D^2\, L/4$$

= (15,000 ft^3/min)(10 s) [q = 10s]

$$= \frac{15{,}000 \times 10}{60} = 2500\text{ ft}^3$$

If L = 1.75 D

$$V = \frac{\pi D^2}{4}(1.75\text{ D})$$

$$D^3 = \frac{2500(4)}{\pi(1.75)} = 1818.9\text{ ft}^3$$

D = 12.2 ft

L = 1.75(12.2) = 21.4 ft

Example 3.138

Incinerator : Stack Emission

Estimate whether the incinerator meets the particulate standard of 0.05 grain/d scf, and also the amount of particulate matter escaping the stack, and the molecular weight of the stack gas (dry and wet basis). Assume the following data:

Data at standard conditions (70°F and 1 atm)

Volume sampled	:	35 dscf
Diameter of stack	:	2 ft
Pressure of stack gas	:	29.6 in. Hg
Stack gas temperature	:	140°F
Mass of particulate collected	:	0.16 g
% Moisture in stack gas	:	7% (by volume)
% O_2 in stack gas (dry)	:	7% (by volume)
% CO_2 in stack gas (dry)	:	14% (by volume)
% N_2 in stack gas (dry)	:	79% (by volume)

Pitot tube measurements made at 8 points across the diameter of the stack provided values :

0.3, 0.35, 0.4, 0.5, 0.5, 0.4, and 0.3 in H_2O

S – pitot tube velocity :

$$v = k(2gH)^{0.5}$$

$$= k\left[2g\frac{\rho_l}{\rho_a}(0.0254)h\right]^{0.5}$$

where g = Gravity contact (9.81 m/s²)

H = Fluid velocity head, H_2O

ρ_l = Density of manometer fluid (1000 kg/m³)

ρ_a = Density of flue gas (1.084 kg/m³)

h = Mean pitot tube reading (in. H_2O)

Solution

1. Required particulate concentration in the stack

$$\text{Particulate concentration} = \frac{0.16 \text{ g collected}}{35 \text{ dscf sampled}}(15.43 \text{ grain/g})$$

$$= 0.0706 \text{ grain/dscf [Not OK]}$$

2. Required stack flow rate

- $$v = 0.85\left[2\left(9.81\ \text{m/s}^2\right)\left(\frac{1000}{1084}\right)0.0254\ h\right]^{0.5}$$

$= 0.85(21.4)h^{0.5}$

$= 0.85\ (21.4)(0.6142) = 11.2\ \text{m/s}\ \ (= 36.75\ \text{ft/s})$

Stack flow rate = vA

$$= \left(36.75\ \text{ft/s}\right)\left(\frac{\pi}{4}\right)\left(2\ \text{ft}\right)^2$$

$= 115.45\ \text{ft}^3/\text{s} = 6.924\ \text{ft}^3/\text{min}$

Dry volumetric flow rate = $(1 - 0.07)\ (6.924\ \text{ft}^3/\text{min})$

$= 6439\ \text{da} - \text{ft}^3/\text{min}$

Standard volumetric flow rate (70°F and 1 atm)

$$= 6439\ \text{d} - \text{a ft}^3/\text{min}\left(\frac{530°\text{R}}{600°\text{R}}\right)\left(\frac{29.6\ \text{psi}}{29.6\ \text{psi}}\right)$$

$= 5631\ \text{d} - \text{a ft}^3/\text{min}$

Particulate emission rate

$= (0.0706\ \text{gr/d} - \text{scf})\ (5631\ \text{d} - \text{a ft}^3/\text{min})$

= 398 grain/min

= (398 grain/min) (1 lb/7000 grain) = 0.0569 lb/min

= 81.9 lb/d

3. Required molecular weight (MW) of stack gas

MW (dry basis : containing O_2 = 7%, CO_2 = 14% and N_2 = 79%)

= 0.07 O_2(32 lb/lb mol) + 0.14CO_2(44 lb/lb mole) + 0.79N_2(28 lb/lb – mole)

= 30.52 lb/lb – mol

MW (Wet basis : 7% water and 93% other components) :

= 0.07 water (18 lb/lb mol) + 0.93 other components (30.52 lb/lb – mole)

= 29.64 lb/lb – mol

Example 3.139

Spray-type Scrubber

Determine the following :

Power loss of the gas passing through the scrubber (P_G)

Power loss of the gas spray liquid during atomization (P_L)

Number of transfer units (N_t).

Assume the following data :

Gas flow rate (q_G)	: 10,000 actual ft^3/min
Water rate (q_L)	: 50 gal/min
Inlet loading	: 5.0 grain/ft^3 (Particulate)
Maximum outlet loading	: 0.05 grain/ft^3 (Particulate, required)
Maximum gas pressure drop across the unit (ΔP)	: 15 in. H_2O
Maximum water pressure drop across the unit (P)	: 100 psi

Vendor's design and operating data :

α : 1.47

β : 1.05

Water pressure drop (P)	: 80 psi
Gas pressure drop across the tower (DP)	: 5.0 in. H_2O

Solution

1. Required necessary expressions

Power loss of the gas passing through the scrubber (P_G)

$$P_G = 0.157\ \Delta P,\ (\text{hp}/1000\ \text{acfm})$$

where ΔP = Pressure drop across the scrubber (in. H_2O)

Power loss of the spray liquid during atomization (P_L)

$$P_L = 0.583\ P(q_L/q_G)\ ;\ (\text{hp}/1000\ \text{acfm})$$

where P = Liquid inlet pressure (psi)

q_L = Liquid feed rate (gal/min)

q_G = Gas flow rate (ft^3/min)

Number of transfer unit (N_t)

$$N_t = \text{In}\left[\frac{1.0}{1-E}\right];\ [\text{Mass transfer anology}]$$

where E = Fractional collection efficiency

– $N_t = \alpha\, P_T^{\beta}$; [Experimental data fit]

where α and β = Parameters for the type of particulate being collected and the scrubber unit

$P_T = P_G + P_L$

2. Required efficiency for the spray-type scrubber

$$P_G = 0.157(\Delta P) = 0.157(5.0) = 0.785 \text{ hp/1000 acfm}$$

$$P_L = 0.583(80)\left(\frac{50}{10{,}000}\right) = 0.233 \text{ hp/1000 acfm}$$

$$P_T = P_G + P_L = (0.785 + 0.233) = 1.018 \text{ hp/1000 acfm}$$

$$N_t = \alpha\, P_T^{\beta}$$

$$= 1.47(1.018)^{1.05} = 1.50$$

$$N_t = \ln\left[\frac{1.0}{1-E}\right]$$

or $$E = (1 - e^{-N_t}) = (1 - e^{-1.5}) = 0.777$$

Required collection efficiency (E_s) is :

$$E_s = \frac{(\text{Inlet loading} - \text{Required outlet loading})}{\text{Inlet loading}}(100)$$

$$= \left(\frac{5.0 - 0.05}{5}\right)(100)$$

$$= 99.0\%$$

Therefore, E_s (99.0%) required is for greater than E (77.7%)

3. Required changes in the operating conditions to meet E_s of 99.0%

- No. of transfer units $(N_t) = \ln\left[\dfrac{1.0}{1-E}\right]$

$$= \ln\left[\frac{1.0}{1-0.99}\right]$$

$$N_t = 4.605$$

Total power loss (P_T) is given by :

$$N_t = \alpha\, P_T^{\beta}$$

$$4.605 = 1.47(P_T)^{1.05}$$

Or, $P_T = 2.96$ hp/1000 acfm

Contacting power based on the gas stream power input (P_G)

$$P_G = 0.157 \text{ DP}$$
$$= 0.157\ (15)$$
$$= 2.355 \text{ hp/1000 acfm}$$

Therefore, $P_L = P_T - P_G = (2.96 - 2.355)$

$$= 0.605 \text{ hp/1000 acfm}$$

- $\dfrac{\text{Liquid flow}}{\text{Gas flow ratio}}\left(\dfrac{q_L}{q_G}\right)$ ratio is

$$\frac{q_L}{q_G} = \frac{P_L}{[(0.583)(P)]}$$

$$= \frac{0.605}{0.583(100)} = 0.0104$$

New water flow rate (q'_L in gal/min) is :

$$q'_L = (q_L/q_G)(10{,}000 \text{ acfm})$$
$$= (0.0104)(10{,}000 \text{ acfm})$$

= 104 gal/min to satisfy the outlet loading (particulate concentration) of 0.05 grain/ft^3.

Example 3.140

Capacity of Secured Landfill for Industrial Solid Waste

Estimate the capacity of a solid waste disposal site, as also the amount of cover material that must be cut from the sides and rear slope of the disposal site to meet the specified cover requirements. Assume the following data :

Total lift height : 3.0 m

Front face of completed landfill will have a slope of 2 to 1 and will be coincident with the front face of the existing site.

Ratio of solid waste to cover material is 5 to 1

Solution

1. Required schematics of the landfill site
   - Plan view

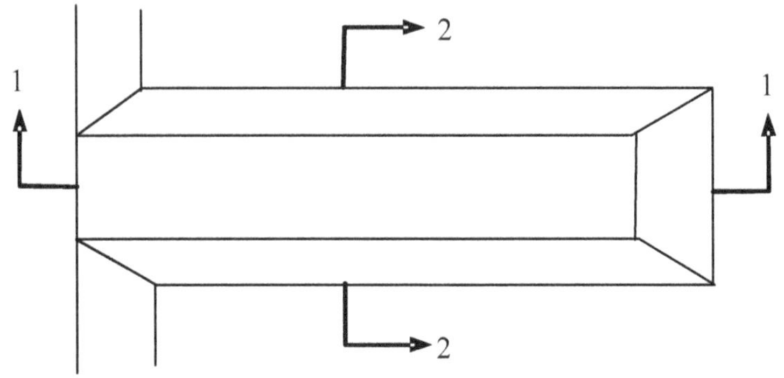

Figure 42 : Schematics of Landfill site.

- Section : 1-1 :

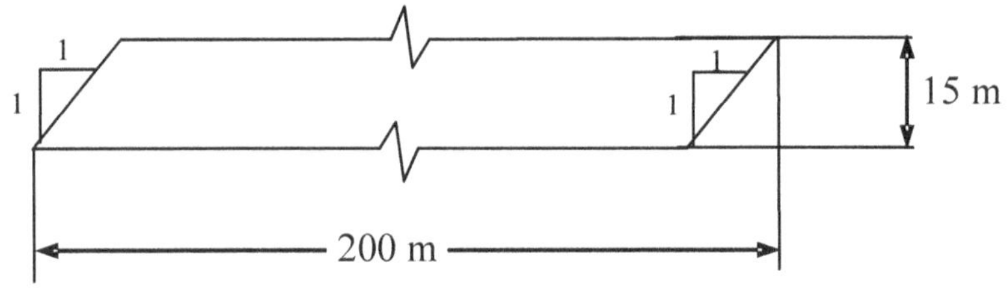

Figure 43 : Schematics of Landfill site.

- Section 2-2 :

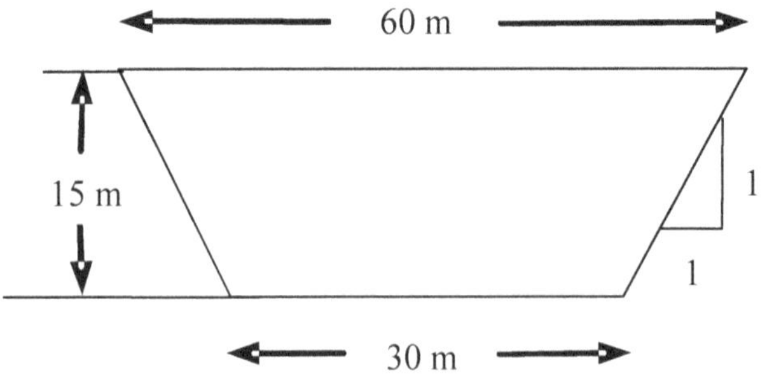

Figure 44 : Schematics of Landfill site.

Table 110 : Required secured landfill capacity

Lift No.	*Elevation (m)*	*Area (m^2)* *At Contour Interval*	*Area (m^2)* *Average between Intervals*	*Capacity between Countours (m^3)*	*Cover Material Required (m^3)*
	0	5500*			
1			6015	18045**	3609+
	3	6480			
2			6945	20835	4167
	6	7410			
3			7875	23625	4725
	9	8340			
4			8805	26415	5283
	12	9270			
5			9735	29205	5841
	15	10200			
	Total capacity (m^3)			118125	
	Total cover (m^3)				23625

* Area at the elevation – 0

[(200 – 15)m] [30 m] = 5550 m^2

** 6015 × 3 m (lift height) = 18045 m^3

+ Cover material = (18045 m^3) [1/5, ratio of cover material to solid waste]

= 3690 m^3

3. Required limits of excavation that be necessary to obtain the required cover material [Figure 44]

Assume that a triangular wedge of material will be excavated all round the disposal site.

Relationship that is used to estimate the volume (V) :

V = 2 [Volume of truncated excavated wedge sections] + L [Area of the continuous triangular wedge]

$$= 2\left[\frac{1}{2}(2x)\right] + L\left[\frac{1}{2}x^2\right]$$

where : L = Wedge length measured along a line drawn on a horizontal plane of a distance (2/3x) from the bottom perimeter of the disposal site

$$\text{Therfore, } L = 2(155\text{ m}) + 30\text{ m} + 4\left(\frac{2}{3}x\right)$$

$$= 340 + 2.67x$$

$$V = 1.33x^3 + (340 + 2.67x)\left(\frac{1}{2}x^2\right)$$
$$= 170x^2 + 2.67x^3$$

It has to be solved by trial and error method, assuming different values of x

Assume x = 110 m

$$23625 = 170\ (11)^2 + 2.67\ (11)^3 = 24123\ \text{[OK]}.$$

Schematics of the landfill is shown in Figure 45.

Note :

Table 111 : Factors that must be considered in evaluating potential landfill sites

Factors	*Remarks*
Available land area	Site should have a useful life greater than 1 yr (mini)
Haul distance	Will have significant impact on operating cost
Soil condition and topography the site	Cover material must be available at or near
Surface water hydrology	Impacts drainage requirements
Geologic and hydrogeologic conditions	Probably most important factors in establishment of landfill site, especially with respect to site preparation
Local environmental conditions	Noise, odour, dust, vector, and aesthetic factors controls requirements
Climatologic conditions	Provision must be made for wet-weather operation
Ultimate-use of site	Affects long-term management for site

2. Landfill gases [NH_3, CO_2, H_2S, CH_4, etc.]

Overall anaerobic reactions :

$$C_aH_bO_cN_d \rightarrow nC_wH_xO_yN_z + mCH_4 + sCO_2 + rH_2O + (d - nz)NH_3$$

where s = a – nw – m

r = c – ny – 2s

Total mineralization of organic compound through anaerobic reaction

$$C_aH_bO_cN_d + \left[\frac{4a - b - 2c + 3d}{4}\right]H_2O \rightarrow \left[\frac{4a + b - 2c - 3d}{8}\right]CH_4$$

$$+\left[\frac{4a - b + 2c + 3d}{8}\right]CO_2 + dNH_3$$

– Rates of decomposition in unmanaged landfills (in terms of gas production) reaches in the first 2 years, and then slowly tapers off (continuing in many cases upto 25 years or more).

Table 112 : Composition of leachate from landfills

Parameters	*Range (mg/L)*	*Typical (mg/L)*
BOD_5	2000 – 30000	10000
TOC	1500 – 20000	6000
COD	3000 – 45000	18000
TSS	200 – 1000	500
Org – N	10 – 600	200
NH_3 – N	10 – 800	200
NO_3^-	5 – 40	25
Total – P	1 – 70	30
Ortho – P	1 – 50	20
Alkalinity (as $CaCO_3$)	1000 – 10000	3000
pH	5.3 – 8.5	6
Total hardness (as $CaCO_3$)	3000 – 10000	3500
Ca^{+2}	200 – 3000	1000
Mg^{+2}	50 – 1500	250
K^+	200 – 2000	300
Na^+	200 – 2000	500
Cl^-	100 – 3000	500
SO_4^{-2}	100 – 1500	300
Total – Fe	50 – 600	60

Table 113 : Landfill sealant for the control of gas and leachate movement

Sealant		*Remarks*
Classification	Type	
Compacted soils	—	Should contain some clay
Compacted clay	Bentonites, illites, kalolinites	Most commonly used sealant for landfills; layer thickness varies from 6 to 48 in.(to cm); layer must be continuous and not allowed to dry out and crack
In-organic chemicals	Sodium carbonate, silicate or pyrophosphate	Use depends on local soil characteristics
Synthetic chemicals	Polymers, rubber latex	Experimental, use not well established
Synthetic membrane liners	PVC, butyl rubber, hypaton, polyethylene, nylon-reinforced	Expensive, may be justified where gas is to be recovered liners
Asphalt	Modified asphalt, asphalt covered polypropylene fabric, asphalt concrete	Layer must be thick enough to maintain continuity under differential settling conditions
Others	Gunite concrete, soil cement, plastic soil cement	—

Table 114 : Generalized rating of suitability of various types of soils for use as landfill cover material

Function	*General Soil Type*					
	Clean gravel	*Clayey–silty gravel*	*Clean sand*	*Clayey–silty sand*	*Silt*	*Clay*
Prevents rodents from burrowing(tunneling)	G	F – G	G	P	P	P
Keeps flies from emerging	P	F	P	G	G	E+
Minimizes moisture entering fill	P	F – G	P	G – E	G – E	E+
Minimizes landfill gas+ entering fill	P	F – G	P	G – E	G – E	E+
Provides planning appearance and controls blowing pipes	E	E	E	E	E	E
Supports vegetation	P	G	P – F	E	G – E	F – G
Vents decomposition gas++	E	P	G	P	P	P

E – excellent; G – good; F – fair; P – poor

+ Except when cracks extend through the entire cover

++ Only if well drained

Table 115 : Important factors in the design and operation of solid waste landfills

Parameter	*Remarks*
[A] Design	
Access	Paved all weather access roads to landfill site; temporary roads to unloading areas
Cell design and construction	Will vary depending on terrain; landfilling method; and whether gas is to be recovered
Cover material	Maximise use of on-site earth material; approximately 1 m^3 of cover material will be required for every 4 to 6 m^3 of solid wastes; mix with sealants to control surface infiltration. In some designs, intermediate cover is not used
Drainage	Install drainage ditches to divert-water run-off; maintain 1 to 2 percent grade on finished fill to prevent ponding
Equipment requirements	Vary with size of landfills
	Fire prevention Water on-site; if non-potable, outlets must be marked clearly, proper cell separation prevents continuous burn through if combustion occurs
Groundwater protection	Divert any under ground springs; if required, install sealants for leachate controls; install wells for gas and groundwater monitoring
Land area	Area should be large enough to hold all wastes for a minimum of 1 year but preferably 5 to 10 years

(Contd..)

Land filling method	Depend on terrain and available cover
Litter control	Use movable fences at unloading areas; crews should pick up litter at least once a month or as required
Operation plan	With or without the co-disposal of treatment plant sludges and the recovery of gas
Spread and compaction	Spread and compact waste in 0.6 m (2 ft) layers
Unloading area	Keep small, generally under 30 m (100 ft)
[B] Operation	
Communication	Telephone for emergencies
Days and hours of operation	Usually 5 to 6 days per week; 8 to 10 h/d
Employee maintenance	Rest rooms and drinking water should be provided
Equipment maintenance	A covered shed to be provided
Operational records	Tonnage, transactions, and billing if a disposal fee is charged
Salvage	No scavenging; salvage should occur away from the unloading area; no salvage storage on site
Scales	Essential for record keeping

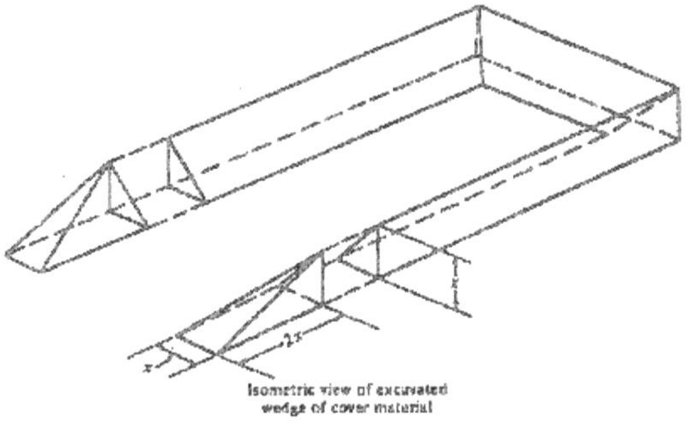

(a)

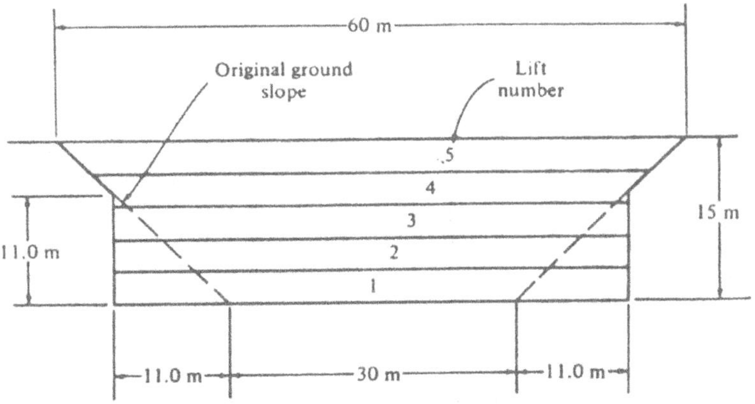

(b)

Figure 45 : Isometric view of excavated wedge of cover material and section through completed landfill

Example 3.141

Leachate Rate

Determine how much leachate in gallons/hour would generate during the operation of landfill in Kanpur. Assume the annual rainfall for the year in question is 45 in. and the affected surface area of the landfill is 1.0 sq. miles. Also assume that the leachate produced in 40% of the estimated rainfall.

Solution

1. Required leachate rate

Effective height (H) of leachate in relation to rainfall is :

$H = (0.40)\ (45 \text{ in.})$

$= 18.0 \text{ in./yr} = 1.5 \text{ ft/yr}$

Cross-sectional area (A) of the landfill is :

$A = 1.0 \text{ mile}^2$

$= 1.0(5280 \text{ ft}^2)$

$= 2.79 \times 10^7 \text{ ft}^2$

Volumetric leachate rate (L) is :

$L = (1.5 \text{ ft/yr})(2.79 \times 10^7 \text{ ft}^2)$

$= 4.18 \times 10^7 \text{ ft}^3/\text{yr}$

Mass based leachate rate (L) is:

$L = (4.18 \text{ ft}^3/\text{yr})(62.4 \text{ lb/ft}^3)$

$= 2.61 \times 10^9 \text{ lb/yr}$

[No evaporation, groundwater infiltration, or biological reactions have been considered].

Example 3.142

Leachate pH in Contact with Landfill Gas

Determine the pH of the Leachate in contact with landfill gas. Assume the composition of the landfill gas in contact with the leachate is 50% CO_2 and 50% CH_4, and the landfill gas is saturated with water vapour at a temperature of 50°C and the pressure within the landfill is one atmosphere. The alkalinity of the leachate is 250 mg/L.

Solution

1. Required pH of the leachate

Required saturation concentration of CO_2 at 50°C and 1 atm is 379 mg/L

Equilibrium equation

$$\frac{[H^+][HCO_3^-]}{[H_2CO_3^*]} = K_1 \quad \text{[First dissociation constant]}$$

where $[H^+][HCO_3^-]$, and $[H_2CO_3^*]$ = Molar concentration of hydrogen ion, bicarbonate ion, and carbonic acid, respectively [mol/L]

$$[H_2CO_3^*] = [CO_2(aq)] + [H_2CO_3]$$

Assume that : $[CO_2(aq)] = [H_2CO_3^*]$

Molar concentration

$$[HCO_3^-] = \frac{250 \text{ mg/L}}{50{,}000 \text{ mg/mol}} = 0.005 \text{ mol/L}$$

[Assume that all alkalinity is due to HCO_3^- ion]

$$[CO_2(aq)] \approx \frac{379 \text{ mg/L}}{44000 \text{ mg/mol}}$$

$$= 0.00861 \text{ mol/L}$$

pH volume of the leachate at 50°C is

$K_1 = 5.07 \times 10^{-7}$ mol/L

$$\frac{[H^+][0.005 \text{ mol/L}]}{[0.00861 \text{ mol/L}]} = 5.07 \times 10^{-7} \text{ mol/L}$$

$$[H^+] = \frac{(5.07 \times 10^{-7})(0.00861)}{0.005}$$

$$= 8.73 \times 10^{-7}$$

$pH = -\log [H^+] = 6.059.$

Example 3.143

Leaching of Soluble Salts from Solid Sludges

1. Multiple leaching of solid wastes in a simple tank (n-stages)

$$E_n = \frac{A + W\left[\left(\frac{R}{n}+1\right)^n - 1\right]}{\left[\left(\frac{R}{n}+1\right)^n\right]}$$

$$R = n\left[\left(\frac{A-W}{E-W}\right)^{1/n} - 1\right]$$

2. Multiple tank, counter-current leaching (n tanks)

$$E_n = \frac{A + W\left(R^n + R^{n-1} + \text{-------} + R\right)}{R^n + R^{n-1} + \text{------} + R + 1}$$

$$= \frac{A(R-1) + WR\left(R^n - 1\right)}{R^{n+1} - 1}$$

$$\left(\frac{R^{n+1}-1}{R-1}\right) = \left(\frac{A-W}{E-W}\right)$$

If total of R volumes of leaching water to with W salt (mg/L) content is added for each volume of moisture in the sludge and salt content of the moisture before leaching is A mg/L, the salt content of the leached moisture, E in mg/L can be found by striking a mass balance between salt content entering and leaving the leaching tank or tanks.

3. Determine the number of stages (n) for counter-current leaching operation. Use the following data :

Waste sludge composition :		
Inerts solids	:	13%
Soluble salt	:	2%
Water	:	85%
Solvent used for leaching	:	Water (with no soluble salt)
Solvent to water ratio	:	2
Required removal of soluble salt	:	99%
Under flow from stages contain water plus soluble salt per lb of inert solids	:	6 lb

Consider 1 lb of dry solid waste :

$$E_n = \frac{A(R-1) + WR\left(R^n - 1\right)}{R^{n+1} - 1}$$

$$\frac{R^{n+1}-1}{R-1}=\frac{A-W}{E-W}$$

$$E_n=\frac{0.02(2-1)+0}{2^{n+1}-1}=0.02\times 0.01$$

$$2\times 10^{-4}\times 2^{n+1}=0.02+0.0002$$

$$=0.0202$$

$$2^{n+1}=\frac{0.0202}{2\times 10^{-4}}$$

$$(n+1)\log(2)=\log\left(\frac{202\times 10^{-4}}{2\times 10^{-4}}\right)$$

$$n+1=\frac{2.004}{0.301}=6.66$$

or alternatively, n = 6.68 – 1 = 5.66 (says 5 or 6)

$$\frac{2^{n+1}}{2-1}=\frac{0.02-0}{0.0002-0}=\frac{2\times 10^{-2}}{2\times 10^{-4}}=10^2$$

$$2^{n+1}-1=100+1=101$$

$$n+1(\log 2)=\log(101)$$

$$n+1=\frac{2.004}{\log 2}=6.6$$

$$n=5.6\text{ (say 5)}$$

A is the concentration of salt in the waste = 0.02 lb/lb of waste.

W is the concentration of salt in the solvent (water) = 0, E is the concentration of salt remaining 0.0002 lb/lb of waste.

BIBLIOGRAPHY

Process Design Manual, Sludge Treatment and Disposal : EPA (Publication No. EPA 625/ 1-79-011, Wash DC, 1979.

Process Design Manual, Sludge Treatment and Disposal, US EPA.

Dick, R.I., Thickening, Ad. Wat. Qual. Improvement : Physical and Chemical process, Ed (Gloyna, E.F, and Eckenfelder, W.W.), Austin, Univ. of Texas Press, p. 358 (1970).

Campbell, H.W., Rush. R. J., and Ten, R., Sludge Dewatering Design Manual, Research Report No. 72, Ontario Ministry of Environment, Toronto, Ontario.

Smith, R., Estimating The Rate of Sludge Production in The Activated Sludge Process, Memo (October 10,1972).

City of Austin and Univ of Texas, Desigin guides line for Biological Wastewater Treatment Process, USEPA : Water Pollution Control Res. Series 110101 ESQ (Aug. 1971).

Kreissl, J.F., Westrick, J.J., Municipal Waste Treatment, by Physical-Chemical Method, Vanderloilt Symp., Application of New Concepts of Physical-Chemical Waste Water Treatment (Sept. 18-22, 1972).

Schaffer, R.B. *et al.*, Application of a Carbon Analyser in Waste Treatment, JWPCF 37(11), 1545 (1965).

Bunch, R.L., *et al.*, Organic Material in Secondary Effluent, JWPCF 33(2), 122 (1961).

Coackley, P. Development in our knowledge of Sludge Dewatering behavior, 8th Public Health Engineering Conference, Civil Eng. Dept., Lough borough Univ. of Tech., (1975).

Rietema, K Stabilising Effects in Compressible Filter Cake, Chem. Eng. Sci. **2**, 88 (1953).

Tchobanoglous, G., Theisen, H., and Eliassen, R., Solid Wastes: Engg. Principles and Management Issues, McGraw Hill, N.Y., 1977.

Coe, H.S., Clevenger, G.H., Methods for Determining the Capacities of Slime Settling Tanks, Trans. AIME, 53, 356, 1916

Yoshioka, N., *et at.*, Continuous Thickening of Homogenous Flocculated Slurries, Chem. Eng. (Tokyo), 21, 1957

Kynch, G.I., A Theory of Sedimentation, Trans. Faraday Society, 48, 166, 1952.

Dick, R.I., Role of Activated Sludge Final Settling Tanks, J. Sanit. Eng. Div., Am. Soc. Civil Engrs., 96(SA-2), 423, 1970.

Dick, R.I., Process Design in Water Quality Engineering, Thackston, E.L., and Eckenfeldor, W. W. (Ed_s), Jenkun Publishing Co., Austin and N. Y., 1972, Chap.17.

US EPA, Process Design Manual for SS Removal, PA-625/1-75-003a, Jan. 1975.

USEPA, Treatability Manual, EPA-600/8-80-042a, 1980.

Talmadge, W. P. and Fitch, E. B., Determining Thickening Unit Areas, Ind. Eng, Chem., 47 (1), 1955.

Roberts, E.J., Thickening? Art or Science?, Mining Eng. 1, 61, 1969.

Schroeder, E.D., Wat and Wastewater Treatment, the McGraw-Hill Book Co., N.Y., 1977.

EPA Process Design Manual for Wastewater Treatment Facilities for Sewered Small Community, US EPA, Washington, D.C.

Gale, R.S., Filtration Theory with Special Reference to Sewage Sludges, J. Inst. Wat. Pollu. Control 66, 622 (1967).

Goodman, B.L., Design Handbook of Wastewater Systems Domestic, Industrial, Commercial, Technomic Publishing Co., Westport, Conn., 1971.

Recommended Standards for Sewage Works, A Report of the Committee of the Great Lakes-Upper Mississipi River Board of State Sanitary Engineers, Health Education Service, Albany, N.Y., 1968.

Krischer, A., The Scientific Principles of Drying Technology, Springer-Verlag, W., Berlin 1963.

Arceivala, S.J. Wastewater Treatment and Disposal, Marcel Dekker, N.Y. (1981).

Barnard, J. and Englande, A. J. Centri-fugation, on Process Design in Water Quantity Engineering, Jenkins Publishing, Austine, Texas, 1972.

Metcalf and Eddy, Inc. Wastewater Engg. Treatment, Disposal; and Reuse, McGraw Hill, N. Delhi, 1979.

Veslind, P.A. and Peirce, J.J., Envn. Engg. Ann Arbor Science, Collingwood, MI, 1982.

Bernard, J. and Englande, B.R.S., Centrifugation in Process Design in Water Quality Engineering, Jenkins, Austin, Texas, 1972.

Kirk and Othmer's Chemical Encyclopedia.

Albertson, O.E. et al., Centrifugation of Waste Sludges, JWPCF, 41(4), 607, April 1979.

USEPA, Design Manual for Dewatering Municipal Wastewater Sludges, EPA-625/1-82-014, Oct., 1982.

USEPA Design Information Report Belt Filter Presses, EPA-600/M-86-011, May, 1986.

Gale, R.S., The Calculation of Theoretical yields of Rotary Vacuum Filters, J. Inst. Water Poll. Control 70, p. 114 (1971).

USEPA, Continued Evaluation of Oxygen use in Conventional Activated Sludge Processing, Report No. 170 50 DNW 02/72, February, 1972.

Friedman, S.J. and Marshell, W.R. Jr., Chem. Eng. Progr. 45(8), 482 (1949).

Miller, C.O., Smith, B.A, and Schuette, W.H., Trans. AIChE, 38, 841, 1942.

Prutton, C.F., Miller, C. O., and Schuette, W.H., Trans. AIChE 38, 123, 1942.

Peck, R.E., Wasan, D.T., Adv. Chem. Eng. 9, 247, 1947.

Saeman, W.C., Mitchell, T.R., Chem. Eng. Progr. 50 (9), 467, 1954.

Badger, W.L., and Bonchero, J. T., Introduction to Chemical Enggs, McGraw-Hill, N. Y. Chap 10, 1955.

McCabe, W. L., and Smith, J.C. Unit Operations of Chemical Engg., McGraw Hill, N.Y., Clap 25, 1976.

The ratio L/D_d is within 4-10, as suggested by Perry, J.H. Ed. Chem. Engr's Handbook, McGraw-Hill, N.Y., 1963.

Crackley and Jones proposed a similar expression for vacuum filtration (Sew. Ind. Wastes 28 (8), 963, 1956).

Blunk, H. Beitary Zur Berechnung von Faulraumen, Gesundheitzing enieur, 48 (4), 37, 1926.

Vater, w., Die Entwasserung, Trocknung and Beseitigang von stadtischen Klarschlan, Doctoral Dissertation, Technische Hochschule, Hanover, 1910.

Jones, B.R.S., Vacuum Sludge Filtration, Prediction of Performance, Sew. Ind. Wastes 28, 1103 (1956).

Baskerville, R.C. and Komorak, J.A. Effect of Operating Variables in Filter Press Performance, J. Indst. Wat. Pollu. Control, 70 400 (1971).

Bartell, R.E., Public Health Engineering : Design in Metric-Wastewater Treatment, Applied Science Publishers, London (1971).

Brovko, N., and Chen, K. Y., Optimising Gas Production, Methane Content, and Buffer Capacity in Digester, Water and Sewage Works, 124 (7), 54, 1977.

Barber, N.R., and Dale, C.W., Increasing Sludge Digester Efficiency, Chemical Engineering 149, July 17, 1978.

Lawerence, A.W. Application of Process Kinetics to Design of Anaerobic Process, Chapter 9 in Anaerobic Biological Treatment Processes, Advances in Chemistry Series 105, Am. Chem. Society, 1971.

Ames Crosta Bobcock Ltd. Publication 100, Simflex Sludge Digestion Plants, Heywod, Lancashire.

Brooks, R.W. Conversion of Sludge into utilizable Gas, Chapter 4 in Dickison (ed), Practical Waste Treatment and Disposal, Applied Science Publishers, London 1974.

Bartlett, R.I. Public Health Engineering, Applied Science Publishers, London, 1971.

Reynolds, T.D., Aerobic digestion of thickened waste activated sludge, Proc. 28th Industrial Waste Conf., Purdue Univ., Ind., 12, 1973.

Randall, C.W., et al, Biological and chemical changes in activated sludges during aerobic digestion., Proc., 18th Southern Wat. Res. and Pollu. Control Conf., North Caroline State Univ., North Carolina, 1969.

Schulze, K.L. Rate of Oxygen Consumption and Respiratory Quotients during Aerobic Decomposition of a Synthetic Garbage, Compost Science 1 (1), 36, 1960.

Haug, R.T., Engg. Principles of Sludge Composting, JWPCF, 51(8), 1979.

Rich, L.G., Unit Processes of Sanitary Engg., John Wiley & Sons, N.Y., 1963.

Burge, W.D., et.al., Criteria for Control of Paths Gens during Sewage Sludge Composting, Proc. Nat. Conf. on Design of Muni. Sludge Composting Facilities, 1978.

Liao, P.B., Fluidized Bed Sludge Incinerator Design, JWPCF 46(8), 1895, 1974.

Gales, R.S. the Sludge Treatment and Disposal Problem, in Helliwell, P. (ed.), Proc. Symp. On Incineration of Refuse and Sludge, South Ampton University Press, 1972.

Hitchock, Solid Waste Disposal: Incineration, Chemical Engrs, May 21, 1997.

Orning, A.A., Principles of Constructions; Principles and Practice of Incineration, Ed. Corey, R.C., Wiley Interscience, N.Y., 1969.

Danielson, J., Air Pollution Engineering Manual, Country of Los Angeles, Air Pollution Control District, Ap-40, May 1973.

Brunner, C.R., Incineration System, Selection and Design, Incinerator Consultants Inc., Virginea, 1984.

Niessen, W.R., Combustion and Incineration Processes: Application in Environmental Engg., Marcel Dekker, Inc., N.Y., 1978.

Techobanoglous, G., Theisen, H., Eliassen, R., Solid Wastes: Engg. Principles and Management Issues., Mc Graw Hill, N.Y., 1977.

Techobanoglous, G., et.al. Solid Wastes: Engg. Principles and Management Issues, McGraw Hill, N.Y., 1977.

CHAPTER 4

NATURAL TREATMENT PROCESSES

Example 4.1

Ecosystem Partitioning and Solute Transport

1. An ecosystem may be thought of as a series of homogenous compartments such as air, water, soil, sediment, and biological material. The concentration of an organic chemical in any compartment is related to the concentration in any other compartment by a specific distribution coefficient :

   Air-water : Henry's law constant

   Solid-water : Distribution constant

   Fish-water : Bio-concentration factor

2. **Liquid-liquid Partitioning :** A distribution coefficient is a rate of the concentration of the solute in each of the two immiscible phases, and is dependent on two factors :

   - Distribution coefficient
   - Volume of each phase present.

   2.1 Immiscible liquid (i) and water

$$K_i = \frac{C_i}{C_w} = \frac{\dfrac{M_i}{V_i}}{\dfrac{M_w}{V_w}} = \frac{M_i \times V_w \,(\mathrm{mg/L})}{M_w \times V_i \,(\mathrm{mg/L})}$$

$$M_i = M_w \left(\frac{V_i}{V_w}\right) K_i$$

If the total mass of solute in the system is the loading, then

$$\text{Loading (LD)} = M_w + M_i = M_w \left[1 + \frac{V_i}{V_w} K_i\right]$$

2.2 Solid(s) and water

$$K_s = \frac{C_s}{C_w} = \frac{\dfrac{M_s}{MASS_s}}{\dfrac{M_w}{V_w}} \frac{mg/kg}{mg/L} = \frac{L}{kg} = \frac{mL}{g}$$

Mass = Density × Volume

$$K_s = \frac{C_s}{C_w} = \frac{\left(\dfrac{M_s}{V_s \times D_s}\right)}{\left(\dfrac{M_w}{V_w}\right)} = \frac{M_s \times 1 \times V_w}{V_s \times D_s \times M_w}$$

$$M_s = M_w\left(\frac{V_s}{V_w}\right)(D_s)(K_s)$$

$$\textit{Loading}\ (LD) = M_w + M_s = M_w\left[1 + \frac{V_s}{V_w}(D_s)K_s\right]$$

C_1	C_2

$C_2 = KC_1$

Phase 1 Conc. = C_1

Phase 2 Conc. = C_2

2.3 Octanol-water partition coefficient (K_{ow})

K_{ow} describes the distribution of a solute between two immiscible phases (that remain separate from one another), octanol and water :

$$K_{ow} = \frac{\text{Concentration in octanol } (C_o)}{\text{Concentration in water } (C_w)}$$

2.4 *Example :* Consider an octanol/water system containing 200 mL octanol and 500 mL water with 10 mg solute distributed between them. The log K_{ow} for solute is 1.0

$\log\ K_{ow} = 1$ or $K_{ow} = 10^1$

or $\dfrac{C_o}{C_w} = 10$

If water contains x mg solute, then octanol contains (10 – x) mg of solute :

$$C_w = \frac{x\ mg}{500\ mL} = 2x\ mg/L$$

$$C_o = \frac{(10 - x)\ mg}{200\ mL} = 5(10 - x)\ mg/L$$

Therefore, $C_o = K_{ow} C_w$

$5(10 - x) = 10 \times 2x$

$50 - 5x = 20x$ or $x = 2$

$C_w = 4$ mg/L, $C_o = 40$ mg/L

- Calculate concentrations for $\log K_{ow} = 2.1$

$K_{ow} = 10^{2.1} = 125.9$

$5(10 - x) = (125.9)(2x)$

$x = 0.195$

$C_w = 0.39$ mg/L

$C_o = 49$ mg/L

- If the amount of solute in each phase is expressed as a percentage of the total amount of solute, and the volume of each phase is expressed as a percentage of the total volume :

$\% M_o$ = % of solute mass in octanol

$\% M_w$ = % of solute mass in water

$\% V_o$ = Volume of octanol expressed as % of the total volume

$\% V_w$ = Volume of water expressed as % of the total volume

Concentration of solute in octanol, $C_o = \dfrac{\% M_o}{\% V_o}$

Concentration of solute in water, $C_w = \dfrac{\% M_w}{\% V_w}$

$$K_{ow} = \frac{\dfrac{\% M_o}{\% V_o}}{\dfrac{\% M_w}{\% V_w}}$$

or $\% M_o/\% V_o = K_{ow} (\% M_w/\% V_w)$

$$\% M_o = K_{ow} \left(\frac{\% V_o}{\% V_w}\right)(\% M_w)$$

$$\% M_o = 100 - \% M_w$$

$$(100 - \% M_w) = K_{ow} \left(\frac{\% V_o}{\% V_w}\right)(\% M_w)$$

or $100 = \% M_w[1 + K_{ow} (\% V_o/\% V_w)]$

or $$\% M_W = \frac{100}{\left(1 + K_{OW} \times \frac{\% V_O}{\% V_W}\right)}$$

and $$M_w = \% M_w \times \text{Loading}$$

$$V_o = 200 \text{ mL}, V_w = 500 \text{ mL}$$

$$\% V_o = \frac{200 \text{ mL}}{(200+500)\text{mL}} \times 100 = 28.6\%$$

$$\% V_w = \frac{500 \text{ mL}}{700 \text{ mL}} \times 100 = 71.4\%$$

$$\% M_w = \frac{100}{\left[1 + 10\left(\frac{28.6}{71.4}\right)\right]} = 20\%$$

$$\% M_o = 10\left(\frac{28.6}{71.4}\right)20 = 80\%$$

$$\text{Loading} = (C_o V_o + C_2 V_2) = 10 \text{ mg}$$

$$M_w = (\% M_w)(\text{Loading})$$

$$= \left(\frac{20}{100}\right)(10) = 2 \text{ mg}$$

$$M_o = (10 - 2) = 8 \text{ mg}.$$

3. Bioconcentration factor (BCF)

BCF describes the distribution of a solute between a fish and water :

$$\text{BCF} = \frac{\text{Concentration in fish } (C_f)}{\text{Concentration in water } (C_w)}$$

$$\text{BCF} = \frac{\left(\frac{M_f}{V_f}\right)}{\left(\frac{M_w}{V_w}\right)} = \frac{M_f \times V_w}{M_w \times V_f}$$

where M is the mass of solute in compartment with volume V

$$M_f = (\text{BCF})\left(\frac{V_f}{V_w}\right)M_w$$

3.1 *Example* : Concentration of a chemical in a pond of 20 mg. Determine the concentration of the chemical in the fish. Assume a log BCF value of 3.2.

$\log (BCF) = 3.2$, $BCF = 10^{3.2}$

$$C_f = BCF\,(C_w)$$
$$= 10^{3.2} \times 20 = 31{,}698\ \mu g = 31.7\ mg/L.$$

- Assume a 1 kg spill of a material in a pond with a capacity of 100 × 100 × 10 m^3. If the fish make-up 1 ppm of the pond and log BCF is 3.2, find the concentration of the material in a pond.

  Pond capacity = $10^5\ m^3$

$$\text{Fish volume} = \frac{\text{pond}}{10^6} = 0.1\,m^3$$

$$BCF = 10^{3.2} = 1585$$

$$\text{Loading (LD)} = M_w + M_f$$
$$= M_w + M_w\,(BCF)\frac{V_f}{V_w}$$
$$= M_w\left[1 + (BCF)\frac{V_f}{V_w}\right]$$
$$1\,kg = M_w\left[1 + \frac{0.1}{10^5} \times 1585\right]$$

or
$$M_w = \frac{1000\ g}{1.001585} = 998.418\ g$$

$$C_w = \frac{M_w}{V_w} = \frac{998.418\ g}{10^5} = 9.984 \times 10^{-3}\ g/m^3$$
$$= 0.01\ mgL$$
$$M_f = (1000 - 998.418) = 1.583\ g$$
$$C_f = \frac{M_f}{V_f} = \frac{1.583\ g}{0.1\ m^3} = 15.83\ g/m^3$$

- BCF-K_{ow} relationship

  $\log\ BCF = 0.935 \ln K_{ow} - 3.443$

  or $\log\ BCF = 0.935 \log K_{ow} - 1.495$

4. **Non-aqueous phase liquid partitioning (NAPL) :** Partitioning of liquid solutes between non aqueous phase liquid (NAPL) and water may be calculated by the relationship :

$$\frac{C_{NAPL}}{C_{water}} = K_{NAPL}$$

C_{NAPL} = mole fraction – x_i

C_{water} = mg/L – S_i^e (effective solubility)

K_{NAPL} = $1/S_i$ where S_i is the pure solubility of compound i

S_i^e = $X_i \times S_i$

where S_i = Maximum solubility of a solute in water

X_i = Mole fraction concentration in non-aqueous phase

S_i^e = Effective solubility of a solute in water partitioning from NAPL in which it is in contact.

4.1 *Example :* If a dense non aqueous phase liquid (DNAPL) contains 0.10 mole fraction of trichloroethene (TCE), then the solubility in water in contact with it is :

$$(0.1)(1000 \text{ mg/L}) = 100 \text{ mg/L}$$

To calculate the partitioning of a solid solute between NAPL and water, the liquid solubility must be obtained, as is calculated from solid solubility by the equation :

$$C_1 = C_s \exp\left[6.8\left(\frac{T_m}{T} - 1\right)\right]$$

where C_l = Liquid solubility

C_s = Solid solubility

T_m = Melting point (°K)

T = System temperature (°K)

Table 1 : General data for various compounds

Compound	*Melting Point(°C)*	*Solubility Solid(mg/L)*	*Solubility Liquid (mg/L) (15°C)*	*Factor* C_l/C_s
Chrysene	256.5	0.0620	0.583	292
Anthracene	216.4	0.073	8.46	116
Pentachlorophenol	191	14	891	64
Pyrene	156	0.135	3.76	28
Phenathrene	161	1.29	9.82	8
Naphthalene	80.5	31.7	149	5
2-Methyl napthalene	34.6	25.4	40.3	2

The liquid phase solubility values for the solutes may be considerably higher than the solid-phase solubility values, and that the difference is greater for the compounds with higher melting points. Thus, an oily waste product containing a small fraction of a high melting point solid may present greater hazard to the groundwater than a concentrated pile of the solid component.

5. Adsorption isotherms equations (solid phase partitioning)

- Linear adsorption isotherms result when the distribution coefficient is independent of concentration :

$$C_s = k_d\, C_w$$

where C_s = Amount of solute adsorbed by the solid phase

C_w = Concentration of solute in the liquid phase

k_d = Distribution coefficient

- Freundlich isotherms

$$C_s = k_d\, C_w^{\,n}$$

[n = 1, linear relationship; n > 1, concave up; n < 1, convex up]

or $$\frac{X}{M} = K\, C_w^n \text{ [Activated carbon partitioning]}$$

where X/M = Concentration in the solid phase (mg/g)

X = Amount of solute adsorbed by the carbon

M = Concentration of carbon in the Solution

K = Activated carbon distribution coefficient (L/g)

C_w = Concentration of solute in the Solution (mg/L)

- A litre of solution containing 10 mg/L solute contains 3 mg/L after adding 2 g of activated carbon, determine the distribution coefficient (K). Assume K = 1

Concentration of solute in water = 3 mg/L

Concentration of carbon = 2 g/L

Mass of solute in carbon = (10 – 3) = 7 mg

$$\frac{X}{M} = K\, C_w$$

or $$K = \frac{(10-3)}{(2)(3)} = 1.17 \text{ L/g}$$

Assume K = 50 and n = 0.7; how much activated carbon would be required to reduce a solute from 15 to 2 mg/L?

Concentration of solute in water = 2 mg/L

Concentration of solute in carbon = (15 – 2) = 13 mg/L

$$\frac{13}{M} = 50(2)^{0.7} \quad \text{or} \quad M = 0.16 \text{ g carbon/L}$$

6. Soil sorption constant (K_d)

It describes the behaviour of a solute in water with respect to solid :

$$K_d\,(\text{mL/g}) = \frac{\text{Conc. in soil}\,(C_s)}{\text{Conc. in water}\,(C_w)} = \frac{\dfrac{M_s}{\text{Mass of soil}}}{\dfrac{M_w}{\text{Mass of water}}}$$

Soil mass = $V_s \times D_s$, where D_s = Bulk density

$$K_d = \frac{\dfrac{M_s}{V_s D_s}}{\dfrac{M_w}{V_w}} = \frac{(M_s)(V_w)(1)}{(M_w)(V_s)(D_s)}$$

$$M_s = (K_d)\left(\frac{V_s}{V_w}\right)(D_s)(M_w)$$

6.1 Normalized K_d

Normalized distribution coefficient obtained from soils to their organic content is expressed as :

$$K_o = K_d\left(\frac{100}{\%\,OC}\right)$$

where K_{oc} = Normalized distribution coefficient

K_d = Distribution coefficient obtained using a soil containing OC percent organic carbon.

6.2 Solute distribution in an aquifer

$K_d = K_{oc} \times$ TOC (expressed as a fraction)

$$100 = M_w\left[1 + \left(\frac{V_s}{V_w}\right)(D_s)(K_d)\right], \qquad \text{[refer Section 2.2]}$$

Assume a porosity *e*,

$$\frac{V_s}{V_w} = \left(\frac{1-e}{e}\right),$$

D_s (bulk density) = 2.65 (1 – e) for a quartz matrix :

$$100 = M_w\left[1 + \left(\frac{1-e}{e}\right)(2.65)(1-e)(K_{oc})(\text{TOC} - \text{fraction})\right]$$

6.3 ***Critical sediment concentration :*** For an acute maximum concentration of a toxic material in water, there exists a maximum critical concentration in the sediment in equilibrium with it :

$$K_{oc} = \frac{K_d}{TOC(\text{fraction})}$$

$$= \frac{C_s}{(C_w \times 100)}$$

$$C_s = K_{oc}(C_w)(TOC)$$

If acute concentration ($C_{w\text{-}ac}$) in water is 0.16 μg/L; K_{oc} = 1950

$$C_{s-ac} = \frac{(1950(0.16)}{(1000\ g/g)} = 0.31\ g/g$$

7. Pond-ecosystem (Water and sediment only)

$\% M_w$ = % Solute in water phase

$\% M_s$ = % Solute in sediment phase

$\% V_w$ = % Total volume occupied by water

$\% V_s$ = % Total volume occupied by sediment

D_s = Bulk density of sediment (2.0)

D_w = Density of water (1.0)

C_w = Concentration of solute in water :

$$C_w = \frac{\% M_w}{\text{Mass of water}} = \frac{\% M_s}{(\% V_w \times D_w)} = \frac{\% M_w}{\% V_w \times 1.0}$$

C_s = Concentration of solute in sediment:

$$= \frac{\% M_s}{\text{Mass of sediment}} = \frac{\% M_w}{(\% V_s \times D_s)} = \frac{\% M_s}{\% V_s \times 2.0}$$

$$\text{Distribution Coefficient } (K_d) = \frac{K_{oc}\,\%OC}{100} = \frac{C_s}{C_w} = \left(\frac{\% M_s}{2.0 V_s}\right)\left(\frac{\% V_w}{\% M_w}\right)$$

$$\% M_s = \frac{(K_d)(2.0)(\% V_s)}{\% V_w} \times \% M_w = X(\% M_w)$$

$$100 = \% M_s + \% M_w = X(\% M_w) + (\% M_w)$$

$$= \% M_w\ (1 + X)$$

$$\% M_w = \frac{100}{X+1}$$

Concentration in ppm (mg/kg) :

Mass of solute in phase X in mg :

$$= \frac{\% \, M_X}{100} \left[\text{loading (kg)} \right] \times 10^6$$

Mass of phase X is kg:

$$= \frac{\% \, V_S}{100} - V_t \left(m^3 \right) \times 10^3 \left(D_x \right)$$

D_x = Density of phase X

Concentration of solute in phase X :

$$= \frac{\left(\% \, M_S \right) \left[LD(kg) \right] \times 10^3}{\left(\% \, V_x \right) \left[V_t \left(M^3 \right) \right] \times D_X}$$

7.1 *Example*

Pond diameter	= 100 m
% OC in sediment	= 5 %
Water depth	= 4 m
K_{oc} solute	= 1000 mg/g
Sediment depth	= 5 cm
Loading of solute	= 1 kg
Volume of water	= π $(50)^2$ (4) = 31,416 m^3
Volume of sediment	= π $(50)^2$ (5/100) = 392.7 m^3

$$\% \, V_W = 98.77 \left(= \frac{31416}{31416 + 392.7} \times 100 \right)$$

$\% \, V_s$ = 100 – 98.77 = 1.23%

$$K_d = K_{oc} \left(\frac{\% \, OC}{100} \right) = 1000 \times \frac{5}{100} = 50$$

$$\left(K_d \right) = (2.0) \left(\frac{\% \, V_S}{\% \, V_W} \right) = \frac{50 \times 2 \times 1.23}{98.77} = 1.254$$

$$\% \, M_W = \frac{100}{1 + 1.254} = 44.4\%$$

$\% \, M_s$ = (100 – 44.4) = 55.6%

$$C_w\,(\text{ppm}) = \frac{(\%\,M_w)(LD)\times 10^3}{(\%\,V_w)(V_t)(D_w)} = \frac{44.4\times 1\text{ kg}\times 10^3}{(98.77)(31808.7)(1)} = 0.014\text{ ppm}$$

$$V_t = (31{,}416 + 392.7)\text{ m}^3 = 31{,}808.7\text{ m}^3$$

$$C_s\,(\text{ppm}) = \frac{(\%\,M_s)(LD)\times 10^3}{(\%\,V_s)(V_t)(D_s)} = \frac{(55.6)\times(1\text{ kg})\times 10^3}{(1.23)(31808.7)\times 2} = 0.71\text{ ppm}$$

8. Air-water partitioning
   - Henry's law constant (H)
     - H (atm-Liter/g)

$$H = \frac{p(\text{atm})}{C_w(\text{g/L})}$$

     - H (atm-Liter/mole)

$$H = \frac{p(\text{atm})}{C_w(\text{mol/L})}$$

     - H (atm/mole fraction)

$$H = \frac{p(\text{atm})}{X_w(\text{mole fraction})}$$

     - H (dimensionless) $H = \dfrac{C_a(\text{mol/L})}{C_w(\text{mol/L})}$
   - Conversion equation
     - Gas law

$$PV = n\,R\,T, \qquad n = \frac{W}{M}$$

$$PV = \frac{WRT}{M}$$

$$P = \frac{WRT}{(M)(V)}$$

$$C_a(\text{mol/L}) = \frac{W}{(M)(V)}$$

$$p = C_a\,R\,T$$

Therefore,

$$H = \frac{P}{C_w} = \frac{C_a\,R\,T}{C_w} \text{ if } C_w \text{ in mol/L}$$

$$R = 0.68205 \text{ atm-litre/mol } ^\circ K$$
$$= 8.2056 \times 10^{-5} \text{ m}^3 \text{ atm/mol } ^\circ K$$

- Approximation for H-

$$C_a\left(g/L\right)=\frac{W\,P\,M}{V\,R\,T}$$

At saturation $C_w = C_{sat}$ (solubility)

For pure solute $P = P_o$

Therefore,

$$\frac{C_a}{C_w}=\frac{P_o\,M}{R\,T\,C_{sat}}$$

If P_o is in mm Hg, then P_o (atm) $= \dfrac{P_o\,(\text{mm Hg})}{760}$

If C_{sat} is in mg/L, then C_{sat} (g/L) $= \dfrac{C_{sat}\,(\text{mg/L})}{1000}$

$R = 0.082054$ litre-atm/mol °K

$T = 25°C = 298.16$ °K

$$\text{Factor}=\frac{1000}{(0.082)(760)}=16.04$$

$$\frac{C_a}{C_w}=\frac{16.04\,P_o\,(\text{mm Hg})M}{T(^\circ K)C_{sat}\,(\text{mg/L})}$$

8.1 ***Example :*** Determine the Henry's law constant for tetra-chloromethane

$M = 153.823$; Solubility (C_{sat}) = 800 ppm;

Vapour pressure (P_o) = 113 mm Hg, T = 25°C

$$H=\frac{(16.04)(113)(153.823)}{(298.15)(800)}=1.169$$

[Larger the Henry's law constant (H), the greater will be the concentration of the contaminant in the air (easily removed by air stripping)].

8.2 Air-water distribution

$$H=\frac{\text{Concentration in air }(C_a)}{\text{Concentration in water }(C_w)}$$

$$= \frac{\frac{M_a}{V_a}}{\frac{M_w}{V_w}} = \frac{M_a\ V_w}{M_w\ V_a}$$

or $M_a = K_H \left(\frac{V_a}{V_w}\right) M_w$

8.3 Aquifer ecosystem (saturated system consisting of rock and water).

Similar to pond ecosystem except that solid phase is dominant and aqueous phase equals the pore space :

$$\frac{V_{sed}}{V_w} = \frac{(1-e)}{e}$$

$$M_{sed} = K_d \left(\frac{V_{sed}}{V_w}\right)(D)(M_w)$$

$$= K_d \left(\frac{1-e}{e}\right)(D)(M_w)$$

9.

Table 2 : Full ecosystem

Phase-I	*Water*
$\frac{M_i}{V_i} = C_i$	$K.C_w = K.\frac{M_w}{V_w}$

$$\frac{M_i}{V_i} = K\frac{M_w}{V_w}$$

$$M_i = K\left(\frac{M_w}{V_w}\right)V_i \text{ or } K\left(\frac{V_i}{V_w}\right)M_w$$

Phase-I	*Water phase*
Octanol	K_{ow}
Fish	BCF
Air	H
Organic carbon	K_{oc}
Soil	K_d

- Fish-water system :

$$BCF = \frac{C_f}{C_w} = \frac{M_f}{V_f}.\frac{V_w}{M_w}$$

$$M_f = M_w \left(\frac{V_s}{V_w}\right) BCF$$

- Sediment-water system :

$$K_d = \frac{C_s}{C_w} = \frac{M_s V_w}{(V_s D_s) M_w} \quad \text{[Ds = Bulk density of sediment (=2)]}$$

$$M_s = M_w \left(\frac{V_s}{V_w}\right)(K_d)(D_s)$$

[Used for both K_d-bed and K_d-soil]

- Air-water system :

$$H = \frac{C_a\,(\text{mol/L})}{C_w\,(\text{mol/L})} = \left(\frac{M_a}{M_w}\right)\left(\frac{V_w}{V_a}\right)$$

$$M_a = M_w \left(\frac{V_a}{V_w}\right) H$$

- *Air-soil-water system :* Equations for air-soil and soil-water system can be combined :

  Air soil water Air-water

$$H = \frac{V_{sw} M_a}{M_{sw} V_a} = \frac{V_w}{M_w} \frac{M_a}{V_a}$$

Therefore,

$$\frac{V_{sw}}{M_{sw}} = \frac{V_w}{M_w} \text{ and } M_{sw} = \frac{M_w V_{sw}}{V_w}$$

Loading (LD) = $M_w + M_f + M_a + M_{sw} + M_{sed} + M_{soil}$

$$= M_w \left[1 + BCF\frac{V_f}{V_w} + H.\frac{V_a}{V_w} + \frac{V_{sw}}{V_w} + k_{d-sed.} \times 2 \times \frac{V_{sed.}}{V_w} + K_{d-soil} \times 2 \times \frac{V_s}{V_w} \right]$$

10. **Partitioning Estimates using Parameter Ranges :** The fundamental parameters for estimating environmental partitioning parameters are solubility and vapour pressure.

    - ***Solubility :*** Many regression equations have been proposed for relating solubility and K_{ow}, solubility and K_{oc}, K_{ow} and K_{oc}, and BCF. As K_{ow} and solubility are actually different expressions of the same property, it is readily apparent that solubility, K_{ow}, K_{oc} and BCF are related. All these parameters have been used to classify organic pollutants into simple categories; however, the divisions used are usually somewhat arbitrary.

In the following discussion, several solubility divisions are used and the related parameter values are calculated using various regression equation such that all are consistent and that as many as possible related organic chemicals fall into the same category. The divisions used are solubilities of 65 ppm and 2000 ppm. The matching parameter ranges are :

Solubility ppm	65	and	2000
log K_{ow}	4.0	and	2.5
log K_{oc}	2.9-3.6	and	2.1-2.4
log BCF	2.2	and	0.8

- ***Vapour Pressure :*** The only common parameter used to estimate solute volatility is vapor pressure. The partition coefficient used to estimate water/air partitioning is Henry's law constant. This parameter has numerous unit choices, but the one most commonly used in distribution calculations is the unitless one. This value is often circulated from solubility and vapour pressure. Molecular weight and temperature are also used in the computation.

  The vapor pressure divisions used are 0.1 mm Hg and 10 mm Hg. In Table 3 below Henry's law values increase from the high-solubility/low-vapour pressure side to the low-solubility/high-vapour pressure side. Several other organic chemicals are shown in Table 4.

- ***Estimates of Partitioning Coefficients***

  – Conversion Factors

1 mm Hg	= 1 Torr (vacuum technology)
	= atmos. * 760
	= kPa (kilo Pascals) * 7.501
	= psia * 51.72
	= bars * 750.1

Note :

psia	= pounds/square inch absolute (measured with respect to zero pressure)
psi	= pounds/square inch (measured with respect to atmospheric pressure)
^oC	= (F – 32) * 5/9
K	= ^oC + 273.15

log x = ln x/ln 10

H (dimensionless) = H′/R * T with R in similar units to H′

R = 0.082054 litre-atmos/deg/mol

Table 3 : Selected organic chemicals in the selected solubility-vapour pressure ranges

	Solubility		
Vapor Pressure	*Low*	*Medium*	*High*
Low	Anthracene H medium	Diethyl phthalate	2-4 Dichlorophenol H low
Medium	Naphthalene	m-Xylene H medium	Aniline
High	Octane H high	Benzene	Vinyl chloride H medium

Table 4 : Organic compounds with a range of solubility and vapour pressures

	Solubility		
Vapor Pressure mm Hg	*Low < 65 mg/L*	*Medium 65-2000 mg/L*	*High> 2000 mg/L*
Low < 0.1 mm	Chlordane DDT PCP Dibutylphthalate Pyrene Anthracene	1.3 Butadene Diethyl phthalate	2,4-Dintrophenol 4-Chloro-m-cresol 2,4-Dichlorophenol
Medium 0.1-10 mm	Naphthalene 1,4 Diethylbenzene 2,2,5,5 Tetramethyl hexene	Ethybenzene	o-Cresol Anline Chlorotoulene 2-Chlorophenol
High > 10 mm	Octane Butane Cyclohexane Isobutane 1-Hexene	Toluene Benzene Tetrachloromethane 1-Pentene	Trichloromethane Dichloromethane Chloroethene 2-Butanone

R = 8.20562E-05 atmos-m^3/deg/mol

K_{oc} = K_{som} * 1.724

D_s = (1 – porosity) * 2.65 (specific gravity of quartz)

K_d = K_{oc} * % organic carbon/100

R_d = 1 + K_d * bulk density/porosity

The equations listed below are used for estimating critical partitioning parameters in the ECOPLUS computer programme. The relationship between the various parameters and the partitioning parameters that they are used to estimate are shown in Figure 1.

– **Estimated Boiling Point-T_b**

$\log T_b = 3 - 4/5 * \sqrt{M}$ For M>200 (Banks equation)

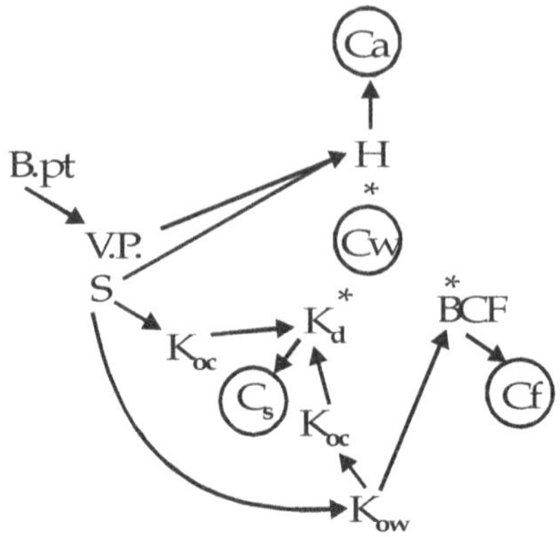

Figure 1 : A plot of the various distribution coefficients and their relationship to solute concentrations in the various ecosystem compartments-air, soil/sediment, fish and aqueous phases

T_b = boiling point in K

M = molecular weight

– **Estimated Melting Point-T_m**

T_m = 0.5839 T_b (Gold and Ogle, 1969)

T_m = melting point in K

– **Vapour Pressure**

Mackay equation :

$$\text{In } P = -(4.4 + \text{In } T_b) * \left[1.803\left(\frac{T_b}{T} - 1\right) - 0.803 * \text{In}\frac{T_b}{T}\right] - 6.8 * \left(\frac{T_m}{T} - 1\right)$$

P = vapour pressure in atmosphere

T_b = boiling point (K)

T_m = melting point (K)

When $T_m > T$ the last term is ignored.

– **Solubility**

$\log S = -\log K_{ow} + 0.76 - 0.01 * T_m$ (Yalkowsky equation)

S = solubility in mole/litre

T_m = melting point in °C

For liquids T_m is set to 25 °C

– **Henry's Law Constant**

Dilling equation :

H (dimensionless) = 16.04 * P_o * M/(T * S)

P_o = vapour pressure of pure compound (mm Hg)

S = solubility in mg/l (maximum solubility 1 mol/L)

M = molecular weight

T = temperature K

This equation cannot be used for miscible solutes

Mackay *et al.*, 1982a

$$H\left(\text{atm m}^3/\text{mol}\right) = 10.6 * \left[1 - \frac{T_b}{T}\right] + 6.8 * \left[1 - \frac{T_m}{T}\right] + 0.0318 *$$

$$(T_b - 273) - 5.15$$

T_b = boiling point (K)

T_m = melting point (K) solids

– ***Estimates of Normalized Distribution Coefficients-K_{oc}*:** Estimates of normalized distribution coefficients are commonly obtained from solubility or octanol/water partition coefficients rather than by the difficult and time-consuming direct determination of adsorption isotherms. Note that some authors use the organic carbon content, whereas others use the percent som. If an assumption is made that 58% of som is organic

$$K_{oc} = \frac{K_{som}}{0.58}$$

K_{oc} from Solubility

Freundlich K is inversely proportional to the solute in question:

$$\log K = A - n * \log S$$

where n = is the Freundlich exponent,

S = is the solubility, and

K = is the Freundlich distribution coefficient

The constants in the equations are largely dependent on the solubility units used and to a lesser extent on the group of compounds investigated.

Kenaga and Goring collected data for 106 organic chemicals, primarily pesticides, and obtained a regression equation:

$\log K_{oc} = 3.64 - 0.55 \log S$ (ppm) [Kenaga and Goring equation]

They estimated the results would be within + 1.23 orders of magnitude from the actual value, assuming 95% confidence limits.

Chiou obtained a relationship between distribution coefficients and solubility for a large number of non-ionic organic compounds. Their relationship covers more than 7 orders of magnitude in S and 4 orders of magnitude in K_{oc}. They found that

$$\log K_{som} = 4.040 - 0.557 \log S \text{ (micro moles/litre)}$$

Assuming that SOM = 58% oc,

$$K_{oc} = 3.80 - 0.557 \log S$$

Karickhoff *et al.* (1979) found the relationship to be

$$\log K_{oc} = 0.44 - 0.54 \log S \text{ (mole fraction).}$$

Only hydrophobic compounds such as aromatic hydrocarbons and chlorinated hydrocarbons were considered.

Means found a relationship :

$$\log K_{oc} = 4.070 - 0.82 \log S \text{ (mg/mL)}$$

Kirchhoff gives :

$$\log K_{oc} = -0.197 - 0.594 \log S \text{ (mole fraction solubility)}$$

Karickoff also recommended the incorporation of a crystal energy term in the equation.

The solubility units used in the expression to calculate the distribution coefficient so far encountered include mg/l, ppm, mmol/L, and mole fraction.

Thus :

$$\text{mg/L} = \text{ppm for dilute solutions}$$

$$\text{Micro mole/L} = \text{mg/L} * \frac{1000}{\text{molecular weight}}$$

$$\text{mole/L} = \text{micro moles}/10^6$$

$$\text{mole fraction} = \frac{\text{moles/L}}{(\text{mole/L}) + 55.51}$$

1. $\log K_{oc} = -0.55 * \log S + 3.64$ (Kenaga and Goring, equation)
   S in mg/L
2. $\log K_{oc} = -0.54 * \log S + 0.44$ (Kirchhoff's equation)
   S in mole fraction
3. $\log K_{oc} = -0.557 * \log S + 4.277$ (Chiou equation)
   S in μmol/L

– K_{oc} from octanol/water partition coefficients

The partitioning of a solute between water and an immiscible organic solvent has been used extensively to estimate biological concentration tendency solutes. Generally, octanol-water partition coefficients are those most commonly measured.

Chiou used octanol/water partition coefficients to obtain better estimates of solubility that is,

$$\log K_{ow} = 5.00 - 0.670 \log S \text{ (mmol/L)}$$

These equations were found to be valid over 6 orders of magnitude K_{oc} (10 to 10^7).

Kirchhoff examined ten hydrophobic pollutants with water solubilities ranging from 1 ppb to 100 ppm and obtained excellent correlations of K_{oc} vs K_{ow}, and poor correlation between K_{oc} and solubility. One of the following equations was suggested :

$$\log K_{oc} = 1.00 \log K_{oc} - 0.21 \text{ or } K_{oc} = 0.63\ K_{oc}$$

Means obtained similar partition coefficients for pyrene and 7, 12-dimethylbenz [a] anthracene (DMBA), respectively :

$$K_{oc} = 0.53\ K_{ow} \text{ and } 0.50\ K_{ow}$$

Banerjee correlated n-octanol/water partition coefficients with solubility and found that

$$\log K_{ow} = 5.2 - 0.68 \log S \text{ (mmol/L)}$$

For solids with known melting points they suggest :

$$\log K_{ow} = 6.5 - 0.89 \log S - 0.015 * T_m$$

where T_m is the melting point in °C, A value of 25°C is used if the solute is a liquid

Chiou and Schmedding state that most inaccurate data on water solubility and partition coefficients are generated with impure compounds or solvents. They discuss methods of ensuring purity of phases and suggest at a minimum that a melting point of a solid phase is minimal and that poor phase separations or persistent emulsions are often an indication of undesirable solvents. Experimentally, they found that for 36 organic chemicals, which ranged over 6 orders of magnitude, that

$$\log K_{ow} = -0.862 \log S \text{ (mol/l)} + 0.710$$

Kenaga and Goring obtained the following relationship for 45 organic chemicals :

$$\log K_{oc} = 1.377 + 0.544 \log K_{ow}$$

Kirchhoff obtained the equation for hydrophobic solutes :

$$\log K_{oc} = 0.989 \log K_{ow} - 0.346$$

The near unity of the coefficient suggests that

$K_{oc} = 0.411\ K_{ow}$ is a good approximation

Briggs gives the relationship

$$\log K_{om} = 0.52 \log K_{ow} + 0.64$$

He presents the data Felsot and Dahm (1979) in the form

$$\log K_{om} = 0.52 \log K_{ow} + 0.78$$

the data of Lord et al (1978) as,

$$\log K_{om} = 0.53 \log K_{ow} + 0.98$$

and his own earlier data as,

$$\log K_{om} = 0.52 \log K_{ow} + 0.62$$

Brown and Flagg found another empirical equation for nine compounds :

$$\log K_{oc} = 0.937 \log K_{ow} - 0.006$$

The equation used in ECOPLUS were those recommended by Lyman. Note that K_{ow} is a unitless number that is often listed in its long form.

1. $\log K_{oc} = 0.544 * \log K_{ow} + 1.377$ (Kenaga and Goring equation)
2. $\log K_{oc} = 0.937 * \log K_{ow} - 0.006$ (Brown and Flagg equation)
3. $\log K_{oc} = 1.00 * \log K_{ow} - 0.21$ (Kirchhoff equation)
4. $\log K_{oc} = 0.94 * \log K_{ow} + 0.02$ (Brown equation)
5. $\log K_{oc} = 1.029 * \log K_{ow} - 0.18$ (Rao and Davidson equation)
6. $\log K_{oc} = 0.524 * \log K_{ow} + 0.855$ (Braggs equation)

Bio-concentration Factor

1. In BCF = 0.935 * in K_{ow} – 3.433 (Kenaga and Goring, 1980)

or

$$\log BCF = 0.935 * \log K_{ow} - 1.495$$

11. Ecosystem distribution formulas

T (K)	= 273.15 + T°C
R	= 8.2054E – 05 atm-m^3/deg/mol
1 atmos	= 760 mm Hg
°C	$= \frac{(F-32)*5}{9}$

$$H \text{ (dimensionless)} = 16.04 * \frac{P_o(\text{mm Hg})^* \text{ M. Wt}}{T(K) * \text{Sol.}(\text{mg/L})}$$

$$H \text{ (dimensionless)} = \frac{H(\text{atm} - m^3/\text{mol})}{R * T(K)}$$

K_{oc} = K_{som}* 1.724

log S = – log K_{ow} + 0.76 – 0.01 T_m, S in mol/L, T_m is melting point (°C) or 25 if a liquid

log K_{oc} = – 0.55* log S (mg/l) + 3.64

log K_{oc} = 0.544* log K_{ow} + 1.377

log BCF = 0.935* log K_{ow} – 1.495

K_d = K_{oc}*f_{oc} (fraction organic carbon)

Bulk density = (1 – porosity)* 2.65 (quartz)

$$R_d = 1 + K_d * \frac{\text{bulk density}}{\text{porosity}}$$

NAPL

S_i^{θ} = X_i * S_i

S is pure phase solubility in water,

X is mole fraction in NAPL,

S^{θ} = partitioning concentration in water

$$\text{Mole fraction} = \frac{\text{moles solute}}{\text{total moles of liquid}}$$

$$C_l = S * \exp\left[6.8\left(\frac{T_m}{T} - 1\right)\right]$$

C_l is liquid solubility,

T_m is melting point (K),

T is system temperature (K)

Half-life

$C_t = C_o * e^{-kt}$ and $t_{1/2} = 0.693/k$

12. Additional examples
    - Parameter estimation
      - Henry's law constant for diazion :

MW = 304.34, solubility = 40 mg/L, vapour pressure (P_o) = 1.4 × 10^{-4} mm Hg at 20°C

Therefore,

$$H = 16.04 \frac{P_o\,(\text{mm Hg})}{T(^{\circ}K)\,C_s\,(\text{ppm})}.M$$

$$= \frac{16.04 \times 1.4 \times 10^{-4} \times 304.34}{293.15 \times 40} = 5.8 \times 10^{-5}$$

- Methylene chloride

  H = 0.003319 atm-m^3/mol

Therefore,

$$H(\text{dimensionless}) = \frac{C_a}{C_w} = \frac{H\left(\text{atm-m}^3/\text{mole}\right)}{R \times T}$$

$$= \frac{0.003319}{293.15 \times 8.20562 \times 10^{-5}}$$

$$= 0.138$$

- Methylene chloride

  MW = 84.93

  Solubility = 16,000 mg/L

  Vapour pressure = 360 mm Hg at 20°C

$$H = \frac{16.04 \times 360 \times 84.93}{293.15 \times 16000} = 0.105$$

- Diazion and methylene chloride (which will be more effectively removed by air stripping)

  Diazion: H = 5.8 × 10^{-5}

  Methylene chloride H = 0.105

Methylene chloride (H) is greater than diazion (H), and therefore, can be easily removed by air stripping.

- Calculate BCF values for DDT and malathion, and determine which would be more highly concentrated in fish.

Assume : DDT – K_{ow} = 960,000

Malathion – K_{ow} = 780

BCF = 0.938 ln K_{ow} – 3.443 [Kenaga and Goring equation]

ln [BCF (DDT)] = 0.938 ln (960,000) – 3.443

= 9.436

$$\text{BCF (DDT)} = 1.2536 \times 10^4 = 12{,}536$$

$$\ln [\text{BCF (Malathion)}] = 0.938 \ln (780) - 3.443$$

$$= 2.783$$

$$\text{BCF (Malathion)} = 1.6175 \times 10^1 = 16.2$$

$$\text{As BCF} = \frac{C_f}{C_w}$$

The compound with a higher value will have concentrated in the fish, that is, the DDT

- Given that 500 mL of water contains 10 mg/L of an organic solute, but after adding 1 g of activated carbon, the concentration of solute is reduced to 2 mg/L. Determine the linear distribution coefficient.

$$\frac{X}{M} = KC_w^n$$

$$n = 1,\ X = (10 - 2)\ \text{mg/L},\ M = \frac{1\,\text{g}}{500\ \text{mL}} = 2\text{g/L}$$

$$C_w = 2\ \text{mg/L}$$

Therefore,

$$K = \frac{(10-2)}{(2)(2)} = 2\ \text{mg/g}$$

- Determine the amount of granulated carbon necessary to reduce 10 mg/L, lindane to 0.1 mg/L. Lindane has K = 256, n = 0.49

$$\frac{X}{M} = KC_w^n$$

$$\frac{(10-0.1)}{M} = 256(0.1)^{0.49}$$

$$M = 0.12\ \text{g} = 120\ \text{mg/L}$$

- Recalculate the amount in two stages of Lindane

10 mg/L to 1 mg/L of Lindane

1 mg/L to 0.1 mg/L of Lindane

$$* \quad \frac{10-1}{M} = (256)(1)^{0.49}$$

$$M = 0.035\ \text{g/L} = 35\ \text{mg/L}$$

$$* \quad \frac{1-0.1}{M} = 256(0.1)^{0.49}$$

M = 0.011 g/L = 11 mg/L

Total carbon required = (35 + 11) = 46 mg/L

[Greater the number of steps, the greater the efficiency]

- Chloroform (K = 2.6, n = 0.75)

  Initial concentration = 10 mg/L

  Final concentration = 0.1 mg/L

$$\frac{(10-0.1)}{M} = 2.6(0.1)^{0.75}$$

M = 21.4 g/L

[Less strongly adsorbed chloroform requires 465 times as much carbon to remove an identical amount of more strongly adsorbed lindane]

- Derive the relationship between bulk density (D_s), and porosity (e) assuming quartz acquifer with a specific gravity of 2.65

  Total volume of sand and air = V

  Volume of sand = V (1 – e)

  Mass of sand = V × S.G.

  V (1 – e)(2.65) [Quartz]

  Total mass = Mass of sand + Mass of air

  = V (1 – e)(2.65)

$$\text{Bulk density} = \frac{\text{Total mass}}{\text{Total volume}}$$

$$= \frac{V(1-e)\times 2.65}{V} = (1-e)(2.65)$$

If e = 0.15

Bulk density = (1 – 0.15)(2.65) = 2.25

If bulk density = 1.85

$$\text{Porosity (e)} = \left[1 - \frac{1.85}{2.65}\right] = 0.30.$$

Example 4.2

Evapo-transpiration System

In evaporation, surface water, soil water, and precipitation falling and collecting on vegatation or other surfaces are converted to atmospheric moisture. The rate of evaporation depends on temperature relative humidity, barometric pressure, and windspread; soil moisture, type of soil, and depth of moisture below the soil surface are also factors.

In transpiration, water is taken in by plant roots, is used to build up plant tissue, moves up through the stem or trunk, and is released as vapour through the leaves.

The quantity of water evaporated and transpired from a cropped area, or the normal loss of water from soil by evaporation and transpiration, is known as the consumptive use or evapo-transpiration.

The evapo-transpiration system can be used where the available soil has no absorptive capacity or where little or no top soil exists over clay, hardpan or bed rock, provided a water balance study shows the transevaporation plus runoff exceeds precipitation infiltration plus inflow. It can also be built where groundwater level is high, provided it is built with a watertight linear and the bottom and sides to exclude the groundwater from transpiration bed. If an impermeable linear is not provided, elevation of the bed or curtain drains may be necessary if seasonal high water is a problem. The evapo-transpiration should be a lost resort.

(A) Evapo-transpiration

The design of evapo-transpiration is based on maintenance of a favourable input-output water balance. The precipitation less runoff (infiltration), plus wastewater flow must be less than evaporation plus transpiration. Tables 1 and 2 gives the consumptive use of crops and approximate evapo-transpiration/ consumptive water respectively.

(B) Capillarity

For evaporation to take place, it is necessary to have upward movement of water (capillarity) in soil to ground surface; for transpiration to take place it is necessary for capillary water (capillary fringe) to react the surface vegetation root system.

Fair, *et al.*, give an example for capillary tube of 0.1 mm in diameter and water temperature 50°F (10°C), resulting in a rise of 30.3 cm based on :

$$h = \frac{\delta}{245\ d} = \frac{74.2}{245 \times 0.01} = 30.3 \text{ cm}$$

where h is the capillary rise of water (cm), d is the tube diameter (cm), and is the surface tension (dyne/cm) [74.9 at 5°C, 74.2 at 10°C, 73.5 at 15°C, and 72.8 at 20°C].

The equation assumes that the weight of air is insignificantly small and the weight of water close to unity. Plants roots can reach a depth of 24 inches in a well developed absorption beds and take up wastewater. Maintenance of permeable soil structure and microbial and microscopic organisms are all essential to minimize system clogging and failure.

(C) Evapo-transpiration Design Basis During Growing Season

Assume a 150-days [5-month] growing season, using a soil cover of meadows or lucern grass, and an evapo-transpiration of 15 in. to 36 in. Consumptive use [soil evaporation + transpiration] dependent on location and vegetation (Table 2). The precipitation during the season is 10 in., of which 8 in. or less infiltrates a well crowned transvap dried. The water than can be disposed of growing season by evapo-transpiration will range from:

$$\frac{(15-8)(0.623)}{5\times 34} = 0.029\ \text{gpd/ft}^2, \quad \text{to}$$

$$\frac{(36-8)(0.623)}{5\times 30} = 0.116\ \text{gpd/ft}^2$$

1 in. of water/ft^2 = 0.623 gal.

(D) Sewage Disposal System in Tight Clay Soil for a Dewelling for Seasonal Occupancy Only

Season	:	6 months
Soil evaporation	:	8 in. in 6 months
Grass transpiration	:	20 in. during 6 month growing season
Precipitation	:	18 in. during 6 month period
Infiltration	:	80 % of precipitation (= 0.8×18) = 14.4 in.
Soil percolation	:	Zero
Daily sewage flow	:	300 gpd

Design trans-vap = [Evaporation + Transpiration – Infiltration) × 0.623/g/ft^2/in.

= [8 + 20 – 14.4] × 0.623 in 6 month period

= 8.473 gal/ft^2 in six month period

= 0.0471 gal/ft^2/day average during 6 month period

$$\text{Required transvap area} = \frac{300}{0.0471} = 6369\ \text{ft}^2$$

- For material specification and design details.
- Adjust design for local rainfall and runoff, and evaporation and transpiration.

- Uniform sand and gravel have 30 to 48% void space for flow storage and equalization.
- Bottom of bed 2 ft above groundwater recommended. Crown bed 1 in. per ft to promote runoff.
- Bed depth D of 12 in. is adequate, providing about 2 months storage. Actually no storage is needed for the 6-month seasonal occupancy.

Or alternatively,

$$A = \frac{Q}{ET + E - 1}$$

where A is the area of the evapo-transpiration bed (ft^2), Q is the sewage flow (gal/yr), ET is the evapo-transpiration from the bed (gal/ft^2/yr), E is the soil evaporation from the bed (gal/ft^2/yr) and I is the infiltration inflow (gal/ft^2/yr). ET = (Evaporation + transpiration) during growing season. E is the soil evaporation during non-growing season.

Therefore,

$$A = \frac{300 \times 6 \times 30}{0.623(20 + 8 - 14.4)} = 6373 \text{ ft}^2$$

$$\begin{aligned} \text{Storage (Y) required} &= Q + (I - EI - E) \times A \\ &= 300 \times 6 \times 30 + (14.4 - 20 - 8) \times 6370 \times 0.623 \\ &= 54{,}000 + (-53{,}972) \\ &= 28 \text{ gal (Negligible)} \end{aligned}$$

(E) Rational Design of Transvap Sewage Disposal System for Year Record Occupancy

Sewage flow	:	100 gpd = 6083 gal/month = 73,000 gal/yr
Bed surface cover	:	Sandy, silty, clayey loam top soil and lawn grass, crowned 1 in./ft. Use gravel bed (40% void space) with sand ridges or with gravel ridges.

Required area of evapo-transpiration bed:

$$\text{Outflow} = \text{Inflow}$$

$$[ET + E]A = Q + I \times A$$

$$A = \frac{Q}{(ET - E - I)}$$

$$A = \frac{200 \times 365}{3.12 \times 5 + 0.934 \times 4 + 12.076} = \frac{73{,}000}{7.26} = 10{,}055 \text{ ft}^2$$

Storage (Y) required during 7-months nongrowing period (J, F, M, S, O, N, D) (or make monthly water balance)

$$Y = Q_1 + I_1 - E_1 = \text{Sewage inflow} + \text{Infiltration} - \text{Soil evaporation}$$

$$= 6083 \times 7 + (0.206 + 0.196 + 0.810 + 0.966 + 0.810 + 0.872 + 0.271)10{,}055 - (0.934 \times 4) \times 10{,}055$$

$$= 42{,}583 + 41{,}537 - 37{,}565$$

$$= 46{,}555 \text{ gal}$$

Bed depth (D) to provide required storage

$$d = \frac{Y}{A \times 7.5 \text{ gal/ft}^2 \times \text{void space}}$$

$$A = \frac{46{,}555}{10{,}055 \times 7.5 \times 0.4}$$

$$= 1.54 \text{ ft (this is within the fine sand capillary range).}$$

Table 5 : Consumptive use of crop in California

Crop	*Evaporation (soil) and Transpiration (plant)*	*Growing Season*	*Average Annual Transvap*[a]
Alfalfa (Lucern grass)	5 in./month or 0.83 $pt/ft^2/day$[b]	6 or 7 months	0.415 $pt/ft^2/day$
Truck garden	4 in./month or 0.66 $pt/ft^2/day$	5 months	0.275 $pt/ft^2/day$
Cotton	4 in./month or 0.66 $pt/ft^2/day$	7 months	0.38 $pt/ft^2/day$
Citrus orchard	3 in./month or 0.50 $pt/ft^2/day$	7 months	0.3 $pt/ft^2/day$
Deciduous orchard	5 in./month or 0.83 $pt/ft^2/day$	6 months	0.415 $pt/ft^2/day$

[a] Soil evaporation for 7 to 5 months outside of growing season not included

[b] Print per square foot per day = $pt/ft^2/day$ (8 pints = 1 gal)

Table 6 : Approximate evapo-transpiration/consumptive uses

Growth [a]	*Water*[b] *(in.)*	*Ref.*
Alfalfa(Lucern grass)		
Brawely, Calif.-annual measure, Lysimeter	80	1
Kimberly, Idaho-April to October	55	4
Kimberly, Idaho-annual	53	1
Kimberly, Idaho-1 May to 30 September	36	1
Upham, N.D.-143 days observed	23	1
Reno, Nev.-124 days	40	1
Arvin, Calif.-annual	50	1
Mesa and Tempe, Ariz-annual	74	1
Swift Current, sask.-annual	25	1
Colorado-annual	26	2
Alberta-annual	22	3

(Contd..)

Trees		
Coniferouse trees[c]	4-9	6
Deciduous trees[c]	7-10	6
Oak-54 ft high, 25 in. dia., 34 gal Summer, 6 gal Winter[c,5]		
Pine-46 ft high, 15 in. dia., 40 gal Summer, 10 gal Winter[c,5]		
Apple-15 ft high, 6 in. dia., 18 gal Summer, 3 gal Winter[c,5]		
Eucalyptus, large, 25 gal[c]		
Fully grown tree, 70 gal[c]		
Grasses		
Seebrook, N. J.-clipped grass, annul	38	1
Lompac, Calip.-hoed grass, annual	41	1
Davis, Calif.-hoed grass, annual	52	1
Copenhagen, Denmark-clipped cover and rey grass, annual	16	1
Aspendale, Australia-clipped cover and rey grass, annual	51	1
Alberta-pasture grass, annual	18	3
Colorada-meadow grass, annual	23	3
Grasses		
Meadow grass-season (bluegrass and fescue)	22 - 60	6
Lucern grass-season	26 – 55	6
Cannery grass-season		
Hay		
Coshocoton, Ohio-grass legume hay, annul	40	1
South park, Colo.-native hay, annual	22	1
Clover		
Aspendale, Australia-annual	51	1
Prooser, Wash.-23 May to 28 October	34	1

1. Marvin E. Jensen, ed., Consumptive Use of Water and Irrigation Water Requirements, a report prepared by the Technical Committee on Irrigation Water Requirements, ASCE, New York, September 1973. See also Eldon L. Johns, ed., Water Use by Natural Occurring Vegetation: An Annotated Bibliograhy, a report prepared by the Task Committee on Water Requirements of Natural Vegetation, Committee on Irrigation Water Requirements, ASCE, New York, 1989, pp. 17-22.
2. Harry F. Blaney, "Water and Our Crops", The Yearbook of Agriculture 1955 Water, USDA, Washington, D.C., 1955.
3. John R. Davis, Evaporation ad Evapo-transpiration Research in the United States and Other Countries, American Society of Agricultural Engineers, December, 1956. Based on report by W.L. Jacobson and L.G. Sonmor, Department of Agriculture Experiment Farms Service, Canada Department of Agriculture.
4. James L. Wright and Marvin E. Jensen, "Peak Water Requirements of Crops in Southern, Idaho," J. Irrig. Drain Div., ASCE, Proc. paper 8940, June 1972.
5. Alfred P. Bernhart, Treatment and Disposal of Waste Water from Homes by Soil Infiltration and Evapo-transpiration, University of Toronto Press, Canada, 1973, p. 146.
6. Leonard C. Urquart, Civil Engineering Handbook, McGraw-Hill, New York, January 1950, p. 796 aWell watered.

(*a*) Obtain more accurate data from local farm bureau or agricultural college, 1 in. of water = 0.623 gal/ft^2. (*b*) Transpiration per day.

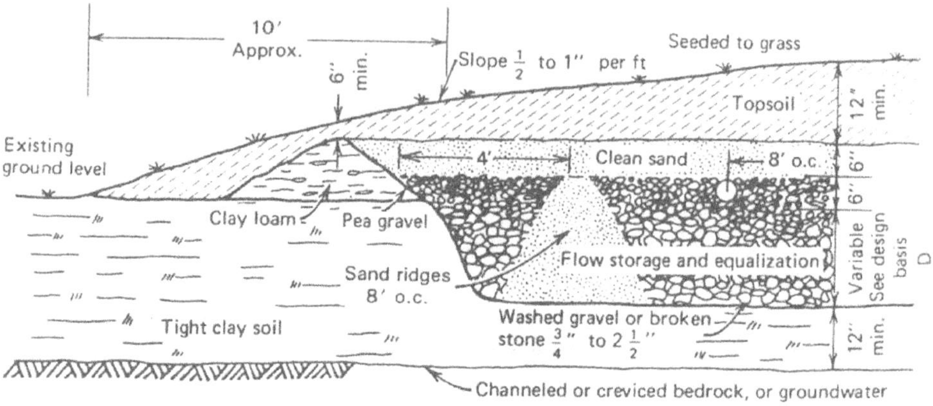

Figure 2 : Transvap sewage disposal system in tight soil (Raise bed as necessary if groundwater or bedrock is a problem). Clean washed sand, 0.1 mm effective size for up to 12 to 16 in. gravel and sand depth, and 0.05 mm sand for up to 24 in. gravel and sand depth. Sand ridges are necessary to obtain capillarity and promote evapo-transpiration. Permeable geotextile fabric is recommended over the sand ridges and in place of 6 in. of sand over the gravel. Add 6-in. or 8-in. diameter perforated risers in and to bottom of gravel bed for inspection and emergency pump out. Pressure distribution is usually required. Silt in sand will increase capillary rise.

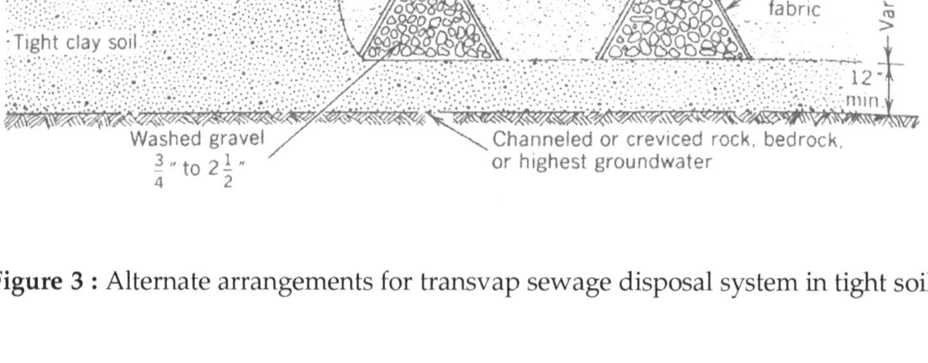

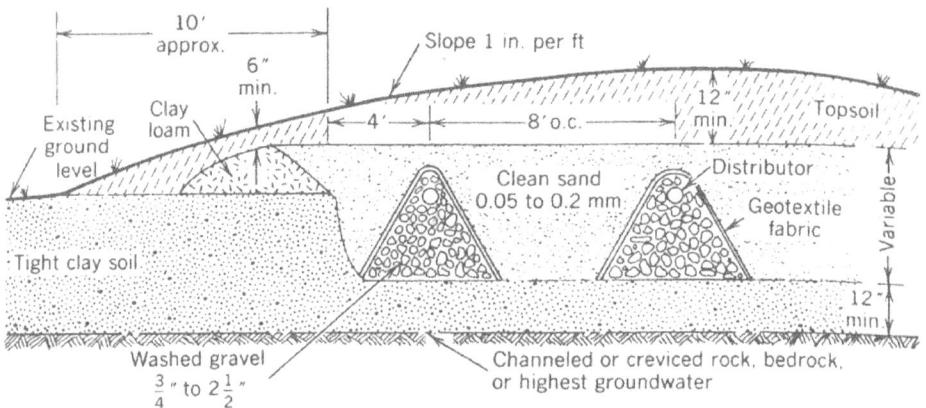

Figure 3 : Alternate arrangements for transvap sewage disposal system in tight soil

Table 7 : Precipitation, infiltration, land evaporation and evapo-transpiration data

S.N.	Item	Months											
		Jan.	*Feb.*	*Mar.*	*April*	*May*	*June*	*July*	*Aug.*	*Sep.*	*Oct.*	*Nov.*	*Dec.*
1.	Precipitation (ppt)(in.)	2.2	2.1	2.6	2.7	3.3	3.3	3.1	2.9	3.1	2.6	2.8	2.9
2.	% ppt infiltration in bed	15	15	50	85	85	85	85	85	50	50	50	15
3.	Infiltration ppt (gal/ft^2)	0.206	0.196	0.810	1.430	1.748	1.589	1.642	1.536	0.966	0.810	0.872	0.271
4.	Land evaporation	0	0	0.934	in ET	in ET	in ET	in ET	in ET	0.934	0.934	0.934	0
5.	Evaporation	0	0	0	3.12	3.12	3.12	3.12	3.12	0	0	0	0

0.623 = gal/ft^2 per in. precipitation

Infiltration = % infiltration × precipitation/month × 0.623(in. gal/ft^2/month)

Land evaporation = 0.6 lake evaporation of 30 in. per year

$= (0.6 \times 30) \div 12 \times 0.623$

= 0.934 gal/ft^2/month

Evapo-transpiration = 20 in. in growing season × 0.623 ÷ 5

= 3.12 gal/ ft^2/month.

Example 4.3

Engineering Considerations for Natural Systems for Wastewater Treatment

1. Lagoon systems (aerobic, facultative, aerated pond anaerobic)

 1.1 Classification based on depth and the biological reactions

 1.1.1 Aerobic lagoons

 Depths ranging from 0.3 to 0.6 m (1 to 2 ft)

 Oxygen supply through photosynthesis and wind aided surface aeration

 Lagoons are often mixed by recirculation to maintain DO throughout their entire depth

 Suitable to warm sunny climates.

 1.1.2 Facultative lagoons (oxidation lagoon)

 Depths varying from 1.5 to 2.5 m (5 to 8 ft)

 Normal detention periods – 25 to 180 days

Depths are kept at 1.5 m or more to avoid the growth of emergent plants

Surface layers are aerobic with an anaerobic layer near the bottom

Oxygen supply through surface aeration and photosynthesis

Designed in series with minimum of 3 – cells to reduce short circuiting

Algae remains in the effluent causing effluent suspended solids to exceed discharge requirements.

1.1.3 Aerated lagoons (partially or completely mixed)

Oxygen supplied by mechanical aerators or diffused aeration

Depths can range from 3 to 6 m (10 to 20 ft) with detention times ranging from 5 to 30 days

Less susceptible to odours and requires less land

Aerated lagoons are followed by a facultative lagoon or settling lagoon (1 – day detention or less) to reduce SS before discharge.

1.1.4 Anaerobic lagoons

Heavily loaded with organic (no aerobic zone)

Depths range from 2.5 to 6 m (8 to 20 ft) with detention times of 20 to 50 days

Used as a pre-treatment to facultative or aerobic lagoons for strong industrial wastewater.

1.2 Classification based on the duration and frequency of effluent discharge

1.2.1 Total containment lagoons

Applies to only climates where evaporation exceeds the precipitation on an annual basis.

1.2.2 Controlled discharge lagoons

Discharge only once or twice a year when the stream conditions are OK.

1.2.3 Hydrograph controlled release lagoons (HCR)

Designed with a discharge rate correlated to the stream flow rate.

[Most controlled discharge and HCR lagoons are facultative. HCR lagoon releases only when the stream flow is higher than minimum acceptable value].

1.2.4 Continuous discharge lagoons

Effluent discharged at the same rate (less evaporation and seepage) as the influent wastewater.

1.3 Facultative lagoons

1.3.1 *Design procedure :* The water surface area for a facultative lagoon is estimated as :

$$A = \frac{(BOD)(Q)}{(1000)(LR)} \tag{1}$$

where A is the area for a facultative lagoon (ha), BOD is the concentration of BOD in wastewater (mg/L), Q is the wastewater flow (m^3/d), 1000 is the conversion factor gms to kg, and LR is the BOD loading rate.

Table 8 : Loading ratio and temperature.

Average Winter Temperature (°C)	*Depth (m)*	*LR (kg BOD/ha.d)*
< 0	1.5 – 2.1	11 – 22
0 – 15	1.2 – 1.8	22 – 45
> 15	1.1	45 – 94

Lagoons can be designed to avoid hydraulic short circuiting by providing manifolds or diffusers on inlet and outlets, keeping inlet and outlet as far as possible (using in lagoon baffles, or providing multiple inlets and outlets)

Multiple lagoons in series (3 or 4) reduce short circuiting

Recirculation from the last lagoon to the first lagoon helps distribution and reduce short circuiting.

Significant amounts of nitrogen and pathogens can be removed as a result of long detention period.

Nitrogen removed ranges from 40 to 95%

Phosphorus removed is small (less than 40%)

Bacteria and viruses are removed as a result of sedimentation, predation, natural die – off, and adsorption.

Ice covered ponds are anaerobic, BOD removal is by sedimentation.

The results of a design example using common methods are presented below: Flow rate – 1900 m^3/d (0.5 HGD), BOD – 200 mg/L and temperature of 0.6°C.

Table 9 : Design Details from various Investigator.

Procedure	*Detention[a] Time (days)*	*Surface Area (ha)[b]*	*Effective Depth (m)*	*LR (kg BOD/ head)[b]*
Areal loading rate	180	22.3	1.4	17
Gloyna	140	26.5	1.0	14
Marais and Shaw	74	5.6	2.4	68
Plug flow	180	22.3	1.4	25
Wehner and Wilhelm	80 – 132	10.8 – 17.9	1.4	30 – 48

[a] – Total system using 4 – ponds in series,

[b] – temp, 5°C, light intensity adequate, and influent TSS – 250 mg/L

1.3.2 Controlled discharge lagoons

Essentially facultative lagoons with detention times of 120 days or more

Lagoons with seasonal discharge have been operated in north central US with following design criteria :

- Overall organic loading of 22 to 28 kg BOD/ha.d
- Water depth of 2 m or less in the first cell and 2.5 m or less in subsequent two or three cells
- Minimum detention time of 6 months above a minimum depth of 0.6 m.

1.3.3 Hydrograph controlled release lagoons

A variation of the controlled discharge lagoons concept to optimise the dilution of lagoon effluent in receiving waters

1.4 Aerated lagoons (partial mix conditions)

The basic design equation is :

$$\frac{S_n}{S_i} = \frac{1}{\left(1+\frac{k\theta}{n}\right)^n}$$

where S_n is the effluent BOD_5 in cell – n (mg/L), S_i is the influent BOD_5 (mg/L), k is the first order reaction rate constant (0.14 to 0.30 d^{-1}), q is the detention time (days), and n is the number of cells in series.

This equation is based on equal – size ponds at the same temperature. The detention time is for each single pond.

No attempt is made to completely mix the lagoons

Typical aeration rate to be provided is 1 to 1.5 kg O_2/kg BOD loading

Anaerobic activity at the bottom of the lagoon due to settling of SS [partial mix lagoons are also referred as facultative aerated lagoons]

Detention times range from 5 to 30 days with water depths ranging from 3 to 6 m

BOD loading rates vary from 100 to 400 kg/ha.d and a settling lagoon is also provided to settle SS (detention time of 0.5 to 1.0 day)

No of less 3 or more

Design equation is :

$$A = \frac{(Q)(\theta)}{(10,000)(d)} \tag{2}$$

where A is the average lagoon area (ha), Q is the wastewater flow (m^3/d), q is the detention time (days, calculated from above referred equation), 10,000 is the conversion factor from m^2 to ha, and d is the lagoon depth (3 to 6 m).

Area should be increased to account for the side slope of the lagoon berm.

Power requirements are based on oxygen transfer (3 W/m^3) and not on mixing requirements.

Permeability (k) of the linear should not exceed the value derived from the following equations:

$$k = FL \tag{3}$$

where k is the permeability (cm/s), $F = 3.0 \times 10^{-9}\ s^{-1}$, and L is the thickness of the scale (cm).

Darcy law permits to determine the seepage rate :

$$q = \frac{(k)(A)(h)}{(L)} \tag{4}$$

where k is the permeability (cm/s), L is the thickness of the scale (cm), q is the flow through the linear (cm^3/s), A is the linear area (cm^2), and h is the hydraulic head over the linear area (cm).

Embankment tops of a minimum of 2.5 m permits access of vehicles for maintenance with outer slopes no steeper than 3:1 allow grass growth and tractor moving. Freeboard of 0.3 m (3 ft) is OK.

1.5 *Dual power multi cellular aerated lagoons (DPMC)* : DPMC aerated lagoon is to combine a series of partial-mixed lagoons.

First cell has a water depth of 3 m and is aerated with a surface aerator (6 kW/1000 m^3) and a detention time of 1.5 to 2 days.

Subsequently three cells are aerated with surface aerators at a power level of 1 kW/1000 m^3

Overall detention time of all four cells is 4.5 to 5 days.

Complete mixing requires a power level of 15 to 30 kW/1000 m^3

Recommended power level of 6 kW/1000 m^3 (first cell) is more than 2 kW/1000 m^3

The combination of two power levels of aeration meets requirements of biological conversion while minimizing algal production by turbulence of mixing.

1.6 Advanced integrated lagoon systems

Combines multiple lagoons with recycle.

System comprises of a deep, primary facultative lagoon followed by a shallow aerobic lagoon.

Primary lagoon has fermentation pits for anaerobic digestion of settled solids. Pits must be unaerated and unmixed and serve as upflow anaerobic digesters.

An example : Three shallow floating aerators have been added to the primary lagoon to supplement the recycled, algae laden lagoon 2 water that serves to provide an aerobic cap on the primary lagoon. A third lagoon is used in series for settling, and the treated effluent is chlorinated before irrigation. The fourth and fifth lagoons are maturation or storage lagoon. There has been little accumulation of solids in the primary lagoon. No solids have been removed from the lagoon. Design factors are given as:

Table 10 : Design factors with values.

Design Factors	*Magnitude*
Design flow (MGD)	0.5
Average flow (MGD)	0.4
Primary lagoon	
Aeration (hp)	5
BOD loading (lb/acre.d)	345
Depth (ft)	10
Area (acre)	3
Detention time (days)	19
High rate aerobic lagoon	
Depth (ft)	3
Area (acre)	5.1

(Contd..)

Design Factors	*Magnitude*
Detention time (days)	10
Settling (maturation) lagoon	
Depth (ft)	9
Area (acre)	2.5
Detention time (days)	15
Influent BOD_5 (mg/L)	250 - 300
Effluent BOD_5 (mg/L)	15 - 40
Influent TSS (mg/L)	200 - 250
Effluent TSS (mg/L)	20 - 40

2. Land application systems (slow rate, overland flow, and rapid infiltration): Slow-rate and rapid infiltration systems, wastewater is treated as it percolates through the soil. Overland flow system : the treatment occurs in a thin film on grassy slopes constructed on slowly permeable soil.

2.1 *Design of slow rate system :* Slow rate systems are classified as :

Slow infiltration (Wastewater treatment).

Crop irrigation (Water and nutrient reuse).

2.1.1 Hydraulic loading rate for slow infiltration and water balance are :

$$L_W = ET - P - W_P \tag{5}$$

where L_W is the wastewater hydraulic loading rate based on soil permeability (m/yr), ET is the evapo-transpiration rate (m/year), P is the precipitation rate (m/yr), and W_P is the percolation rate (m/yr). ET should be measured in the field using a cylinder infiltrometer, sprinkler infiltrometer, or basin flooding technique. The limiting design factors are soil permeability, allowable loading rate for particular wastewater constituent (such as Nitrogen). Slow infiltration systems are suitable in humid areas. Compatible crops having high nitrogen uptake capacity, high ET rates, and tolerance to moisture and wastewater constituents (perennial forage grasses, turf grasses, some tree species, and some field crops) are necessary conditions for this type of system.

2.1.2 Hydraulic loading rate and water balance for crop irrigation :

$$L_W = (ET - P)\left(1 + \frac{LR}{100}\right)\left(\frac{100}{E_u}\right) \tag{6}$$

where L_W is the annual wastewater loading (m/yr), LR is the leaching requirement (15 to 25%), and E_u is the irrigation efficiency (65 to 85%).

The specific crop and its sensitivity to wastewater TDS determines the leaching requirement. The irrigation efficiency for sprinklers ranges from 70 to 80%; surface irrigation efficiencies range from 65 to 85%. The total percolation is as combination of the leaching fraction and the irrigation in efficiency fraction.

$$\left(1-\frac{E_u}{100}\right)$$

2.1.3 *Nitrogen loading rate* : The nitrogen balance is :

$$L_n = \frac{\left(U + 0.001 C_p P_w\right)}{(1-f)} \quad (7)$$

where L_n is the nitrogen loading rate (kg/ha.yr), U is the crop uptake rate of nitrogen (kg/ha.yr), f is the fraction of applied nitrogen lost (denitrification, volatilization and soil storage), C_p is the percolate nitrate nitrogen concentration (mg/L), and P_w is the percolate flow (m/yr).

The value of 'f' depends on BOD/Nitrogen ratio in the wastewater and the air temperature during application period [high strength wastewater] (BOD/N) has the highest f value. Lower f values apply to cold climates.

Combining water and nitrogen balance equations, the hydraulic loading rate based on nitrogen limits can be estimated as :

$$L_{wn} = \frac{C_p (P - ET) + 0.1U}{(1-f)C_n - C_p} \quad (8)$$

where L_{wn} is the hydraulic loading rate based on nitrogen limits (m/yr), and C_n is wastewater nitrogen concentration (mg/L).

The design limiting loading rate is the lowest of the two estimated values (L_W or L_{wn}) for the slow rate.

Removal of BOD, nitrogen and phosphorus in slow rate systems

Table 11 : Nitrogen and Phosphorus Balances.

Site	*BOD (mg/L)*		*Total Nitrogen (mg/L)*		*Total Phosphorus (mg/L)*	
	Applied	*Percolate*	*Applied*	*Percolate*	*Applied*	*Percolate*
North Dakota	42	< 1	11.8	3.9	6.9	0.05
Hanover						
PST	101	1.4	28.0	9.5	7.1	0.03
SST	36	1.2	26.9	7.3	7.1	0.03
Michigan	34	1.3	8.2	2.5	3.8	0.10
New Mexico	43	< 1	66.2	10.7	8.0	0.39
Massachusetts	85	< 2	30.8	1.8	12.0	0.04

Nitrogen uptake of selected crops

Table 12 : Nitrogen Liptake for various crops.

Crop	*Nitrogen Uptake kg-N/ha.yr*
Fi rage Crops	
Alfalfa	225 – 675
Bermudagrass	400 – 675
Sweet clover	175 – 300
Field Crops	
Barley	125 – 160
Corn	175 – 250
Cotton	75 – 180
Grain sorghum	135 – 250
Oats	115
Sugar beets	255
Wheat	160 – 175
Tree Crops	
Mixed hard wood	225
Mixed hard wood	340

Legumes take minimum amount of nitrogen from the atmosphere under nitrogen fertilization.

Electrical conductivity (E_{CW}) values resulting in reductions in crop yields.

Table 13 : Crop yield reduction with Electrical conductivity.

Crop	*EC_W Values of Applied Wastewater (m mhos/cm)*		
	Crop Yield Reduction (%)		
	0	25	100
Forage crops			
Alfalfa	1.3	3.6	10.3
Bermuda grass	4.6	7.2	15.0
Clover	1.0	2.4	6.7
Field crops			
Barley	5.3	8.7	18.8
Corn	1.1	2.5	6.7
Cotton	5.1	8.7	18.0
Wheat	4.0	6.3	13.3

Barley and wheat are less tolerant during germination and seeding stage EC_W should not exceed 2.7 m mhos/cm.

Ranges of f values for municipal wastewater :

Table 14 : Magnitude of 'f' for Municipal Wastewater.

Wastewater Type	*Magnitude of f*
High strength	0.5 – 0.8
Primary effluent	0.25 – 0.5
Secondary effluent	0.15 – 0.25
Advanced treatment effluent	0.10 – 0.15

The percolate nitrogen quality is limited to 10 mg/L or less ($NO_3 - N$)

3. **Overland flow systems :** The design factors are :

Application rate (0.03 – 0.37 m^3/m.h)

Slope (terrace) length (30 – 60 m)

Slope grade (1 – 12%)

Application period (6 – 12 hours/d, 5 to 7 days/week)

Loading rate (total daily flow/total field area; 11 – 50 mm/d).

3.1 Relationship between BOD removal and application rate

$$\left(\frac{C_z - C}{C_i}\right) = A \exp\left(\frac{Kz}{q^n}\right) \tag{9}$$

where C_z is the effluent BOD concentration at point z (mg/L), C is the residual BOD at the end of the slope (mg/L), A is the constant, C_o is the effluent BOD (mg/L), z is the slope length (m), q is the application rate (m^3/m.h), and K and n are empirical constant.

This equation has been used for screened raw wastewater and primary effluent but not for high strength wastewater. It is necessary to reduce the application rate by dividing by a safety factor of 1.5 before calculating field area (Figure 1). BOD loading upto 100 kg BOD/ha.d have been successfully used in overland flow system. Oxygen transfer capacity of the system becomes limiting when BOD is greater than 800 mg/L and, therefore, pre treatment or effluent recycling may be necessary to obtain successful operation of the system.

3.2 **Suspended solids (SS) loadings :** SS are effectively removed on overland flow (low velocity and shallow depth of flow except for algae. Sprinkler application is suitable for high strength and high solids concentration (for uniform distribution of SS over the upper 65% of the sludge).

Algae removal by overland flow :

Depends on application rates, types, and concentration.

Removal rates range from 45 to 83%

Buoyant (motile) algae resist removal by sedimentation or filtration.

Loading rate not to exceed 0.09 $m^3/m.h$ for wastewater treatment system using facultative lagoons as pre-treatment devices.

3.3 **Land requirement for overland flow system :** Area (without storage of wastewater) required is expressed by :

$$A = \frac{(Q)(Z)}{(q)(p)(10,000)} \tag{10}$$

Area (with wastewater storage) required is expressed by :

$$A = \frac{(365Q + V_s)Z}{(q)(p)(10,000)(D)} \tag{11}$$

where A is the field area (ha), Q is the wastewater flow (m^3/d), Z is the slope length (m), q is the application rate ($m^3/m.h$), P is the period of application (h/d), V_s is the net loss or gain from evaporation, seepage, or precipitation on the storage lagoon (m^3/yr), and D is the operating time (d/yr).

To summarise, the overland flow system can be designed to achieve secondary treatment, advanced treatment, or nitrogen removal. Phosphorus removal requires either pre or post treatment. The treated runoff concentrations of BOD and SS differ little among raw, primary, and secondary effluent applications. Algae in lagoon effluent are not effectively removed in overland flow systems.

BOD/N ratio is important for nitrogen removal. Secondary effluent application to overland flow system removes nitrogen effectively (low BOD/N ratio) in comparison to primary treated effluent (large BOD/N ratio).

The primary mechanism for nitrogen removal is nitrification-denitrification (BOD/ N = 3:1 optional value).

4. **Rapid infiltration :** This system requires deep, permeable soil for wastewater treatment, and can effectively remove BOD and SS through filtration, and bacterial decomposition. SS are removed to low levels (approaching 1 mg/L).

4.1 Nitrogen removal

Nitrogen removals vary from 40 to 90% (because of nitrification and denitrification processes).

Important design factors are : BOD/N ratio, hydraulic loading rate, and ratio of flooding period to dry period.

BOD/N should be greater than 3:1 (for effective denitrification).

Loading rates within 15 to 30 m/yr (provides adequate detention time within soil profile for nitrogen removal.

Soil profile depth to be effective is 3 m or deeper to ensure adequate retention time at a loading rate of 30 m/yr.

4.2 Phosphorus removal :

Mechanisms include absorption and chemical precipitation.

Detention time (critical for chemical precipitation) is a function of the percolation rate through the soil and the aquifer, and the distance to the monitoring point.

Phosphorus removal declines with time but removal rate may remain high for several years.

4.3 Rapid infiltration systems are effective in removing metals, pathogens, and trace organics

4.3.1 Design objectives:

Treatment and avoidance of direct discharge to receiving water bodies by discharging to groundwater.

Treatment and groundwater recharge.

Treatment and recharge of streams by interception of groundwater.

Treatment and recovery of treated water by wells or under drains for reuse.

Treatment and temporary storage of water on the aquifer.

4.3.2 *Design procedure :* General

Determine the field infiltration rate.

Predict the hydraulic pathway of treated water.

Determine overall treatment requirements.

Determine the appropriate pre-application treatment.

Estimate hydraulic rate (on yearly basis).

Required field area.

Estimate the potential for groundwater mounding.

Selection of the final hydraulic loading cycle.

Estimate the application rate.

Estimate the number of individual basins required.

Establishment of the monitoring wells.

Additional requirements for nitrogen:

Estimate the mass of NH_3–N that can be absorbed on the cation exchange sites in the soil.

Estimate the loading period based on NH_3–N concentration and daily application rate without exceeding the mass loading for step 1.

Compare the NH_3–N and org–N loading rate to the maximum nitrification rate of 70 kg/ha.yr (complete nitrification).

Consider limiting the infiltration rate (30 to 45 m/yr).

Pilot studies may be required if NO_3–N concentration required is less than 5 mg/L or removals greater than 80%

Moderate infiltration rates are conductive to higher nitrogen removal rates.

4.3.3 *Hydraulic loading rate* : The design of hydraulic loading rate is based on the soil infiltration rate, subsurface flow rate, or loading of BOD or nitrogen. Each of these loading rates may be calculated and the lowest value is selected for design purposes.

Convert hourly infiltration rate to a yearly rate (×8760 h/yr)

Multiply 'this' by a factor to account for the wetting and drying cycle, variability of the soils, and type of infiltration rate field test.

4.3.4 Annual design loading rate is :

$$L_W = (a)(l) \tag{12}$$

where L_W is the yearly design loading rate (m/yr), a is the design factor (ranging from 0.015 to 0.15), and l is the measured steady – state infiltration rate (m/yr).

a = 0.02 – 0.04 [Small scale tests]

a = 0.07 – 0.15 [Large scale basin flooding]

[The design factor must not exceed the fraction of the loading cycle during which the basins are flooded. If application period is 1 day and the drying period is 9 days [(a total cycle period of 10 days, the design factor must be less than 0.10)].

4.3.5 Organic loading rate

BOD loading rates range from 10 to 200 kg BOD/ha.d

Suggested BOD loading rate is 670 kg BOD_5/ha.d.

4.3.6 Area (A) requirements : Basin bottom area is given by :

$$\text{Net area}\,(A) = \frac{365\,Q}{(10{,}000)(L_W)} \tag{13}$$

where A is the net field area (ha), Q is the wastewater flow (m^3/d), and L_W is the limiting loading rate (m/yr)

(A) BOD loading and removals in rapid infiltration systems :

Applied BOD loading (kg/ha.d)	:	13 – 177
Applied BOD (mg/L)	:	15 – 220
Percolate BOD (mg/L)	:	0 – 12
% BOD removal	:	89 – 100

(B) Total nitrogen removals at rapid infiltration systems :

Applied BOD loading (kg/ha.d)	:	13 – 177
Total N – loading applied (kg/ha.yr)	:	1330 – 16710
Applied nitrogen (mg/L)	:	10.9 – 50.0
Percolate nitrogen (mg/L)	:	2.8 – 20.0
BOD/N ratio	:	1:1 – 5.5:1
% N – removal	:	38 – 80

(C) Phosphorus removals at rapid infiltration system

Years of operation	:	5 – 88
Applied concentration (mg/L)	:	2.1 – 9.0
Distance to sample point (m)	:	0.8 – 1700
Percolate concentration (mg/L)	:	0.03 – 0.45
% removal	:	85 – 99

(D) Loading cycles for rapid infiltration system

Table 15 : Application period and drying period with applied wastewater in different seasons.

Objectives	*Applied Wastewater*	*Season*	*Application Period (d)*	*Drying Period (d)*
Maximise infiltration rate	PE	Summer	1 – 2	6 – 7
	Winter	1	7 – 12	
	SE	Summer	1 – 3	4 – 5
	Winter	1 – 3	5 – 10	
Maximise nitrification rate	PE	Summer	1 – 2	6 – 7
	Winter	1	7 – 12	
	SE	Summer	1 – 3	4 – 5
	Winter	1 – 2	7 – 10	
Maximise nitrogen removal	PE	Summer	1 – 2	10 – 14
	Winter	1 – 2	12 – 16	
	SE			
	Summer	7 – 9	10 – 15	
	Winter	9 – 12	12 – 16	

PE : Primary effluent; SE : Secondary effluent.

5. **Summary of land application systems :** This covers features of land treatment systems, pre-application for land treatment systems, site requirements, and removal of BOD, nitrogen and phosphorus in slow rate systems.

5.1

Table 16 : Features of land treatment systems

Features	*Slow Rate*	*Overland Flow*	*Rapid Infiltration*
Treatment goods	ST or AWWT or zero discharge	ST, NR	ST, AWWT, GWR, Zero discharge
Vegetation	Yes, various crops	Yes, water tolerant	Only for soil stabilization
Climate restriction	Storage required for cold and heavy rain fall	Storage needed for cold weather	No storage required for proper designed and operated system
Hydraulic loading (m/yr)	0.5 – 6	3 – 20	6 – 100
Area needed (ha)	23 – 280	7 – 46	1.4 – 23
Mechanism	Plant uptake, denitrification, soil storage nitrogen, phosphorus removal (adsorption + chemical precipitation), metals removal (adsorption + chemical precipitation), pathogen removal by soil filtration, adsorption, dessication, radiation, prediation and exposure to other environmental conditions.		Described in relevant sections
	Trace organics removal by photo – decomposition, volatilization, sorption and biological degradation		Described in relevant sections

ST : Secondary treatment; AWWT : Advanced wastewater treatment.

5.2

Table 17: Preapplication treatment for land application systems

Land Treatment System	*Suggested Pre-treatment*
Slow rate system	• Primary treatment is acceptable (restricted public access). Crops grown not for human consumption. • Lagoons with control of faecal coliforms (1000 MPN/100 mL) is acceptable for controlled irrigation (not for crops to be eaten raw). • BWT by ponds + disinfection (Faecal coliform 200 MPN/100 mL) for parks and golf courses.
Overland flow systems	• Screening or comminution is acceptable with no public access. • Screening or comminution + aeration (odour ontrol) acceptable for urban location with no public access
Rapid infiltration system	• Primary treatment with restricted public access. • Biology treatment (pond system) is acceptable for urban location with controlled public access.

5.3

Table 18 : Site requirements for land treatment processes

Characteristics	*Slow Rate*	*Overland Flow*	*Rapid Infiltration*
Soil depth (m)	> 0.6	> 0.3	> 1.5
Soil permeability class range	Slow to moderately rapid	Very slow to moderately slow	Rapid
Soil permeability (mm/h)	1.5 – 500	< 5.0	> 50
Depth to ground water (m)	0.6 – 1	Not critical[a]	1 during flood[b] cycle 1.5–3 during drying period
Slope (%)	< 20 or cultivated land;< 40 on non cultivated land	0 – 15; finished slope 2 to 8[c]	< 10; excessive slopes require much earth work

[a] : Permeable soils require to determine effect on groundwater

[b] : Underdrains can be used to maintain this level at sites with high groundwater table

[c] : Slopes as low as 1% and as high as 10% may be considered.

5.4 BOD, N, and P removals in slow rate systems

BOD applied (mg/L)	:	34 – 101
Percolate BOD (mg/L)	:	< 1 – 2
Total – N applied (mg/L)	:	8.2 – 66.2
Percolate – N (mg/L)	:	1.8 – 10.7
Total – P applied (mg/L)	:	3.8 – 12.0
Percolate – P (mg/L)	:	0.03 – 0.39

6. **Floating aquatic plant systems (Water hyacinths and duckweed) :** An attempt to control SS in aerobic and facultative lagoon discharges. The floating plants shield the water from sunlight and reduce the growth of algae (as also BOD, nitrogen, metal, and trace organics).

6.1

Table 19 : Design criteria for water hyacinth systems

Factors	*Secondary Treatment (Non-aerated)*	*Advanced Secondary Treatment (Aerated)*	*Nutrient Removal (Non-aerated)*
Design criteria Influent wastewater Source	Screened, settled or lagoon effluent	Screened or settled	Secondary
Influent BOD_5 (mg/L)	130 – 180	130 – 180	30
BOD_5 loading rate (kg/ha.d)	40 – 80	150 – 300	10 – 40
Detention time (days)	10 – 36	4 – 8	6 – 18

(Contd.)

Factors	*Secondary Treatment (Non-aerated)*	*Advanced Secondary Treatment (Aerated)*	*Nutrient Removal (Non-aerated)*
Hydraulic loading (m^3/ha.d)	200 – 800	550 – 1000	< 800
Harvest schedule	Annual	Twice monthly to continuous	Twice monthly to continuous
Water depth (m)	0.5 – 0.8	0.9 – 1.2	0.6 – 0.9
Effluent quality			
BOD_5	< 30	< 15	< 10
SS	< 30	< 15	< 10
T – KN	< 15	< 15	< 5
Total phosphorus	< 6	< 6	< 1 – 2
Mechanism	Rapid growth rate coupled with excellent biological sulphate medium for bacteria are their extensive root system. Sensitive to cold climates		

6.2

Table 20 : Design criteria and effluent quality for final polishing with duckweed treatment systems

Item	*Magnitude*
Design criteria Wastewater input	Facultative lagoon effluent
BOD_5 loading (kg/ha.d)	22 – 28
Hydraulic loading (m^3/ha.d)	22 – 28
Water depth (m)	1.5 – 2.0
HRT (days)	20 – 25
Water temperature (°C)	> 7
Harvest schedule	Mostly for secondary treatment, weekly for nutrient removals
Expected effluent quality Secondary (mg/L)	
BOD_5	< 30
SS	< 30
TKN	< 15
Total phosphorus	< 6
Nutrient removal (mg/L)	
BOD_5	< 10
SS	< 10
TKN	< 5
Total phosphorus	< 1 – 2

Effluent from duckweed systems is often deficient in DO and may require aeration of treated water. Duckweeds have shallow root system and are sensitive to wind movements. They show lower sensitivity to cold weather as compared to water hyacinths.

The land – area (A) requirement can be estimated from :

$$A = \frac{(Q)(\theta)}{(10,000)(d)} \tag{14}$$

Select the HRT (θ), and depth (d) from the design criteria given in the tables. Q is the wastewater flow (m^3/d) and 10,000 is a conversion factor for m^2 to ha. To minimise development of anaerobic conditions (sulphate reduction to sulphide); recycle, step feed, and warparound, configuration alongwith supplemental aeration may be necessitated (to overcome anaerobic conditions). Mosquito control through sprinklers can be a useful system.

7. **Constructed wetlands :** An emerging technology for sewage and industrial wastewater are considered for the treatment of wastewater using emergent plants such cattails, reeds, and rushes.

The three principal types of constructed wetlands are :

Free water surface (FWS); the flowpath of applied wastewater is above the soil surface.

Subsurface flow (SF); the flow path is lateral through the root zone and the medium which ranges from sand to coarse gravel to rocks.

Vertical flow (VF); the application is either by spray or surface flooding, and the flow path is down the medium and out through the underdrains.

7.1 **Land requirement :** Area requirement for the FWS wetlands is expressed by :

$$A_{fw} = \frac{(Q)(\theta)}{(10,000)(d)} \tag{15}$$

where A_{fw} is the FWS wetland area (ha), Q is the wastewater flow (m^3/d), q is the detention time (days), d is the depth (m)

Area requirement for the SF wetland is given by :

$$A_{sf} = \frac{Q(\ln C_i - \ln C_e)}{(k)(d)(n)(10,000)} \tag{16}$$

where A_{sf} is the SF wetland area (ha), Q is the flow rate (m^3/d), C_i is the influent BOD (mg/L), C_e is the effluent BOD (mg/L), k is the first order rate constant (d^{-1}), d is the depth of media (m) and n is the drainable voids.

7.2 **Important considerations :** The typical value of k_{20} and n are 1.1 d^{-1} and 0.38, respectively for a bed of medium to coarse gravel. Typical bed depths for SF wetlands range from 0.5 to 0.75 m. The FWS wetland is 1.52 times larger than the SF wetlands for the same temperature and the performance and flow. FWS wetlands can provide significant wildlife habitat. Alternating shallow (less than 0.6 m) and deep (greater than 1 m) water areas can provide supplemental oxygen for aerobic treatment, provide open water fowl, and reduce the need for planting and harvesting of emergent plants.

Distribution of wastewater in an FWS wetlands can be through manifolds or weirs to allow variation in water depth.

Mosquito control

FWS wetlands; chemical and biological controls, water level management and encouragement of predators.

SF wetlands; water is not exposed to adult mosquitoes and, therefore, mosquito control is not necessary.

Range of BOD and TSS removals observed in FWS wetlands :

Table 21 : BOD and TSS Values.

BOD (mg/L)	*TSS (mg/L)*
Influent : 23.9 – 140	380 – 8.9
Effluent : 6.5 – 19	8.0 – 53

Table 22 : BOD removals observed in SF wet lands including pre-treatment.

Pretreatment	*BOD*			
	Influent	*Effluent*	*% Removal*	*Nominal Detention Time (days)*
Oxidation Pond	23	8	65	5
Oxidation pond	78	25	68	3.3
Primary	118	1.7	88	6
Secondary	33	4.6	86	7

8. Soil absorption systems (Wastewater flows less than 0.05 MGD)

8.1 Typical absorption systems

General filled trenches preceded by a septic tank; the effluent flows from septic tank into trenches (leach or drain lines).

Basis for the system design is the ability of applied wastewater to infiltrate and then percolate through the soil profile.

Conventional percolation test measures this ability–both horizontal and vertical percolation rates, therefore, with over estimate the actual vertical percolation rate.

Most common natural system for wastewater management is the sub-surface absorption system.

8.2

Table 23 : Alternative soil absorption system

Type	Applicability under Otherwise Restrictive Conditions
Mounds	High groundwater; shallow imperious layer; low permeability soil; shallow Soil profile
At grade	Shallow fractured bed rock; high groundwater
Deep trench, bed or seepage pit	Shallow (but rippable) impermeable layer in soil profile, with more permeable soil below (must not have high groundwater)
Sand lined beds and fill systems	Shallow fractured bed rocks; high permeability soil
Evapo transpiration beds	Net positive annual evaporation; high groundwater; low permeability soil; shallow fractured bedrock; percolation isallowed by regulatory agency.

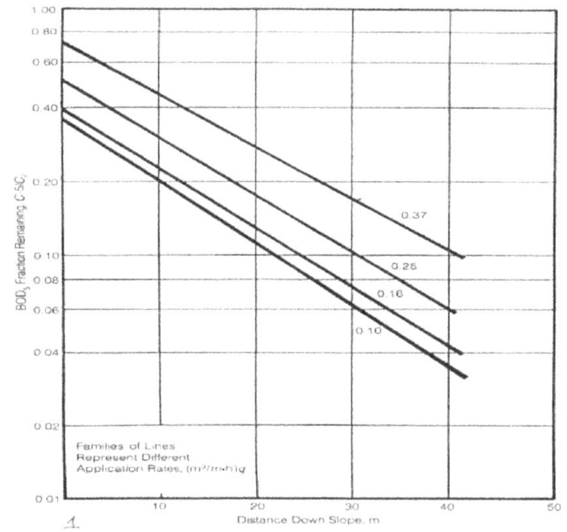

Figure 4 : Biochemical oxygen demand fraction remaining versus distance downslope for overload flow of primary effluent

Example **4.4**

Ecological Engineering Methods for Wetland Application

1. An overview of the ecotechnological methods that are used in lake restoration is presented in Table 24. The table comprises names of the methods, along with the problems they are used to solve and whether the methods change the forcing functions of the lake or attempt to reach the natural balance faster than otherwise would have been the case. In other words, the methods will directly change the mass balance of the lake or the methods will change the structure (the state variables) of the system. Indirectly, all methods will change the mass balance and the structure

more or less. The methods that change the state variables of the systems but do it by use of a new forcing function (e.g., aeration) are indicated in Table 24 by F(+ S).

2. Another overview of ecotechnological methods applied in lake restoration can be obtained by the use of models. From eutrophication models we know that algae growth (A) is a function of light (L), soluble inorganic nitrogen (N_S), soluble inorganic phosphorus (P_S), soluble inorganic carbon (C_S), the settling rate (S), the retention time (R), and the grazing rate (G):

$$A = f(L, N_S, P_S, C_S, S, R, G)$$

Table 24 : Ecotechnological Methods Applied in Lake Restoration

Method	*Problem to Solve*	*Direct Change in Forcing Function (F) or State Variables (S)*
Aeration	Oxygen depletion in hypolimnion. Release of P and Fe from sediment (eutrophication)	F (+S)
Siphon of hypolimnion water	Removal of P and oxygen–poorwater (eutrophication)	F
Nitrate to sediment	Reduce release of P from sediment (eutrophication)	F (+S)
Aeration with circulation	Eutrophication: depression of algal growth	S
Removal of uppersediment layer	Reduce release of P from sediment or remove toxic substances (eutrophication and toxic substance pollution)	S
Removal of P of algae from water	Eutrophication (reduce P/ algaalin water)	S
Wetland	Removal of N and P from non–point sources (eutrophication)	F
Decreased retention time	Reduce P and toxic substanceconcentration (eutrophication and toxic substances pollution)	F
Precipitation of P inlake	Eutrophication (reduce P in water)	S
Precipitation of P in inflowing water	Eutrophication (reduce P input)	F
Addition of $Ca(OH)_2$	Change of pH	F (+S)
Coverage of sediment	Eutrophication	F (+S)
Preimpoundment	Eutrophication (reduce P input)	F
Plastic beads or Sheet, dyes or soot	Reduce high penetration (eutrophication)	F (+S)
pH – modification shock inorganiccarbon in water)	Eutrophication (reduce	S

3. Restoration methods may also be classified according to which factor they change either in the lake or by decreasing input or increasing output, that is, by changing forcing functions (Table 25). The principle of this classification may also be applied to toxic substance and acidification problems.

Table 25 : Classification of Ecotechnological Methods for Lake Restoration According to Factor Controlled

Factor Controlled	*In Lake*	*By Changing Input/Output*
Light	Plastic beads, sheets, dyes or soot	Shading the shores
Soluble inorganic nutrient	Removal of sediment algae or P	Precipitation of P in water, NO_3 to sediment Pre-impoundment, wetlands. Precipitation of P in inflow,siphon of hypolimnion water
Soluble inorganic carbon	Change of pH	–
Retention time	–	Reduction of retention time
Grazing rate	Biomanipulation	–

4. Artificial re-aeration of lake water is widely used in lake restoration to obtain better oxygen conditions. The same is the case with the other methods, which improve the redox conditions – for example, addition of nitrate to the sediment.

5. Anaerobic sediments will generally release more phosphorus than aerobic sediments, for the latter contain iron in oxidation state three, whereas the former contain iron in the oxidation state two. Iron (III) has a higher adsorption capacity to phosphorus than iron (II), and iron (III) phosphate is much more insoluble than iron (II) phosphate. These processes may cause a "runaway" effect because increased eutrophication will mean a higher production of phytoplankton, which will settle and thereby give a high input of organic matter to the sediment. It therefore becomes anaerobic, which will mean enlarged release of phosphorus, which will cause more eutrophication, and so on. If the sediment becomes sufficiently anaerobic (low redox potential), sulfide formation may occur and, because iron (II) sulphide is highly insoluble, this will promote further the release of phosphorus. There is a closed relation between the redox conditions and the nutrients cycling in lakes.

6. Aeration may be used to depress eutrophication. It is achieved if the aerator circulates the water as shown in Figure 5. The water of the bottom layer is lifted by aeration and is replaced by the upper layer of water, which may be supersaturated with oxygen produced by the photosynthesis. The algae will be brought down to the dark hypolimnion, where light conditions are unfavourable for algae growth. Figure 6 shows the profile of phytoplankton and oxygen concentration before and after

aeration with water circulation. Table 26 shows the effects on the eutrophication measured in reservoirs with different depths. If the depth is more than 5 – 10 m, the effect is very pronounced.

Table 26 : Effects of Aeration on Eutrophication as Function of Lake Depth

Depth (m)	*% Reduction in Eutrophication*
2 – 3	0.5
5 – 7	10 – 20
10	45 – 50
> 10	> 50

6.1 Siphoning of hypolimnetic water serves two purposes – to improve the oxygen condition and to change the concentration of nutrients in the lake water. This is the case when hypolimnetic water is siphoned downstream, because hypolimnetic water generally has lower oxygen and higher nutrients concentrations.

6.2 Removal of the upper layer of the sediments may be used when the upper layer has higher nutrients concentrations than the deeper layers, which is most often the case. Figure 7 shows the phosphorus profile in a typical lake sediment. The phosphorus of the upper layer is either nonexchangeable because it is bound in insoluble chemical compounds or it is exchangeable because it is organic bound and after decomposition of the organic matter it is released. Consequently, it will be an advantage to remove the upper layer of the sediment and thereby reduce the release of phosphorus from the sediment. If the upper layer of the sediment contains toxic substances the method must be considered rather attractive, but because the method is expensive, it seems applicable only for small and very valuable lakes.

6.3 Reduction of the retention time may also have a positive effect on the reduction of eutrophication, Table 28, and it also implies a smaller concentration of toxic substances, if that is considered a pollution problem of the lake. The method is realized by the discharge of unpolluted water to the lake. For reservoirs it is possible to control the retention time by the dam. The use of dams enables changes not only in the retention time, but also in the water depth and water volume. In general, a longer retention time means increased eutrophication and a larger volume means a decreased eutrophication. It is therefore necessary to use models to make predictions of the overall effects of these changes in the hydrologic balance of a lake or reservoir. A further complication is related to the soil exposed to the lake water at increased depth. It will be in an oxidized state and therefore in general have a great capacity for non-exchangeable phosphorus. If this new sediment surface is significant in area, an increased water depth may cause an essential change in the phosphorus balance of the entire lake.

6.4 Removal of phosphorus directly from the lake water is also possible. The water is either pumped through an activated aluminum oxide column, which removes the phosphorus very effectively, or is filtered for removal of phytoplankton. The treated lake water is returned to the lake with a reduced concentration of soluble phosphorus or of phytoplankton, thereby reducing the total concentration of nitrogen and phosphorus as a result.

6.5 Precipitation directly in the water by use of aluminum sulphate or iron(III) chloride is also applicable. The method has, however, given less promising results in shallow lakes, because wind may stir up the flocs formed by the precipitation. Precipitation of phosphorus in inflowing water has also been used.

The use of wetlands is often a workable method, which can be applied for removal of non point pollutants. If the sources of lake pollution are diffuse, the problem is in general more difficult to solve. Wetlands often have an aerobic sediment and therefore a high denitrification rate. In addition, the wetland soil has a high capacity for uptake of phosphorus, pesticides, and other pollutants. Finally, the plants of the wetlands take up phosphorus and nitrogen and even other pollutants, and by harvest of the plants the pollutants are removed.

6.6 The use of pre-impoundments is based on a similar idea. Lakes and reservoirs have a phosphorus retention capacity, which implies that the phosphorus concentration of the water leaving the lake is lower than in the water entering the water body. Phosphorus retention increases with decreasing hydraulic retention time and mean depth and the phosphorus retention is also less for a lake or reservoir with an anoxic hypolimnion. Consequently, construction of a man made lake (called a pre-impoundment) in front of another lake may have a highly positive effect on the eutrophication. Models have been developed to evaluate the phosphorus retention of pre impoundments.

6.7 Reduction of light penetration by change of the extinction coefficient has also been used to depress phytoplankton growth (Jorgensen, 1980; Los *et al.*, 1982). The presence of humic substances in lake water may have a pronounced effect on the eutrophication owing to a significant change in the extinction coefficient of the water. Humic substances are natural, colored organic matter, but also human-induced changes may alter the extinction coefficient, particularly for pollution from pulp and paper mills. It has been studied that the use of yellow tetrazin of 1 mg/L was needed to increase the extinction coefficient 0.08. Although yellow tetrazin may be harmless, it is not considered a sound ecological method to add artificial compounds to lake water. Other possibilities are the shading of shores by trees or floating vegetation. The use of plastic non-transparent sheets, plastic beds, or soot has been tried rather successfully for small reservoirs and tanks.

6.8 Biomanipulation has been studied quite intensively. The idea is to increase grazing on phytoplankton, but it is still not clear whether the method can be used by itself or can be used only to bring the ecosystem faster to a new steady state after the forcing functions have been changed.

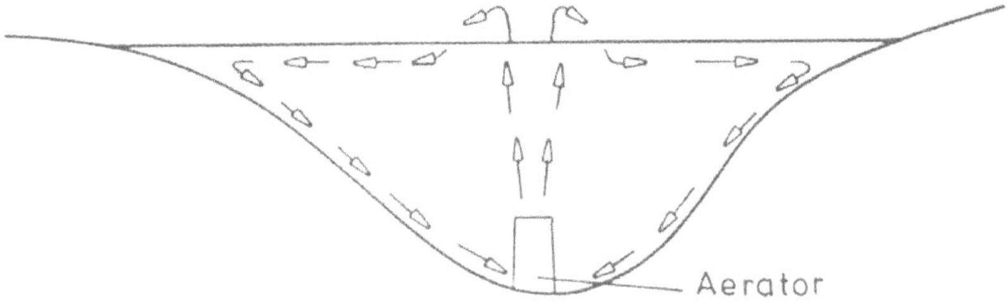

Figure 5 : The circulation of a lake by use of an aerator.

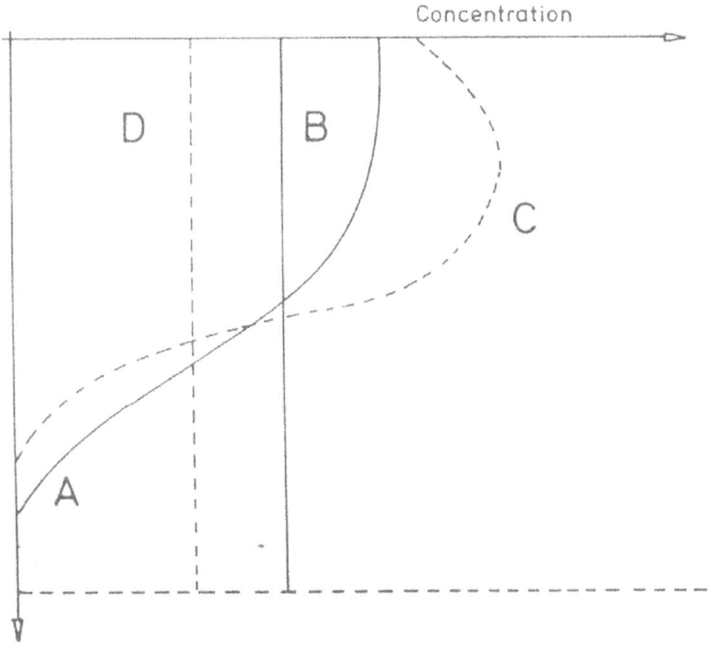

Figure 6 : The oxygen profiles before (A) and after (B) aeration. The phytoplankton profiles are also shown before (C) and after (D) aeration.

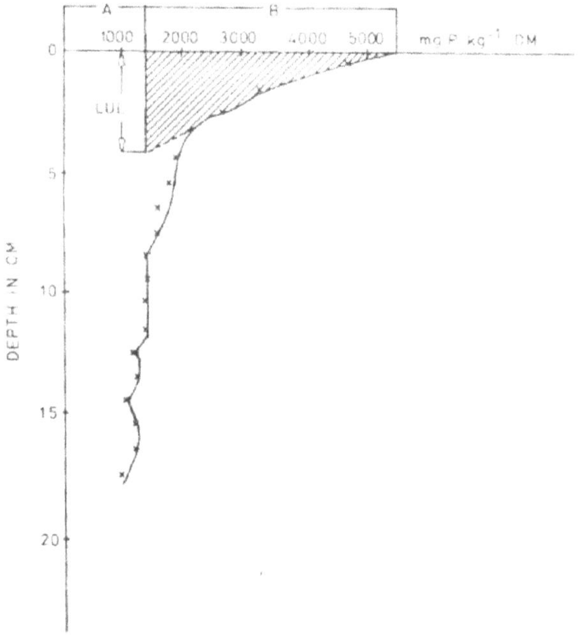

Figure 7 : Phosphorus profile of sediment core a typical case. The hatched area represents exchangeable phosphorus. LUL is the unstabilized area.

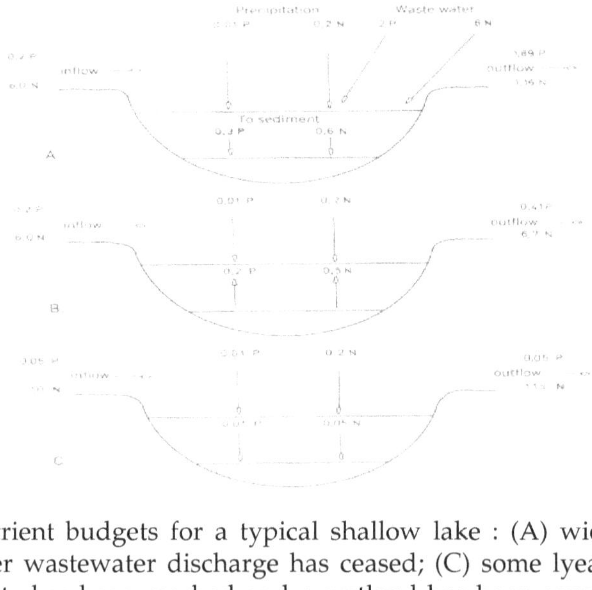

Figure 8 : Nutrient budgets for a typical shallow lake : (A) wioth wastewater inflow; (B) after wastewater discharge has ceased; (C) some lyears later after a new steady state has been reached and a wetland has been constructed to cope with runoff. Lake (20 ha, 5 m deep) has a retention time of 1 yr. and a catchment area of 10 am^2. All figures are × 1000 kg/yr.

Example 4.5

Redox Effects on Chemical and Biological Processes in Wetlands

1. Figure 9 shows a correlation of redox potential with reactions taking place in waters and sediments, including nitrification and denitrification. Some of the important relationships between chemical and biological processes and the redox potential are given below :

1.1 The rate of nitrification depends highly on the redox potential, as can be seen from the chemical reaction scheme :

$$NH_4^+ + 2O_2 \rightleftarrows NO_3^- + H_2O + 2H^+ \quad (1)$$

1.2 If the oxygen concentration is low, denitrification may occur :

$$nH^+ + nNO_3^- + n(CH) \rightleftarrows \frac{n}{2}N_2 + nH_2O + nCO_2 \quad (2)$$

1.3 The oxidation state of sulphur is determined by the redox potential. Nitrate is able to oxidize sulphide, because

$$3H^+ + 5HS^- + 8NO_3^- \rightleftarrows 5SO_4^{2-} + 4N_2 + 4H_2O \quad (3)$$

1.4 The oxidation states of iron and manganese are related to the redox potential :

$$Fe^{3+} + e^- \rightleftarrows Fe^{2+} \quad (4)$$

$$MnO_2 + 2e^- + 4H^+ \rightleftarrows Mn^2 + 2H_2O \quad (5)$$

1.5 The redox potential determine the composition of the micro-organisms culture (species and abundance).

1.6 A part of the phosphorus bound in the sediment may be released at lower redox potential owing to the following process :

$$2H_2O + FePO_4 + (CH) + 5HS^- \rightleftarrows 5FeS + CO_2 + 5HPO_4 + 5H^+ \quad (6)$$

This implies that anaerobic conditions in the sediment may influence eutrophication significantly.

1.7 Species composition and diversity depend on the oxygen concentration, as can be seen in Figure 11.

1.8 Anaerobic decomposition of organic matter is different from aerobic decomposition. The complete oxidation under aerobic condition can be expressed as :

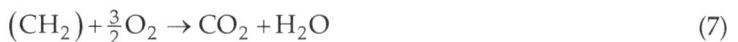

$$(CH_2) + \tfrac{3}{2}O_2 \rightarrow CO_2 + H_2O \quad (7)$$

and the partial decomposition of organic matter under anaerobic conditions can be represented by the following reaction scheme :

$$\text{Organic matter} \rightarrow CO_2 + CH_4 + CH_3(CH_2)_{\underline{n}}COOH \quad (8)$$

1.9 Many aquatic organisms utilize gills, whereby dissolved oxygen is passed from the water into the circulatory fluid of the organisms. Low oxygen concentration in water lead to lower oxygen uptake. This can partly be compensated for by fish and other aquatic organisms by pumping the water more rapidly over the gills. Moderately reduced oxygen concentrations result in reduced physiological activity and in sufficient muscle activity. Food consumption, growth, and swimming velocity will all decrease for fish at dissolved oxygen concentrations less than about 8 mg/L. A slightly lower oxygen concentrations, oxygen becomes the limiting factor, as shown in Table 27.

Table 27 : Limiting Oxygen Concentrations for Selected Organisms

Organism	*Limiting O_2 Concentration (mg/L)*	*Temperature (°C)*
Brown trout	2.9	6 – 24
Coho salmon	2.0	16 – 24
Rainbow trout	3.7	11 – 20
Amphipods	0.7	–

2. The redox potential is related to the oxygen concentration, as also indicated in Figure 9. The oxygen concentration is, however, a dynamic state variable. Oxygen is consumed by oxidation processes, and it is formed by the photosynthesis. Furthermore, oxygen can be exchanged with air. Reaeration is very important for the maintenance of a suitable oxygen concentration in many polluted rivers. The relations among consumption, formation, and exchange of oxygen are shown in conceptual model in Figure 10 and 20.

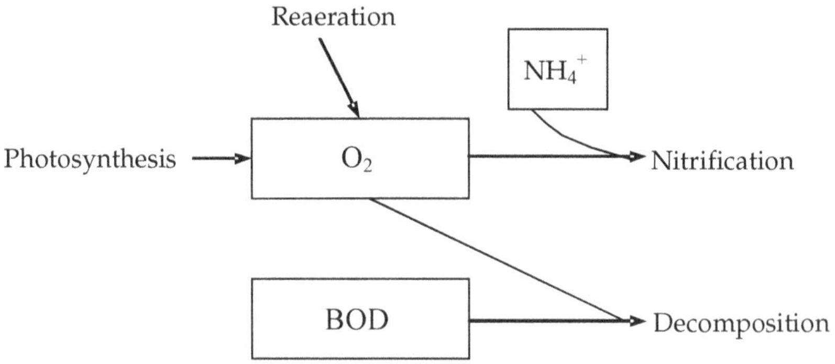

Figure 10 : Conceptual diagram of a dissolved oxygen model

3. The solubility of oxygen in water depends on the temperature and the salinity of water. High temperature means lower solubility, but the rate of oxygen consumption is higher at higher temperature. The relation between the rate constant and the temperature for oxidation of organic matter as well as nitrification may be formulated as follows :

$$K_T = K_{20}Q^{(T-20)} \tag{9}$$

where K_T and K_{20} are the rate constants at T°C and 20°C, respectively, and Q is a constant. These relations imply that the lowest oxygen concentrations are recorded in a warm summer before sunrise, for oxygen formation by photosynthesis depends on solar radiation (Figure 13).

A further complication of the oxygen exchange processes is caused by the formation of a thermocline or a halocline, which reduces considerably the rate of exchange between epilimnion and hypolimnion. Therefore anaerobic conditions are often observed in temperature lakes during the summer in a hypolimnion (Figure 14).

It is understandable that the problem of oxygen depletion is so complex that modelling has been used rather extensively to find a good management Solution.

The redox conditions are further complicated by the influence of pH. Redox processes depend on pH, as can be seen from Equation. Hydrogen ions are formed or consumed by these processes and therefore a changed hydrogen ion concentration (changed pH) will change the concentration of the other components.

Figure 15 shows how the oxidation state of iron depends on the redox potential as well as pH. As seen, the redox potential that determines the transformation from one oxidation state to another depends on pH. The rate constants for oxidation processes also depend on pH. This is illustrated in Figure 16, where the oxidation rate of sulphide versus pH is shown.

3.1 ***Application of ecotechnological methods :*** The redox potential of water is increased most easily by use of artificial aeration, which may be used in streams or lakes. The direct effect is, of course, a higher oxygen concentration, but in addition the aeration may also affect the sediment – water exchange processes, because the release of phosphorus from sediments is affected by the redox potential.

The most effective method of solving the problem of oxygen depletion in streams is by treatment of the wastewater. The biological treatment should preferably take place at wastewater plants, not in the streams. However, if the streams have an accumulation of organic sludge, aeration may be needed to accelerate the decomposition (oxidation) of the sludge.

For lakes the problem is more complex. First of all, the oxygen depletion is most often caused by eutrophication, not by discharge of biological oxygen demand. Secondly, the high concentration of nutrients cannot be reduced rapidly, owing to the long retention time of water in lakes. Therefore artificial aeration has been used more for lakes than for streams and it has been used mainly to prevent anaerobic conditions at the sediment surface.

Four different aeration methods have been applied. They are shown in Figure 17. Table 28 gives the characteristic features of the methods.

Lake restoration will also have the effect of increasing oxygen and decreasing BOD concentration. But addition of nitrate for stabilization of the sediment should be included here, because the effect mainly is on the redox condition in the sediment. The nitrate will oxidize the organic matter (denitrification), giving a higher redox potential at the sediment surface. This method has some pronounced advantages – quick, permanent, and effective results – but also some pronounced disadvantages – the risk for wrong dose is high, control of sediment after the addition is required, and it involves addition of a nutrient (nitrate), which may cause side effects with eutrophication. The method is often combined with the addition of iron (III) chloride, which increases the binding capacity of the sediments for phosphorus.

Table 28 : Features of Four Different Aeration Methods

Method	*Is Thermocline Maintained?*	*Advantages*	*Disadvantages*
Pressurized air in perforated tubes	No	Simple installation, moderate costs	Daily control required, clogging of perforated tubes, low efficiency, difficult to move
Use of diffusions (similar aeration to an activated sludge plant)	No	Good efficiency, movable, relatively simple installation	Daily control required noisy
Aeration with turbines	No	Simple installation, good efficiency, movable	Some fish may be killed in the turbine noisy
Mammut pumps	Yes	Good efficiency, movable	Costly, daily control required outboard motor

4. **Need for increasing the redox potential :** For streams it may be relevant to use aeration for a shorter period to oxidize soluble BOD_5, if conventional biological treatment of wastewater is not operating. Aeration should not replace the biological treatment, at least not permanently.

   Streams with slow and/or little flow may show low oxygen concentrations inspite of sufficient biological treatment of wastewater discharged to the streams owing to a high accumulation of organic sludge. The accumulation of sludge is often caused by a previously insufficient wastewater treatment. Because the water flow is low, it does not contain sufficient oxygen to oxidize the organic sludge. Aeration may here solve the problem caused by previous omissions of wastewater treatment.

   Many BOD – DO stream models do not include the BOD originated from bottom sediment, but it is obviously necessary to include BOD in sediments as a state variable, if a stream model should be able to give any predictions on the effects of aeration. Such models show clearly and not surprisingly that the lower the water flow, the more significant the organic matter in the sediment and the more pronounced the effect of the aeration.

Aeration or addition of nitrate to lake is closely related to internal loading of phosphorus. If the sediments have a high content of phosphorus that may be released as the phosphorus input to the lake is reduced, internal loading may be significant. If, in addition, the retention time is long, it will be advantageous to reduce the internal loading by aeration, thereby achieving a more rapid decrease of the phosphorus concentration in the lake water.

Figure 11 shows two different phosphorus profiles for lake sediments. As demonstrated, the profile may give an indication of the expected internal loading.

Lakes covered by ice during the water have a particular problem. The microbiological activity (and oxygen demand) is of course low during the winter, but because the lake water is separated from the atmosphere by ice even the very low decomposition rate may cause oxygen depletion. Figure 11 illustrates the concentration profiles for oxygen, methane, carbon dioxide, and hydrogen sulfide for a lake in southern Sweden after a severe winter (1970). The formation of hydrogen sulfide shows that the redox potential is at a very low and very critical level.

Ammonia (NH_3) is toxic to fish (but ammonium NH_4^+, the ionized form, is harmless). It is therefore important for the redox potential to be sufficiently high to allow nitrification to occur. pH is high in eutrophic lakes during summer, which implies that the ratio NH_3/NH_4^+ increases. An ammonia concentration of 0.025 mg/L is considered an upper limit, for most fish excrete ammonia through the gills at this level. As seen in Table 29, this limit does not allow a high total ammonia + ammonia concentration at high pH values such 8.5 – 9.5, which often may be found in shallow eutrophic lakes in the summer. Aeration may be urgently needed in such cases to accelerate nitrification.

In general it can be concluded that the following features may increase the advantage of aeration or nitrate addition :

1. high phosphorus concentration in sediment;
2. relatively large amount of phosphorus in the sediment in the exchangeable form (compare with Figure 11);
3. long retention time;
4. formation of thermocline (anaerobic hypolimmion);
5. ice cover during winter, and
6. high ammonium concentration in lake water.

Table 29 : Concentration of Ammonia (NH_3 + NH_4^+) Containing an Unionized Ammonia Concentration of 0.025 mg NH_3/L

Temp.(ºC)	*pH*					
	7.0	7.5	8.0	8.5	9.0	9.5
5	19.6	6.3	2.0	0.65	0.22	0.088
10	12.4	4.3	1.37	0.45	0.16	0.068
15	9.4	5.9	0.93	0.31	0.12	0.054
20	6.3	2.0	0.65	0.22	0.088	0.045
25	4.4	1.43	0.47	0.17	0.069	0.039
30	3.1	1.0	0.33	0.12	0.056	0.035

Note :

Table 30 : Classification of models (Pairs of model types)

Types of Models	Characterization
Research models	Used as a research tool
Management models	Used as a management tool
Deterministic models	Predicted values are computed exactly
Stochastic models	Predicted values depend on probability distribution
Compartment models	Variables defining the system are quantified by means of time-dependent differential equations.
Matrix models	Matrices are used in the mathematical formulation
Reductionistic models	As many relevant details are included as possible
Holistic models	General principles are used
Static models	Variables defining the system do not depend on time
Dynamic models perhaps of space)	Variables defining the system are a function of time (or
Distributed models	Parameters are considered functions of timeand space
Lumped models	Parameters are within certain prescribed spatial locations and time, considered as constants
Linear models	First–degree equations are used consecutively
Nonlinear models	One or more of the equations are not firstdegree
Casual models	Inputs, states, and outputs are interrelated byuse of casual relations
Black box models	Input disturbances affect only the output responses. No causality is required
Autonomous models	Derivatives do not explicitly depend on the independent variable (time).

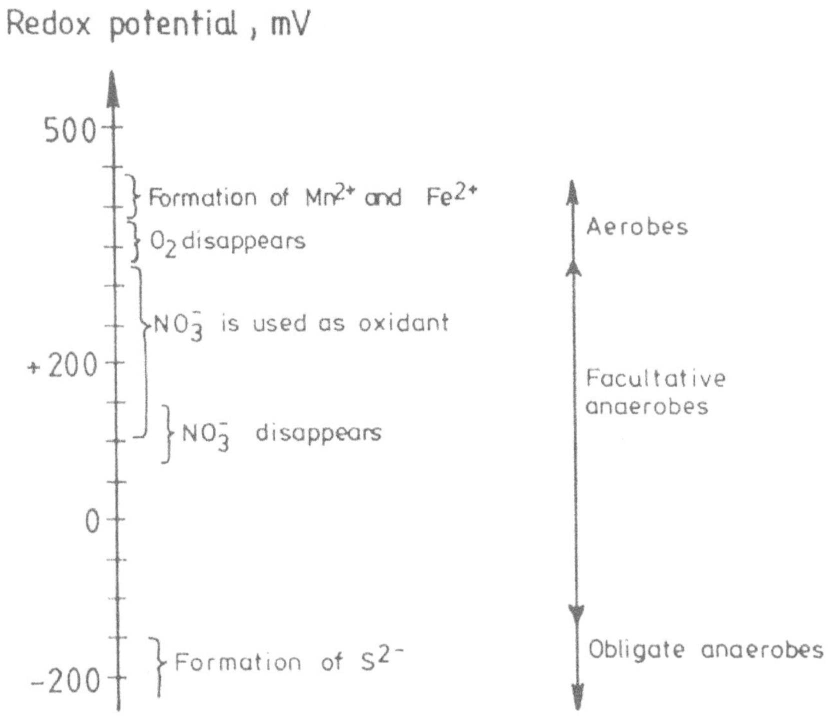

Figure 10 : Rexdox potential relations and associated biological communities.

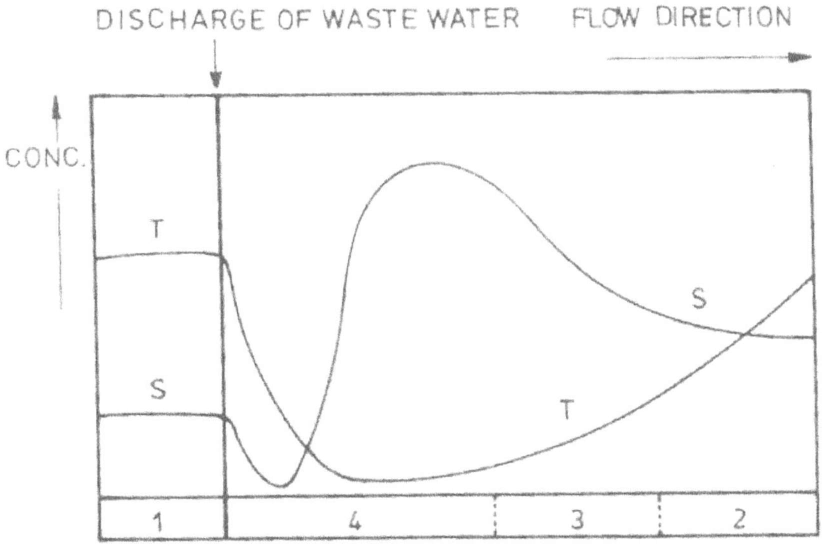

Figure 11 : Patterns of pollution in a river showing number of higher organisms (S) and number of species (T) versus distance in running water. Classification with numbers in accordance with saprobic system.

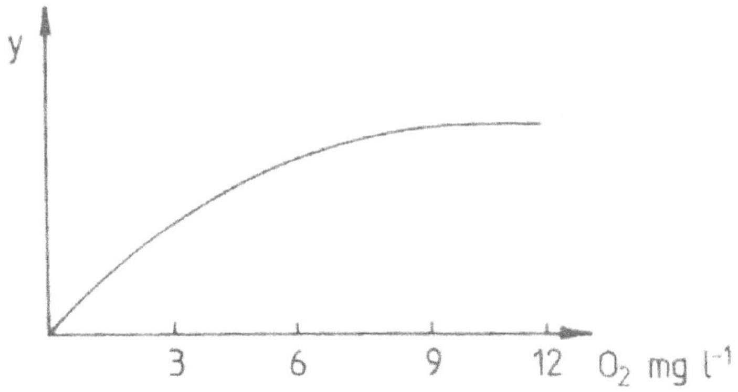

Figure 12 : Activity of fish (food consumption, growth, or swimming activity) versus dissolved oxygen concentration. The relationship varies from species to species and can therefore be considered only very approximate.

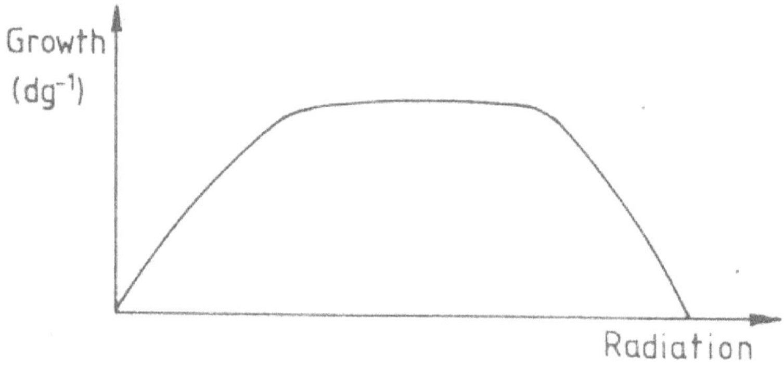

Figure 13 : Growth of phytoplankton (photosynthesis) versus solar radiation.

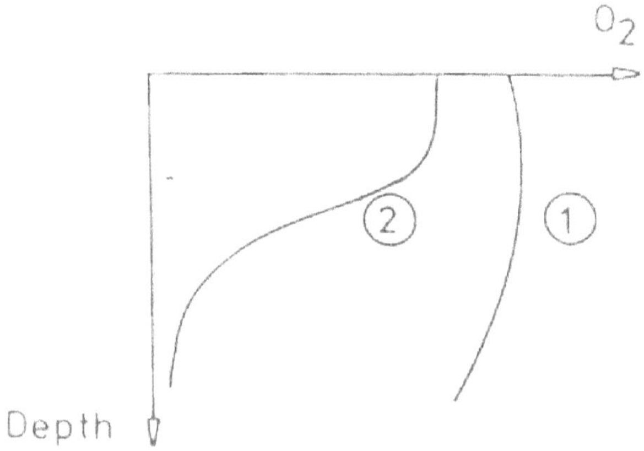

Figure 14 : Dissolved oxygen profiles in a stratified lake (1) winter conditions; (2) summer condition.

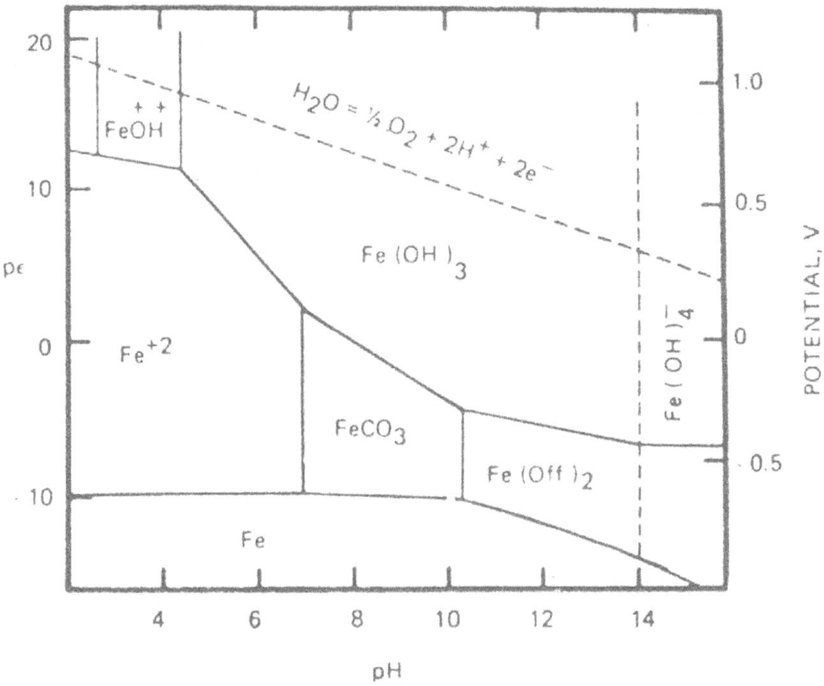

Figure 15 : pi-pH diagram for iron.

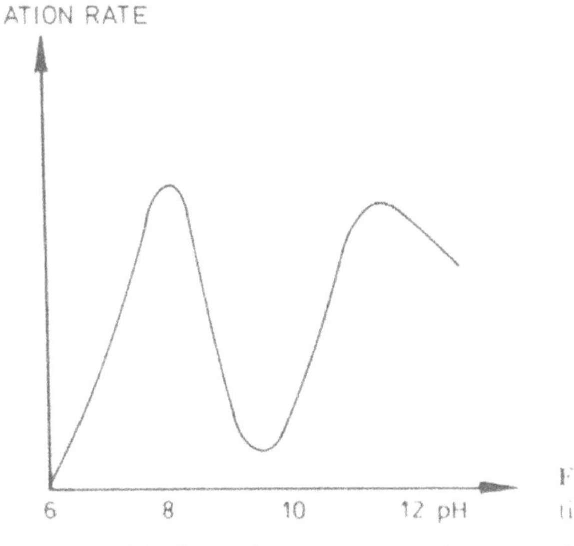

Figure 16 : Sulphide oxidation rate as a function of pH.

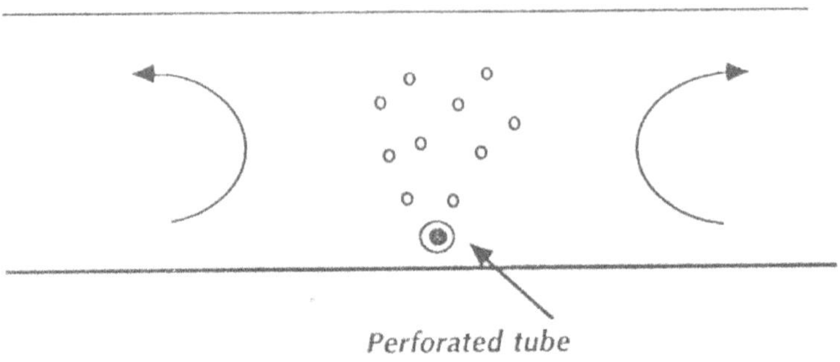

(a)

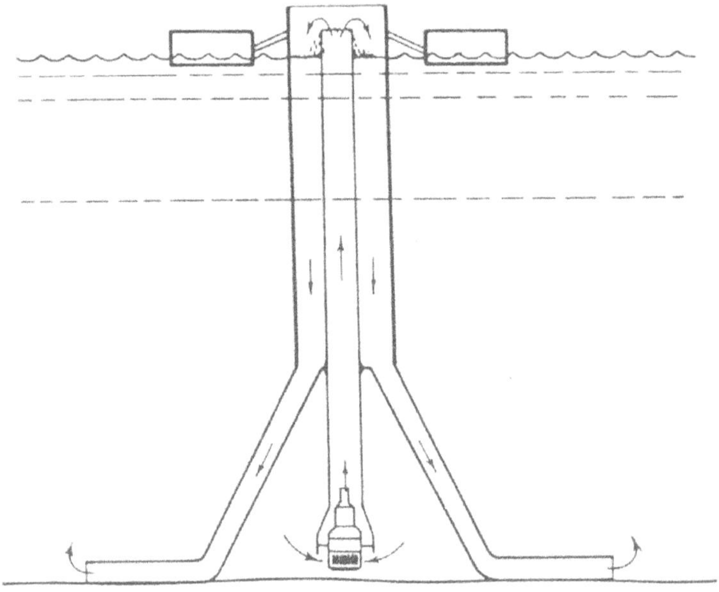

(b)

Figure 17 : Various methods for artificial aeration of lakes: (a) pressurized air in perforated tubes; (b) pumping hypolimnetic water with submerged pump; (c) pumping hypolimnetic water with surface outboard motor; (d) aeration directly into hypolimnion.

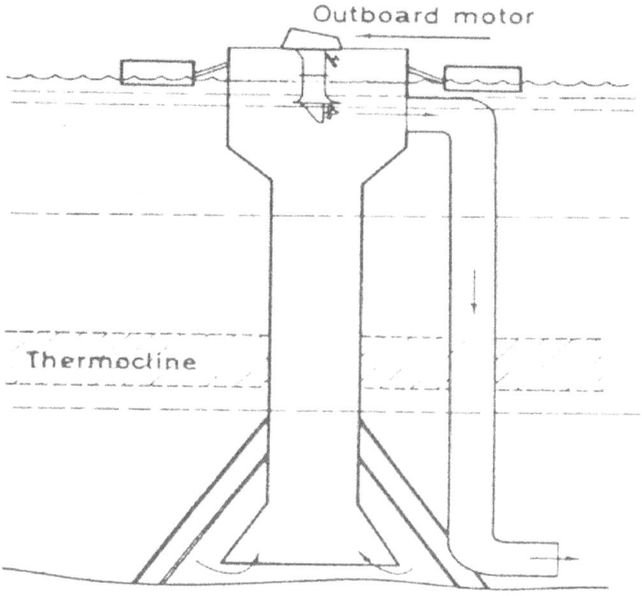

(c)

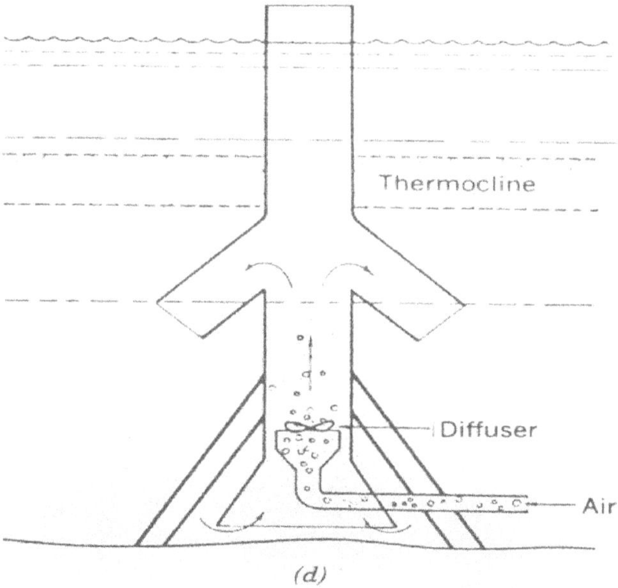

(d)

Figure 18 : Various methods for artificial aeration of lakes.

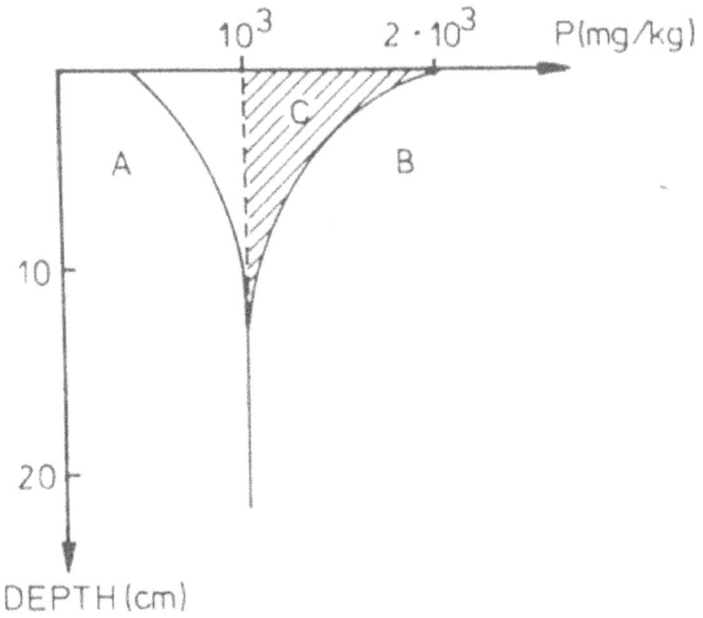

Figure 19 : Possible profiles of phosphorus in lake sediments. A profile will give only a small release of phosphorus, whereas profile B most probably will exchange phosphorus corresponding to area C.

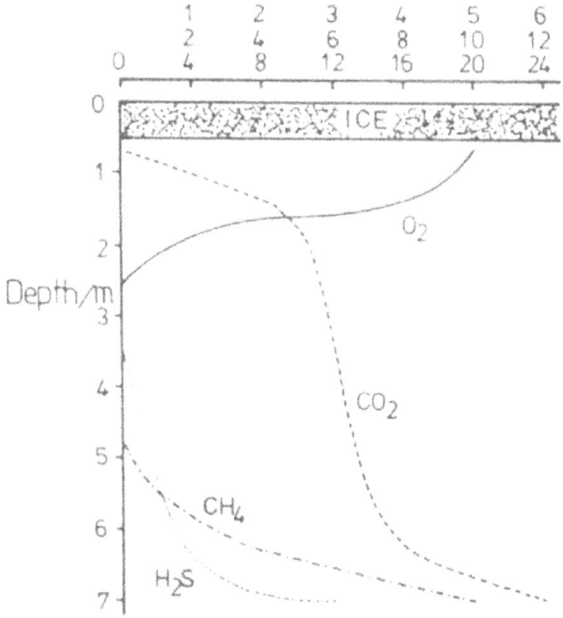

Figure 20 : Profiles of oxygen, methane, hydrogen sulphide, and carbon dioxide in an ice-covered lake in Skaane, Sweden, April 1970 during a severe winter.

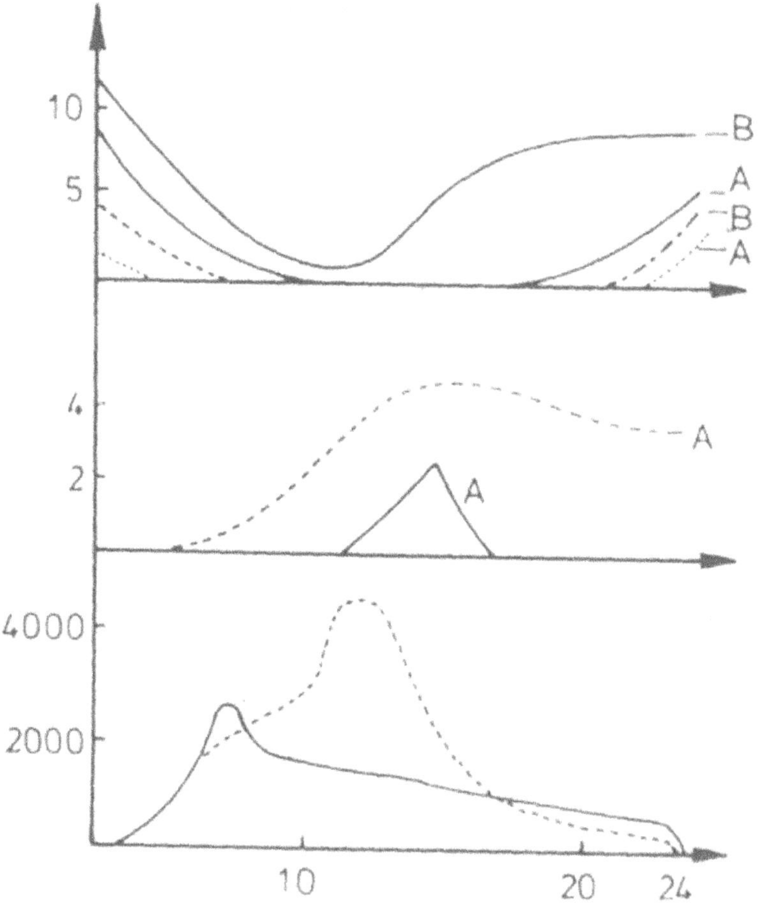

Figure 21 : Results of aeration of Bolts Lake, Kentucky, Solid lines test lake; dotted lines, control lake (A) 15 m depth; (B) 10 m depth. Upeer plot; mg O_2/L, middle plot: S^{-2}; mg/L; Lower plot, algae counts per millimeter.

Example 4.6

Design Considerations for Constructed Wetlands

1. **Wetland classification (Natural and constructed wetlands) :** Natural wetlands occur throughout the landscape as traditional areas between aquatic and uplands (savana, pampas, champaign or moors), while man-made (constructed wetlands) are planned, designed and constructed.

   Natural wetlands include a variety of plants such as cattails (Typha), rushes (Scirpus), and reeds (Phragmites). They can also contain some floating plants such as water hyacinths (*Eichhornia crassipes*), duck weeds

(*Lemna* and *Spirodela* sp.), pennywort (*Hydrocotyle umbellata*), and water ferns. The submerged plants found in natural wetlands include aquatic plants such as water weeds, water milfoil, and water cress. Willows, cypress, and ash trees tend to be found in swamps.

Most constructed wetlands contain predominately herbaceous plants. Free water surface (FWS) wetlands: Water level is above the ground surface; vegetation is rooted and emergent above the water surface; water flow is primarily above the ground; vegetation may be planted or allowed to colonize voluntarily (majority of the water flows over the sediments and through the above ground plant zone). Vegetated submerged bed (VSB) wetlands, designed to conduct water through the bed of the system to make contact with the plant roots (water level is below ground, water flow is through soil or gravel bed; root penetration is to the bottom of bed; wetland plants are generally common reed, bulrush or Cattail). The VSB wetland system is also often referred to as sub-surface flow system (SFS).

2. **Water balance and hydrologic factors :** A simple equation expressing the water balance for such a system is as follows :

$$Q_i - Q_o + P - E = \frac{dV}{dt}$$

where Q_i is the influent wastewater flow (m^3/d), Q_o is the effluent wastewater flow (m^3/d), P is the precipitation (m^3/d), E is the evapo – transpiration rate (m^3/d), and V is the volume of water (m^3).

Wetlands that are continuously inundated, E can generally be estimated as being equal to lake evaporation or approximately 70 to 80% of pan-evaporation values. If the wetland system operates at relatively constant depth, [dV/dt] = 0.

Hydraulic loading rate (HLR) vary for natural and constructed wetlands (may also very with climate). To minimize vegetative changes and maximize treatment efficiencies, HLRs in natural wetlands should generally not exceed 1 to 2 cm/d. The maximum acceptable HLRs for constructed wetlands should generally not exceed 2.5 to 5 cm/d for FWS wetlands and 6 to 8 cm/d for VSB wetlands.

Water depths in constructed wetlands (FWS) are related to the type of plants growing in the system. Water depths can vary from 0.15 to 1.0 m. Normal wetland design water depths are generally less or equal to 0.5 m.

The hydraulic detention time (t) can be calculated by dividing wetland volume (V) by the average flow rate (Q) :

$$\text{Detention time (t)} = \frac{V}{Q}, \text{ and}$$

$$\text{Wetland volume (V)} = A_r d$$

In FWS wetland, a portion of the available volume is occupied by vegetation, therefore, the actual detention time is a function of porosity (voids), e :

$$t = \frac{(A_r)(d)(e)}{Q}$$

where A_r is the wetland area (m^2), d is the water depth (m), and e is the porosity or void fraction.

3. **Organic loading rates, nutrient and contaminant removal :** Organic loading rates for FWS can vary from 1 kg BOD_5/ha.d to 125 kg BOD_5/ha.d and have removal rates that range from 49 to 95%. A maximum BOD_5 loading rate of about 100 kg/ha.d has been recommended to prevent the occurrence of mosquito populations. Others have recommended a maximum BOD_5 loading rate of 110 to 120 kg BOD_5/ha.d for both FWS and VSB systems.

   While wetland systems exhibit seasonal variability, they have achieved high removal rates for BOD_5, SS, trace organics, nitrogen, and heavy metals. Table 31 describes the removal mechanisms.

   The BOD_5 removal in a wetland is described by a first order reaction:

$$\left(\frac{S_e}{S_i}\right) = \exp(-kt)$$

$$K(T) = (20)\theta^{T-20}$$

$$q = 1.056 \text{ to } 1.135 \text{ depending upon temperature}$$

$$\theta\ (20) = 1.1$$

$$\left(\frac{S_e}{S_i}\right) = A\exp\left[-0.7\,k_T\,\frac{(A_v)^{1.75}(A_r)(d)(e)}{Q}\right]$$

   where S_e is the effluent BOD_5 (mg/L), S_i is the influent BOD_5 (mg/L), k_T is the temperature dependent first order reaction rate constant (d^{-1}), t is the detention time (d), A is the fraction of BOD_5 not removed as settle able solids near head works of the system, A_v is the specific area for microbial activity (m^2/m^3), A_r is the wetland area (m^2), d is the design depth, e is the porosity of the system (0.75), and Q is the average hydraulic loading on the system (m^3/d). Assume the following data :

$A = 0.52$

$k(20) = 0.0057\ d^{-1}$

$A_v = 15.7\ m^2/m^3$

$e = 0.75$

Therefore, $A_r = \dfrac{Q(\ln S_i - \ln S_e - 0.6539)}{(65)(k_T)(d)}$

If $A = 0.52$, $\ln(0.52) = -0.6539$, and

$[-0.7\ k_T (15.7\ m^2/m^3)^{1.75}\ A_r(d)(0.75)]$

$= [(0.7)(123.82)(0.75)(k_T)(A_r)(d)] = 65\ k_T\ A_r d$

or $t = \dfrac{[\ln(S_i) - \ln(S_e)] - 0.6539}{65\ k_T}$

4. Determine the wetland area requirement as well as the hydraulic and organic loading rates. Assume the following data :

Effluent BOD_5	:	20 mg/L
Influent BOD_5	:	150 mg/L
Mean Summer temperature	:	25°C
Mean Winter temperature	:	15°C
Design flow (Q)	:	1500 m^3/d
Depth of flow (d)	:	0.5 m (Constant)
Wetland slope	:	Greater than 1%
Average precipitation	:	120 cm/year
Evapo-transpiration	:	120 cm/year
k (20)	:	0.0057 d^{-1}
θ	:	1.1

4.1 Temperature connection for Summer and Winter

$$k(25) = k(20)\theta^{T-20}$$

$$= 0.0057(1.1)^5 \quad \text{[Summer]}$$

$$= 0.0092\ d^{-1}$$

$$k(15) = 0.0057(1.1)^{-5} \quad \text{[Winter]}$$

$$= 0.00354\ d^{-1}$$

4.2 Wetland area (A_r) requirement for reducing BOD_5 from 120 to 20 mg/L

Summer :

$$A_r = \frac{Q\left[\ln(S_i) - \ln(S_e) - 0.6539\right]}{65\,(k_T)(d)}$$

$$= \frac{1500\left[\ln(120) - \ln(20) - 0.6539\right]}{(65)\,(0.0092)(0.5)}$$

$$= \frac{1500\left[4.79 - 2.99 - 0.6539\right]}{(65)\,(0.0092)(0.5)} = 5737\ m^2$$

$$= 0.5737\ \text{ha}$$

Winter :

$$= \frac{1500\left[4.79 - 2.99 - (0.6539)\right]}{65\,(0.00354)(0.5)} = 14,943\ m^2$$

$$= 1.5\ \text{ha.}$$

4.3 Detention time (t)

$$t = \frac{\left[\ln(S_i) - \ln(S_e)\right] - 0.6539}{65\ k_T}$$

Summer :

$$t = \frac{\ln(120) - \ln(20) - 0.6539}{(65)(0.0092)} = 1.9\ \text{days (Possible not included)}$$

Winter :

$$t = \frac{\ln(120) - \ln(20) - 0.6539}{(65)(0.00354)} = 489\ \text{day (Possible not included)}$$

With porosity :

$$t = \frac{(A_r)(d)(e)}{Q}$$

$$= \frac{(5737)(0.5)(0.75)}{1500} \quad \text{[Summer]}$$

$$= 1.43\ \text{days}$$

$$t = \frac{(14,943)(0.57)(0.75)}{1500} \qquad \text{[Winter]}$$

= 3.74 days

[Winter temperature controls, A_r = 14,943 m^2 and t = 3.74 days]

4.4 Water balance (dV/dt = 0)

$$Q_i - Q_e + (P - E) = 0$$

Therefore, $Q_i = Q_e = Q$

4.5 Hydraulic loading (HLR)

$$HLR = \frac{Q}{A_r}$$

$$= \frac{1500}{14,943} = 0.1 \text{ m/d}$$

$$= 10 \text{cm/d}$$

Practical aspect : Therefore, we must increase the area of wetland by a factor of at least 2 to reach the maximum acceptable HLR_s for constructed wetland (= 5 cm/d), A_r = 29,886 m^2

$$HLR = \frac{1500}{29,886} = 0.05 \text{ m/d} = 5 \text{cm/d}.$$

4.6 Organic loading rate (BOD_5)

$$\text{Organic Loading Rate (OLR)} = \frac{Q S_i}{A_r}$$

$$= \frac{1500 \times 120}{1.4943 \times 1000}$$

= 120 kg BOD_5/ha.d (OK)

Practical aspect : As HLR dictates (2.5 to 5.0 cm/d), therefore, the designed

$$OLR = \frac{150 \times 120}{2.9886 \times 1000} = 60 \text{ kg } BOD_5/\text{ha.d}$$

[This loading rate of 60 kg BOD_5/ha.d is well below the suggested loading rate of 120 kg/ha.d based on practical data on operation of the system].

Table 31 : Removal mechanisms in wetlands for the contaminants in wastewater

Mechanism	Contaminant Effected[a] *Settleable Solids*	*Colloidal Solids*	***BOD***	***N***	***P***	*Heavy Metals*	*Refractory Organics*	*Bacteria and Virus*	*Description*
Physical Sedimentation	P[a]	S[a]	I[a]	I	I	I	I	I	Gravitational settling of solids (and constituent contaminants) in pond/marsh settings.
Filtration	S	S							Particulates filtered mechanically as water passes through substrate, root masses, or fish.
Adsorption		S							Interparticle attractive forces (van der Waals force).
Chemical Precipitation				P	P				Formation of or co-precipitation with insolube compounds.
Adsorption				P	P	S			Adsorption on substrate and plant surfaces.
Decomposition						P		P	Decomposition or alteration of less stable compounds by phenomena such as UV irradiation, oxidation, and reduction.
Biological Bacterial metabolism[b]			P	P	P			P	Removal of colloidal solids and soluble organics by suspended, benthic, and plant-supported bacteria. Bacterial nitrification/denitrification.
Plant							S	S	Uptake and metabolism of tabolism[b] organics by plants. Root excretions may be toxic to organisms of enteric origin.
Plant adsorption				S	S	S	S		Under proper conditions, significant quantities of these contaminants will be taken up by plants.
Natural die-off								P	Natural decay of organisms in an unfavourable environment.

[a]P = Primary effect; S = secondary effect; I = incremental effect (effect occurring incidental to removal of another contaminant)
[b]The term metabolism includes both biosynthesis and catabolic reactions

Example 4.7

Wetland Design Parameters

The use of both artificial and natural wetlands for water quality enhancement requires the development of several design parameters:

1. **Wetland retention time :** The total amount of water or chemicals flowing into a wetland (Q), divided by the area of the study wetland (A), gives an estimate of the loading rate of a wetland :

$$L = \frac{Q}{A} \tag{1}$$

When we divide the loading rate by the average depth (d) of the wetland, we calculate the turnover rate (t^{-1}) as

$$t^{-1} = \frac{L}{d} \tag{2}$$

or alternately express the retention or residence time of the wetland (t) as the reciprocal of the turnover rate. The retention time (t), expressed then as :

$$t = \frac{A \times d}{Q} \tag{3}$$

is one of the most important variable in the use of wetlands as wastewater treatment systems. The longer the retention time of the wetland, the more time the water is in contact with the biological active sediments, and the greater the rates of physical processes such as sedimentation. But longer retention times require either a larger wetland area or a deeper wetland. When average depth becomes too great, the system may no longer function as a wetland and the role of vegetation and sediments is diminished.

2. **Chemical loading rates :** In addition to the hydrologic detention time, the amount of chemicals entering the wetland per unit time is a very important variable. There is an inverse relationship between chemical loading rate and percent removal efficiency. The retention of phosphorus, can be expressed as :

$$R_{\%p} = f\left(\frac{1}{L_P}\right) \tag{4}$$

where $R_{\%P}$ is the percent retention of phosphorus in a wetland and L_P is the phosphorous loading in g P/m^2 yr.

Likewise, nitrogen retention is described by :

$$R_{\%N} = f\left(\frac{1}{L_N}\right) \tag{5}$$

where $R_{\%N}$ is the percent retention or loss of nitrogen in a wetland and L_N is the nitrogen loading in g N/m^2 – yr.

3. **Example :** If a wetland system is being designed for removal of phosphorus, then highest efficiencies (60 – 90%) have been measured in the ranges of loading rates of 1 – 5 g P/m^2 – yr, but efficiencies reduce to about 30% removal at loading rates of 15 g/m^2 – yr. If total amount of chemical retained is more important than efficiency (remember, nature does not maximize for efficiency !) then the lower efficiency case above retains 4.5 g/m^2 – yr, and the higher efficiency system retains from 0.6 to 4.5 g P/m^2 – yr. The chemical loading rate, then, depends on whether one selects for efficiency or total retention of chemicals.

4. **Seasonal patterns :** Wetlands often display seasonal patterns of chemical uptake and release, often in synchronization with the needs to control certain chemicals. The design should be made for the limiting situations, either when the receiving body of water is most vulnerable (perhaps at low flow), or when the runoff from uplands is the greatest (possibly in the spring).

5. **Aging :** It has been observed that some wetlands have received artificial loadings of chemicals for many years and are still functioning as nutrient sinks. Studies of this type have included freshwater marshes in Wisconsin and in forested wetlands in Florida. Other investigators have suggested that continual chemical loading of wetlands will lead to a reduction of the wetland's efficiency in a process now called aging. Such a "saturation" in a wetland after a number of years of receiving chemicals can sometimes result in a reduced capacity for pollutant retention. It would therefore be prudent in the design of a wetland treatment system either to design for a larger wetland than is really needed (perhaps twice the size) or to have a second wetland designed to take over after a few years and give the original wetland a chance to recover to a state of seminatural conditions. The aging of natural wetlands receiving low loadings of nutrients, such as from non-point sources, is not as well understood.

6. **Ecological Engineering Role of the Lagoon :** Nishinoko Lagoon plays an important role as a whole ecosystem in removing nutrients from the inflowing wastewater into Lake Biwa, as if it were a natural ecological wastewater treatment system. A simplified model of the system is shown schematically in Figure 22. This is a first estimate of an ecological engineering model applied to the environs of Lake Biwa from the point of view of nutrient removal by utilizing natural ecosystems. This model must be studied further in other types of lagoons and applied to more lake environments suffering from eutrophication.

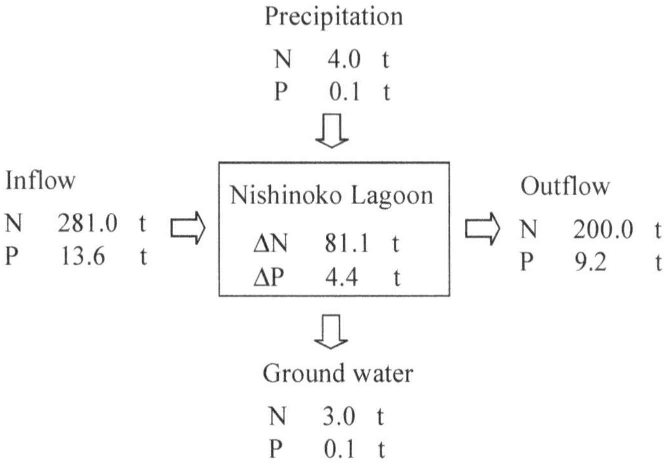

Figure 22 : Nitrogen and Phosphorus Budgets for Nishinoko Lagoon Adjacent to Lake Biwa, Japan

Note :

Key ecotechniques for treatment and utilization of wastewater

Interrelations among main elements of semi-artificial ecosystem, six ecotechniques for increasing the purification efficiency are suggested :

1. ***Ensure sufficient planting area of water hyacinths for the absorption of nutrients and for supplying matrices for sessile micro-organisms :*** The area may vary and depends on (1) the expected degree of purification of wastewater discharged into the semi-artificial ecosystem, (2) the discharge volume and pollutants of the inflow, and (3) growth rates and production of water hyacinths including the rate of absorption of nutrients in the local conditions. A model for estimating the planting area of water hyacinths in this ecological engineering application has been established as follows :

$$A_w = \frac{(C_3 - C_1)Q - (C_3V + C_4P_n)A}{C_2V_1 + C_4P_n - \left(BGE + BRA\left(MM_a + ZZ_a + C_5\right)\right)} \quad (1)$$

where A is the total area of water surface in the semi-artificial ecosystem; A_w is the area required for planting water hyacinth to attain at least the expected level of purification; B is the expected average biomass of water hyacinth; C_1 is the concentration of a given pollutant in the discharged water at the inlet; C_2 is the concentration of a given pollutant in the water discharged at the output before establishment of this ecological engineering; C_3 is the expected concentration of a given pollutant in the output; C_4 is the concentration of organic substances produced by phytoplankton; C_5 is the coagulation and adsorption rate of organic pollutants by the water hyacinth; E is the absorption rate of nitrogen, phosphorus, or other pollutants by water hyacinths; G is the growth rate of water hyacinth; M is the average biomass of sessile micro-organisms on the roots of water hyacinth; M_a is degradation rate of organic substances by those micro-organisms; P_n is the mean net production (mg/l) of phytoplankton; Q is the discharge per unit of time at the input; R is the proportion of roots to whole body of water hyacinth; V is the evapo-transpiration in the local region; Z is the average standing crop of periphyton and micro-invertebrates attached to the roots of the water hyacinths; and Z_a is the degradation rate of organic pollutants by those sessile periphyton.

2. ***Adjust the density of the water hyacinth population to maintain its rapid population growth rate and net production :*** More nitrogen, phosphorus, sulphur, and other pollutants are absorbed and more organic pollutants are degraded with a more rapid growth rate and higher net production of water hyacinth. It is one of the important conditions for the use of water hyacinths in this ecological engineering application that the plant have a rapid growth rate and high production. However, these may vary with the density of the plant. For example, when the density approaches to the carrying capacity, its growth rate becomes lower. In order to maintain and promote a rapid growth rate and high production of the water hyacinth, its population density should be adjusted at the time of the change in its density during its growth period. This is in accordance with the equation

$$\frac{dW}{dt} = r - \frac{2r\,W}{K} \tag{2}$$

which is derived from

$$\frac{dW}{dt} = \frac{r\,W(K - W)}{K} \tag{3}$$

when the first-order derivative value equals zero. W is water hyacinth biomass; r is the intrinsic growth rate, and K is the carrying capacity. dW/dt has a maximal value near 1/4rK where W = 1/2K. In other words, when the population density of the water hyacinth is nearly half of the carrying capacity, its net production is highest. The results of our experiments and field surveys show that the carrying capacity of the water hyacinth population is about 24 kg/m^2, with its highest net production occurring in a density of 10 – 15 kg/m^2 during July to September in our semiartificial ecosystem. It requires 7 – 10 days for water hyacinths to increase in density from 10 to more than 15 kg/m^2. During this period, the net production of this plant is about 0.5 – 0.8 kg/m^2.day.

3. ***Harvest the water hyacinth on time :*** It is necessary to adjust the plant density and transfer some nutrients absorbed by the water hyacinth from this semi-artificial ecosystem to another place to avoid secondary population by saprophytism and decomposition. This requires that the

plant harvest be carried out on time. Generally, a 7 – 10 day interval between harvests is proper for maintaining its density approaching 1/2K, nearly half of the value of its carrying capacity, in summer. In other seasons a slightly longer interval is needed. The area of planting water hyacinth is divided into seven to ten subregions, and they are harvested in turn. One-third to one-half of the standing crop is harvested and removed from this semiartificial ecosystem every 7 – 10 days, one after another in rotation.

4. ***Utilize the water hyacinth fully :*** Full utilization of this floating vegetation, the prerequisite for removing them, is another key for ecological engineering. One of the ways that water hyacinth is used is as feed for fishes, ducks, oxen, pigs, and snails. As mentioned above, this results in economic benefit and environmental efficiency.

5. ***Add artificial matrices into this semiartificial ecosystem :*** One of the purifying mechanisms for organic pollutants in this ecological engineering is the enlarged matrices for sessile microorganisms that can degrade the organic pollutants provided by the root system of the water hyacinth. The species of strains, density, and production of microorganisms increase and result in higher purification efficiency for organic pollutants. However, the more fertile the water is, the smaller the volume and surface of the roots. Generally, the roots of this floating vegetation are only about one-fourth or one-fifth of the whole body in relatively fertile water, such as most sewage. It is imperative to add some artificial matrices in the semiartificial ecosystem if the water is very fertile, in order to increase the number of sessile micro-organisms and to raise the purification efficiency for organic pollutant removal.

6. ***Inoculate bacteria to accelerate biodegradation of special organic pollutants such as phenol and petroleum oil and its products :*** The strains of bacteria that are able to grow on phenol as sole carbon sources were identified as *Bacillus meraterium*, isolated from the roots of the water hyacinth. Through our experiments with bacteria cultured in various media with different contents of phenol, results show that these strains of bacteria have a great capability for utilizing and biodegrading phenol. If water polluted by phenol is put into a semiartificial ecosystem, it is better to inoculate these strains of bacteria on the roots of water hyacinth or artificial matrices. In the same way, if the water is polluted by another special organic pollutant, such as petroleum oil or its products, organophosphates, or BHC, then another strain of bacteria that can rapidly biodegrade that special kind of pollutant should be inoculated.

Example 4.8

Infiltration and Inflow

Determine the infiltration and compare this quantity to the average daily and peak hourly domestic wastewater flows. Use the following data :

Sewered population	:	25,000 persons
Average domestic flow	:	100 gpcd
Peak hourly domestic flow	:	250 gpcd
Infiltration rate	:	450 gpd/mile/in.pipe diameter
Sanitary sewer system	:	
4 in. building sewer	:	40 miles
6 in. street laterals	:	20 miles
8 in. sub mains	:	5 miles
12 in. trunk sewers	:	5 miles

Solution

1. Required infiltration rate (I)

$$I = \left(\text{Rate}, \frac{\text{gal}}{\text{day} \times \text{miles} \times \text{in.}}\right) \times \{\Sigma[\text{dia (in.)} \times \text{length (miles)}]\}$$

2. Required average domestic flow

$$= 25{,}000 \times 100 = 25{,}00{,}000 \text{ gal}$$

$$\text{Ratio of } \frac{\text{Infiltration}}{\text{Average domestic flow}} = \frac{126{,}000}{25{,}00{,}000} = 0.05 \text{ (or 5\%)}$$

Peak domestic flow = (25,000) (250) = 62,50,000 gal/d

$$\text{Ratio of } \frac{\text{Infiltration}}{\text{Peak flow}} = \frac{126{,}000}{62{,}50{,}000} = 0.02 \text{ (2\%)}.$$

Example 4.9

Water Flow System and Infiltration

Determine the concentration profile of the chemical if both the upstream and infiltrating flows do not contain the chemical. The first order reaction constant rate is 0.2 d^{-1}.

Assume the following :

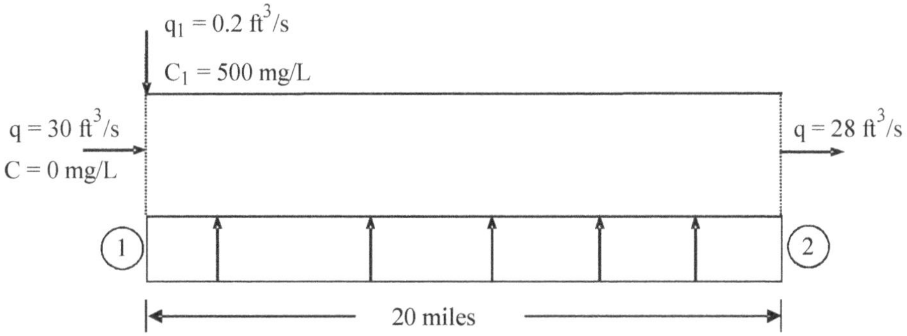

Figure 23 : Cross-sectional area : 40 ft^2 (Constant throughout the flow).

Solution

1. Required concentration profile

- Inlet velocity $(v) = \dfrac{(20+0.2)\,ft^3/s}{40\ ft^2}$

$$= 0.505\ ft/s$$

$$= 8.28\ miles/d$$

Velocity at exit point (2) :

$$v = \frac{28\ ft^3/s}{40\ ft^2} = 0.7\ ft/s$$

$$= 11.5\ miles/d$$

Plug flow system

$$\frac{dC}{dx} = -\frac{kC}{v}$$

$$C_o = \frac{(0.2\ ft^3/s)(500\ mg/L)}{(20+0.2)\,ft^3/s} = 4.95\ mg/L$$

Using constant velocity v = 8.28 miles/d

Therefore, $C = C_o \exp[-k(x/v)]$

$$= 4.95 \exp\left[\frac{-0.2(x)}{8.28}\right]$$

$$= 4.95 \exp[-0.0241x]$$

[Inlet velocity is constant and concentration variations arising due to the infiltration is not significant].

Using average velocity v = 9.89 miles/d

Using average velocity = 1/2[8.28 + 11.5] miles/d

= 9.89 miles/d

Therefore, C = 4.95 exp [– 0.0202x]

[Average velocity is constant and concentration variation arising due to the infiltration is not significant].

Using velocity profile as $v = 8.28(1 + 0.161x)$

[Linear profile for the velocity]

$v \propto (1 + 0.161x)$ [v increases with x]

$$C \;\alpha\; (1 + 0.161x)^{-1} \text{ [c decreases with } x\text{]}$$

$$\frac{dC}{dx} = -k\left(\frac{C}{1+0.161x}\right)\left(\frac{1}{1+0.161x}\right)$$

$$\text{or} \quad \frac{dC}{C} = -\frac{k\,dx}{(1+0.161x)^2}$$

$$\ln[C/C_o] = k\left[-\frac{1}{(0.161)(1+0.161x)}\right]_0^x$$

$$C = C_o \exp\left\{-k\left[\frac{1}{(0.161)(1+0.161x)} - \frac{1}{0.161}\right]\right\}$$

$$= 4.95 \exp\left(1.242 - \frac{1.242}{1+0.161x}\right)$$

[Both velocity and concentration variation account for the infiltration].

Example 4.10

Sub-Soil System

It is desirable to control release of nitrogen to protect ground and surface water resources. Nitrogen control techniques are expensive [air supply, pump and (or) chemicals], and need some skill of operation. Based on laboratory findings and literature search, the field model was developed and designed on the following basis :

Black water (containing most of the household wastewater nitrogen), and represents 40% of the total flow.

Grey water (household flow), and represents 60% of the total flow. The carbon in gray water is mostly organic (soluble form), and biodegrades faster than black water carbon.

Separation of the black and gray water system has an added advantage. Black water septic tank accumulates solids 8-times faster than a gray water tank, septage pumping costs could be reduced. Black water (containing most of the bio-resistant or soil-clogging components (feces and tissuse paper), can be pre-treated to protect the seepage field, resulting in reduced nutrient and solids load.

Sub-surface sand filter bed (black water nitrification) must be aerobic. Sand could be ES (effective size) = 0.16 mm and uniformity coefficient (US) less than 5, and should not be loaded over 30 $mm/m^2.d$. The back fill material must be porous or have peagravel vents. Nitrification requires the ratio of alkalinity to nitrogen be in the ratio of 7:1, and the air flow be 1.5 m^3/m^3 of wastewater. The characteristics of the sand filter (12 m^3, 0.5 m – deep) effluent must be measured :

Sub-surface Sand Filter for Black water Nitrification

Table 32

Parameters	*Input*	*Output*
TKN (mg/L)	92	5
$NO_3^- - N$ (mg/L)	0	30 – 80
Alkalinity (mg/L)	370	0
pH	8	3.5 – 4.1
TOC (mg/L)	140	14
O_2 uptake rate (mg/hr)	0.7	0.2
Load ($L/m^2.d$)	25	–

Anoxic 3.8 m^3 denitrification reactor was based on laboratory studies (5-day liquid detention time with no deliberate cell wastage). The specific fixed surface area needed was estimated to be 600 m^2 (2 in. rock). The characteristics of anoxic denitrification reactor are:

Table 33 : Characteristics of input and utput in an anoxic denitrification.

	Input		*Output*
Parameter	Grey water	Sand filter	
TKN (mg/L)	12	5	8
$NO_3^- - N$ (mg/L)	0	30 – 80	0.2 – 0.5
Alkalinity (mg $CaCO_3$/L)	200	0	100
pH 7.2	3.5 – 4.1	6.8	
O_2 uptake rate (mg/hr)	1.2 to 2	0.2	–
Flow (L/d)	100	300	–
TOC 100	14	–	

$R_x = 2 \times 10^{-3}$ kg NO_3 – N/kg VSS.d at 8°C

Determine the nitrogen removal using different treatment systems.

Solution

1. Required nitrogen balance for the sub-surface treatment system

Total estimated nitrogen excreted	= 12 g/Cap.d
Design flow	= 240 L/Cap.d
Flow of grey water using septic tank	= 140 L/Cap.d
Flow of black water from septic tank	= 100 L/Cap.d
$NO_4^+ - N$ effluent	= 11 mg/L [Grey water from septic tank]

$NO_4^+ - N$ effluent = 18 mg/L [Black water from septic tank and sand filter]

$NO_4^+ - N$ effluent = 60 mg/L [Black water from septic tank and sand filter]

Denitrification process in Rock – filter :

$NO_3^- - N$ removal efficiency is = 95%

$NO_3^- - N$ in effluent (240 L/Cap.d)

$$= \frac{60\,mg/L(0.05) \times 100 + 0.0(140)}{240}$$

= 1.3 mg/L [prior to land application]

* NH^+ – N will be tied up in required cell synthesis to remove 60 mg/L × 100 L/L = 6 g $NO_3^- - N$ /Cap.d

* Required cell yield ($C_5H_7NO_2$) is 0.6 mg/mg

* Cells required = 3.6 g/d

* $$\text{Excess } NH_4^+ - N = \left[\frac{11(40)+18(100)}{1000}\right] - 0.4\,g\,\frac{NH_4^+ - N}{day}$$

[0.4 g NH_4^+ – N/d is required by cells]

$$= 2.8\,g\ NH_4^+ - N/Cap.d \left(= \frac{2.8\,g\,NH_4^+ - N/Cap.d}{240\ L/Cap.d} \right.$$

= 12 mg/L prior to land application)

* Overall nitrogen – removal efficiency prior to land application

$$= \frac{\left[12\,g\,NH_4^+ - N/Cap.d - 2.8\,g\,NH^+ - N/Cap.d\right] - \left[\dfrac{1.3\,mg/L \times 240\ L/Cap.d}{1000}\right]}{12\,g\,NH_4^+ - N/Cap.d}$$

= 0.75 (75% removal efficiency)

* $NO_3^- - N$ concentration is 1.3 mg/L

* $NH_4^+ - N$ concentration is 12 mg/L.

2. Required nitrification using sand-filter

Hydraulic load on sand filter = 1 to 1.5 gal/ft^2.d [Septic tank effluent]

Effective sand size (ES) = 0.24 mm

Uniformity coefficient (EU) = less than or equal to 5.5

Nutrient load (BOD_s) on sand filter

* Black water, grey water and combined wastewater have a similar BOD_5 of about 250 mg/L
* Grey water has greater BOD_5 soluble fraction, and is less efficiently removed in a septic tank as compared to Black water
* Black water BOD_5 of septic tank effluent is 90 mg/L
* Grey water BOD_5 of septic tank effluent is 150 mg/L
* Solids load on sand filter
  - SS in black waters (450 mg/L) are significantly higher than in grey waters (160 mg/L)
  - SS concentration in the effluent treating balck water is 80 mg/L [Typical value of SS = 64 mg/L]
  - Grey water septic tank effluent contains 160 mg/L of SS [Typical value of SS = 100 mg/L]
  - Sludge accumulation in black water septic tank is about 8 times greater than in a grey water septic tank

Table 34 : Total septic tank capacity for Black water and Grey water.

Parameters	*Black Water*	*Grey Water*
Flow (Q)	40%	60%
Solids accumulation	80%	20% (Grease)
Sludge storage for 3 year cleaning period	1 (9) 3	$0.3\ (9)^3$
@ $1 ft^3$/Cap.yr and 9 person and 0.3 ft^3/Cap.d	= 27 ft^3	= 5.4 ft^3
Liquid detention time @ 24 hours or a peak day	Q_{avg} = 60 gal/Cap.d (40%) = 24 gal/Cap.dQ_{peak} = 24 gal/Cap.d (9) (5 × 0.134) = 145 ft^3	Q_{avg} = 60 gal/Cap.d (60%) = 36 gal/Cap.d Q_{peak} = 36 gal/Cap.d × 9 × 5 × 0.134 = 217 ft^3
Total septic tank capacity	27 + 142 = 172 ft^3	5.4 + 217 = 222 ft
Capacity	[1270 gal; use 1500 gal]	[1700 gal; use 2000 gal]

Sand filter for black water (40% of the total flow)

* Average maximum monthly flow [persons = 9]

  Q = 9 persons (75 gal/Cap.d) 40%

  = 675 gal(0.4) = 270 gal/d

* Sand filter load (1 $gal/ft^2.d$) ES = 0.24 mm; UC < 5.5 mm; Thickness = 24 in.; Cover = 12 in. with pea gravel]

$$\text{Sand filter area}(A) = \frac{270\,\text{gal/d}}{1\,\text{gal/ft}^2.\text{d}} = 270\ \text{ft}^2.$$

Nitrification on sand filter

Q_{avg} = 100 L/Cap.d (24 gal/Cap.d)

Daily excreted nitrogen = 12 g/Cap.d [Total removal from septic tank and sand

= 12(0.4) filter = 40%]

= 4.8 g/Cap.d

Effluent N from, sand filter = $\frac{(12-4.8)}{100\,\text{L/Cap.d}}$ [Requirement for alkalinity to convert $NH_4^+ - N$ to $NO_3^- - N$]

$$= \frac{(12-4.8)}{100} - \frac{420}{7} = 18\ \text{mg/L}$$

[Alkalinity to $NH_4^+ - N$ ratio = 7:1, Alkalinity = 420 mg/L.

Lift station [4 pumpings per day]

Lift station comprises of sump, pump, and force main,

$$\text{Sump volume}(V) = \frac{Q_{avg}}{4};\ [4\ \text{pumping/day}]$$

= (9 persons)(60 gal/Cap.d)(0.25)

= 135 gal

Head required [max H] = 2 ft + friction losses

Hourly peak flow = 12 (Q_{avg})

$$= \frac{(9)(60 \times 12)}{24} = 240\ \text{gal/hr}$$

Flow = Q(A) [Pumps 8 gal/min]

$$\text{Pipe area}(A) = \frac{Q}{V} = \frac{0.0022}{1};[V = 1\ \text{ft/s}]$$

$$\text{Pipe diameter}(D) = \left[\frac{4A}{\pi}\right]^{0.5} = \left[\frac{4(0.0022)}{3.14}\right]$$

= 0.2 in.

It is desired to use a pump of 20 gal/min capacity will total developed head (TDH) of 50 ft.

4. Required denitrification

Reactions

$$NO_3^- + \text{Organic carbon} \rightarrow N_2 + CO_2 + H_2O + OH^-$$

Ammonia (or NO_3^-) nitrogen + H_2CO_3 + Organic carbon → cells ($C_5H_7NO_2$) + H_2O + HCO_3^-

Carbon source

* Use of volatile acids (C_1 to C_5) increased to 0.36 kg $NO_3^- - N$ /kg MLVSS.d as compared to 0.18 kg $NO_3^- - N$ /kg MLVSS.d with methanol at 20°C
* Volatile acids are also produced from wastewater organics in a grey water septic tank.
* VSS from the grey water septic tank are about 128 mg/L (80% of SS in the septic tank effluent are VSS).
* Removal rate of nitrogas by VSS for a grey water flow of 140 L/Cap.d is :

$$\frac{0.36(128)140}{1000} = 6.4 \text{ g } NO_3^- - N/\text{Cap.d}$$

* Denitrification reactor size (upflow rock filter) :

  Liquid detention time = 2 days

  Size of stones = 2 to 4 in. (specific area = 200 m^2/m^3)

  Average flow (Q_{avg}) = Average maximum monthly flow for total flow

$$\text{Tank volume} = \frac{(9 \text{ persons})(75 \text{ gal/Cap.d})(2 \text{ days})}{0.33}$$

  [Void fraction = 0.33]

* Sand filter effluent contains about 60 mg/L $NO_3^- - N$ $\left(\frac{60 \times 140}{1000}\right)$ = 8.4 g / $NO_3^- - N$ /Cap.d. The balance (8.4 – 6.4) = 2g $NO_3^- - N$ / Cap.d can be denitrified using the remainder of the carbon source in the grey water effluent.
* Cell residence time in rock filter (attached growth) is longer than in suspended growth reactor [detention time is 2 days with 33% voids].

* Sufficient biomass will grow on stone (200 m^2/m^3) to remove 52 g N removed/m^3.d
* Rate of denitrification is 52 g × 15 = 780 g N removed/d
* Denitrification will add alkalinity.
* Required methanol to $NO_3^- - N$ ratio is 3:1 or carbon to $NO_3^- - N$ ratio is 1:1
* Carbon required to denitrify 6.0 g $NO_3^- - N$ /Cap.d can be supplied by grey water [140 L/Cap.d × minimum of 80 mg/L of total organic carbon = 11.2 g/Cap.d resulting in C to N ratio of = 11.2/6.0 ≈ 2:1]
* Upflow anaerobic flooded packed bed reactor is suitable to avoid the presence of dissolved oxygen.
* Sand filter effluent is expected to develop a DO level of 1 to 3 mg/L which will be consumed by the grey water tank effluent having a BOD_5 of 160 mg/L with a decay constant (k_d) of 0.05/d at 10°C.
* The effective yield coefficient for denitrifier is :

$$\overline{Y} = \frac{Y}{1 + k_d \theta_c}$$

Net cell yield should be equal to the cell loss in the stone filter effluent, and would not exceed 130 mg/L. The grey water tank effluent input is 128 mg/L which limited the net cell yield per day to (130 – 128) = 2 mg/L, or

$$\overline{Y} = \frac{2\,\text{mg/L}(2.40)}{6\,\text{g}(1000)} = 0.08$$

* Cell residence time (q_c)

$$0.08 = \frac{0.6}{1 + 0.05\,\theta_c}; \ [Y = 0.6]$$

Therefore, θ_c= 130 days

* The reactor has enough space to store more than 300 kg of VSS (Cells) to remove less than 1 kg $NO_3^- - N/d$
* Rock filter will not remove more than 60% of the incoming BOD that will be used for denitrification.

5. Required clogging of the rock filter

$\frac{(9\ \text{persons})(240\ \text{L/Cap.d})}{15\ \text{m}^3}$ m^3.d (≈ 5 gal/ft^2.d). This load is approximately 150 times less than all of the denitrification packed media

Loading at minimum rates of 0.5 gal/ft^2.min, proper cleaning due to clogging or cell yield is necessary.

Substrate or BOD load is from the grey water septic tank effluent which is diluted by sand filter effluent. The combined BOD_5 load is:

$$= \frac{(160\ \text{mg/L})(140\ \text{L/Cap.d}) + 10\,\text{mg/L}(1000\ \text{L/Cap.d})}{240\ \text{L/Cap.d}}$$

= 100 mg/L (1.6 g BOD_5/m^3.d or 2.4 g BOD_5/m^2. d)

[Over one-half of this carbon is required for denitrification]

Rock filter will not remove more than 60% of applied BOD which is used for denitrification.

[Yield = 0.6 g VSS/g $NO_3^- - N$ removed and a $NO_3^- - N$ removed of 6 g × 9 person = 54 g/d]

No cleaning is required as the cleaning rate is of the order of 2 mg/L of VSS/d

Clogging limit (Cl) is :

$$Cl = 5k - \frac{1.2}{\log k};\left[k = 5 \times 10^{-3}\ \text{ft/min}\right]$$

$$Cl = 5\left(5 \times 10^{-3}\right) - \frac{1.2}{\log\left(5 \times 10^{-3}\right)}$$

$$= 0.43\ \text{gal/ft}^2\text{.d}$$

Interface area (IA) based on Q_{max} is :

$$IA = \frac{(9\ \text{persons})(75\ \text{gal/Cap.d})}{0.43\ \text{gal/ft}^2\text{.d}} = 1570\ \text{ft}^2$$

Assume total bed rooms = 44

Total persons = (1.5 persons/bed room) (44) = 66

$$\text{Interface area required (IA)} = \frac{1570\ \text{ft}^2}{9}(66\ \text{persons})$$

$$= 11600\ \text{ft}^2$$

$$\text{Linear feet} = \frac{11600\ \text{ft}^2}{6} = 1933$$

[Interface per it is 6 ft^2]

www.ingramcontent.com/pod-product-compliance
Lightning Source LLC
Chambersburg PA
CBHW020216290326
41948CB00001B/68
9789351308959